Nelson MindTap + You = Learning amplified

"I love that everything is interconnected, relevant and that there is a clear learning sequence. I have the tools to create a learning experience that meets the needs of all my students and can easily see how they're progressing."

— Sarah, Secondary School Teacher

NELSON

WAmaths

UNITS ③ + ④

Dion Alfonsi
Amanda Pettitt
Michael Loh
Greg Neal
Tal Ellinson
Dirk Strasser
Contributing author
Neale Woods

mathematics
applications

12

Nelson WAmaths Mathematics Applications 12
1st Edition
Dion Alfonsi
Amanda Pettitt
Michael Loh
Greg Neal
Tal Ellinson
Dirk Strasser
ISBN 9780170476959

Publisher: Dirk Strasser
Associate product manager: Cindy Huang
Project editor: Tanya Smith
Series text design: Alba Design (Rina Gargano)
Series cover design: Nikita Bansal
Series designer: Nikita Bansal
Permissions researcher: Corrina Gilbert
Content developer: Rachael Pictor, Katrina Stavridis
Content manager: Alice Kane
Typeset by: Nikki M Group Pty Ltd

Any URLs contained in this publication were checked for currency during the production process. Note, however, that the publisher cannot vouch for the ongoing currency of URLs.

Acknowledgements
TI-Nspire: Images used with permission by Texas Instruments, Inc.

Casio ClassPad screenshots Images used with permission by Shriro Australia Pty. Ltd.

School Curriculum and Standards Authority. Adapted use of 2016–2021 Mathematics Applications and Mathematics Methods examinations, marking keys, and summary examination reports, ATAR 11 and 12 Mathematics Applications and Mathematics Methods syllabus. The School Curriculum and Standards Authority does not endorse this publication or product.

Selected VCE Examination questions are copyright Victorian Curriculum and Assessment Authority (VCAA), reproduced by permission. VCE ® is a registered trademark of the VCAA. The VCAA does not endorse this product and makes no warranties regarding the correctness or accuracy of this study resource. To the extent permitted by law, the VCAA excludes all liability for any loss or damage suffered or incurred as a result of accessing, using or relying on the content. Current VCE Study Designs, past VCE exams and related content can be accessed directly at www.vcaa.edu.au.

For product information and technology assistance,
in Australia call **1300 790 853**;
in New Zealand call **0800 449 725**

For permission to use material from this text or product, please email
aust.permissions@cengage.com

National Library of Australia Cataloguing-in-Publication Data
A catalogue record for this book is available from the National Library of Australia.

Cengage Learning Australia
Level 5, 80 Dorcas Street
Southbank VIC 3006 Australia

For learning solutions, visit **cengage.com.au**

Printed in China by 1010 Printing International Limited.
1 2 3 4 5 6 7 27 26 25 24 23

Contents

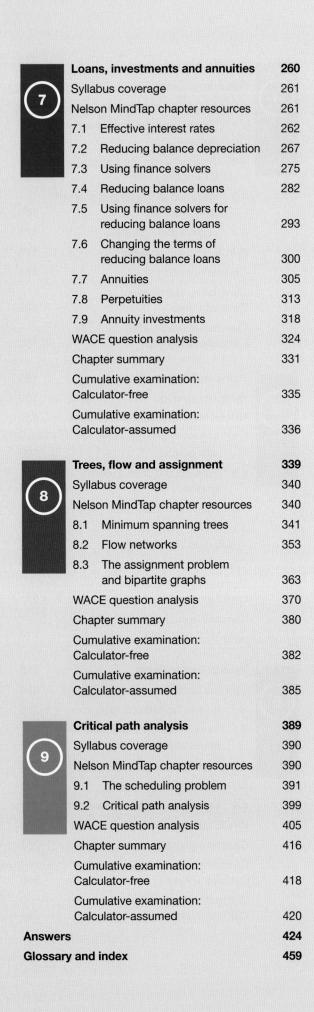

To the teacher

Now there's a better way to WACE maths mastery.

Nelson WAmaths 11–12 is a new WACE mathematics series that is backed by research into the science of learning. The design and structure of the series have been informed by teacher advice and evidence-based pedagogy, with the focus on preparing WACE students for their exams and maximising their learning achievement.

- Using **backwards learning design**, this series has been built by analysing past WACE exam questions and ensuring that all theory and examples are precisely mapped to the SCSA syllabus.

- To reduce the **cognitive load** for learners, explanations are clear and concise, using the technique of **chunking** text with accompanying diagrams and infographics.

- The student book has been designed for **mastery** of the learning content.

- The exercise structure of **Recap**, **Mastery**, **Calculator-free** and **Calculator-assumed** leads students from procedural fluency to **higher-order thinking** using the learning technique of **interleaving**.

- **Calculator-free** and **Calculator-assumed** sections include exam-style questions and past SCSA exam questions.

- The cumulative structure of Exercise **Recaps** and chapter-based **Cumulative examinations** is built on the learning and memory techniques of **spacing** and **retrieval**.

About the authors

Dion Alfonsi is the Head of Mathematics and a Secondary Mathematics Teacher at Shenton College. In the past, he has also had the roles of Year 9 & 10 Mathematics Curriculum Leader and Gifted & Talented/Academic Programs Coordinator. Dion has been a Board Member of MAWA, is a frequent presenter at the MAWA Secondary Conference and a teacher of the MAWA Problem Solving Program.

Amanda Pettitt is a Secondary Mathematics teacher at Lesmurdie Senior High School. She teaches ATAR Mathematics in senior school and lower school classes, including extension classes at the lower school level. Prior to teaching Amanda worked as an Analytical Chemist, working locally and in remote parts of Western Australia.

Michael Loh is a Senior Mathematics Teacher at Shenton College, where he also co-ordinates the Gifted and Talented Program. He has taught in the secondary and tertiary sectors for many years and is a regular presenter at various conferences including MAWA, the Google Education Conference and EdTech Summit.

Greg Neal has taught in regional schools for over 40 years and has co-written several senior textbooks for Cengage Nelson. He has been an examination assessor, presents at conferences and has expertise with CAS technology.

Tal Ellinson is an experienced teacher across all senior secondary maths subjects. He was the Learning Specialist for Numeracy, leading maths teacher training and numeracy intervention at his previous school. He is a maths presenter for Edrolo, has consulted for education companies and presented at national conferences. Recently he has been working overseas as a data scientist in the field of artificial intelligence.

Dirk Strasser is an experienced teacher, a former Head of Mathematics, and a lead author and senior publisher of mathematics series for over 30 years. He has published and co-written eight best-selling mathematics series and won several Australian Educational Publishing Awards. He is the Manager of Secondary Mathematics at Nelson Cengage.

Syllabus grid

Topic		*Nelson WAmaths Mathematics Applications 12* chapter
Topic 3.1: Bivariate data analysis (20 hours)		
The statistical investigation process	1	Bivariate data introduction
Identifying and describing associations between two categorical variables	1	Bivariate data introduction
Identifying and describing associations between two numerical variables	1	Bivariate data introduction
Fitting a linear model to numerical data	2	Bivariate data analysis
Association and causation	1	Bivariate data introduction
The data investigation process	1	Bivariate data introduction
Topic 3.2: Growth and decay in sequences (15 hours)		
The arithmetic sequence	3	Growth and decay in sequences
The geometric sequence	3	Growth and decay in sequences
Sequences generated by first-order linear recurrence relations	3	Growth and decay in sequences
Topic 3.3: Graphs and networks (20 hours)		
The definition of a graph and associated terminology	4	Graphs and networks
Planar graphs	4	Graphs and networks
Paths and cycles	5	Paths and cycles
Topic 4.1: Time series analysis (15 hours)		
Describing and interpreting patterns in time series data	6	Time series analysis
Analysing time series data	6	Time series analysis
The data investigation process	6	Time series analysis
Topic 4.2: Loans, investments and annuities (20 hours)		
Compound interest loans and investments	7	Loans, investments and annuities
Reducing balance loans (compound interest loans with periodic repayments)	7	Loans, investments and annuities
Annuities and perpetuities (compound interest investments with periodic payments made from the investment)	7	Loans, investments and annuities
Topic 4.3: Networks and decision mathematics (20 hours)		
Trees and minimum connector problems	8	Trees, flow and assignment
Project planning and scheduling using critical path analysis (CPA)	9	Critical path analysis
Flow networks	8	Trees, flow and assignment
Assignment problems	8	Trees, flow and assignment

In each chapter

Syllabus coverage and extracts are shown at the start of the chapter, along with a listing of **Nelson MindTap chapter resources**.

Important words and phrases are printed in blue and listed in the **Glossary and index** at the back of the book.

Important facts and formulas are highlighted in a shaded box.

Worked examples are explained clearly step-by-step, with the mathematical working shown on the right-hand side.

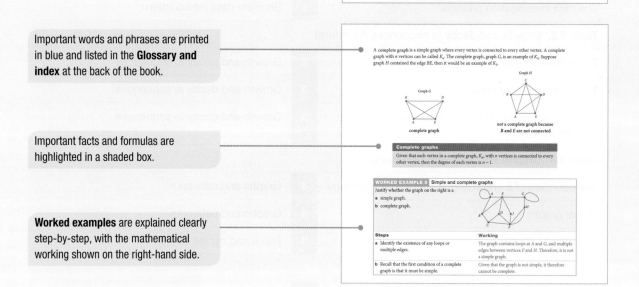

Using CAS provides clear instructions and screenshots for Casio ClassPad and TI-Nspire calculators.

Exam hacks highlight valuable exam hints and common student errors.

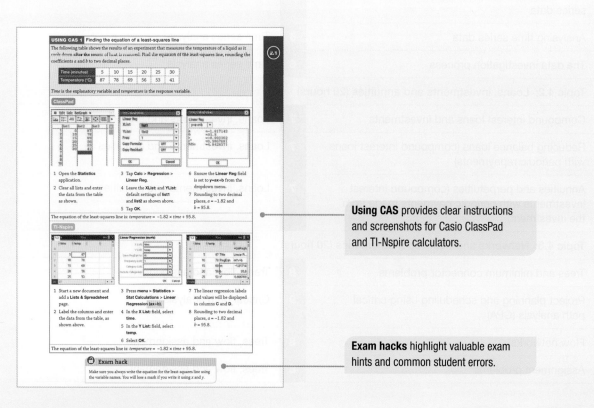

Graded exercises include **Recap**, **Mastery**, **Calculator-free** and **Calculator-assumed** questions. **Recap** questions revise skills from the previous exercise and function as lesson starters.

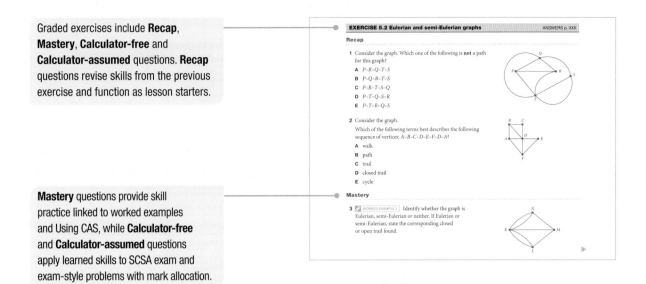

EXERCISE 5.2 Eulerian and semi-Eulerian graphs   ANSWERS p. XXX

Recap

1 Consider the graph. Which one of the following is **not** a path for this graph?
   A $P–R–Q–T–S$
   B $P–Q–R–T–S$
   C $P–R–T–S–Q$
   D $P–T–Q–S–R$
   E $P–T–R–Q–S$

2 Consider the graph.
   Which of the following terms best describes the following sequence of vertices: $A–B–C–D–E–F–D–A$?
   A walk
   B path
   C trail
   D closed trail
   E cycle

Mastery

3 WORKED EXAMPLE 3   Identify whether the graph is Eulerian, semi-Eulerian or neither. If Eulerian or semi-Eulerian, state the corresponding closed or open trail found.

Mastery questions provide skill practice linked to worked examples and Using CAS, while **Calculator-free** and **Calculator-assumed** questions apply learned skills to SCSA exam and exam-style problems with mark allocation.

© SCSA   MA2020 Q16

Past SCSA exam questions are clearly tagged:
MA   Mathematics Applications
MM   Mathematics Methods
2020   2020 exam year
Q16   Question 16

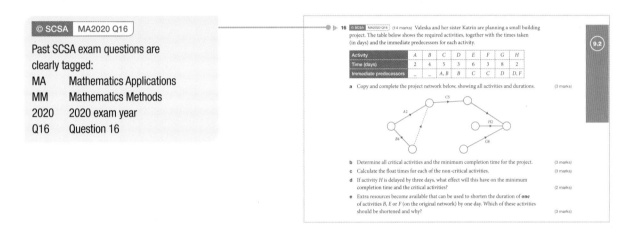

16 © SCSA   MA2020 Q16   (14 marks) Valeska and her sister Katrin are planning a small building project. The table below shows the required activities, together with the times taken (in days) and the immediate predecessors for each activity.

Activity	A	B	C	D	E	F	G	H
Time (days)	2	4	5	3	6	3	8	2
Immediate predecessors	–	–	A, B	B	C	C	D	D, F

a Copy and complete the project network below, showing all activities and durations.   (3 marks)

b Determine all critical activities and the minimum completion time for the project.   (3 marks)
c Calculate the float times for each of the non-critical activities.   (3 marks)
d If activity H is delayed by three days, what effect will this have on the minimum completion time and the critical activities?   (2 marks)
e Extra resources become available that can be used to shorten the duration of **one** of activities B, E or F (on the original network) by one day. Which of these activities should be shortened and why?   (3 marks)

At the end of each chapter

WACE question analysis leads students through a past WACE exam question that exemplifies the chapter, discussing how to approach the question, providing advice on interpreting the question, common student errors and a full worked solution with a marking key.

WACE QUESTION ANALYSIS

© SCSA   MA2021 Q4   Calculator-free (9 marks)

Video
WACE
question
analysis:
Bivariate data
introduction

A public opinion survey was conducted on the statement 'our overwhelming dependence on computers is a good thing', with partial results being shown in the table below.

Age	Opinion			Total
	Agree	Disagree	Undecided	
20–39 years	40	28		80
40–59 years	38		20	100
60–79 years	20		18	
Total				230

a Copy and complete the table above.   (3 marks)
b Identify the response variable.   (1 mark)
c Use the template below to construct a percentaged two-way frequency table showing either column or row percentages as appropriate, to investigate if there is an association between age and opinion.   (4 marks)

Age	Opinion			Total
	Agree	Disagree	Undecided	
20–39 years				
40–59 years				
60–79 years				
Total				

d State an association that can be observed from the percentaged two-way frequency table.   (1 mark)

Chapter 1 : Bivariate data introduction   27

Chapter summary for easy reference.

Cumulative examination: Calculator-free and **Cumulative examination: Calculator-assumed** are mini-exams based on the format of the WACE examinations, revising work from the chapters in which they appear, as well as previous chapters.

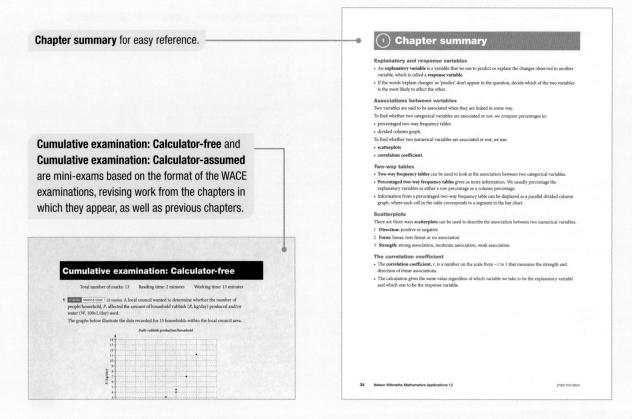

At the end of the book

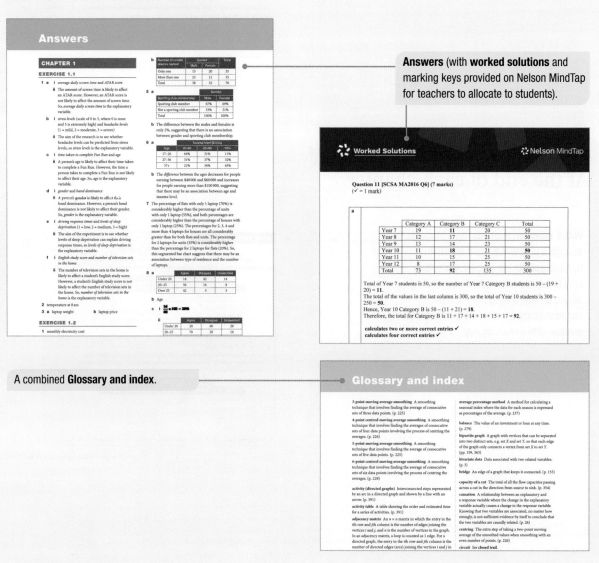

Answers (with **worked solutions** and marking keys provided on Nelson MindTap for teachers to allocate to students).

A combined **Glossary and index**.

Nelson MindTap

Nelson MindTap is an online learning space that provides students with tailored learning experiences. Access tools and content that make learning simpler yet smarter to help you achieve WACE maths mastery.

Nelson MindTap includes an eText with integrated interactives and online assessment.

Margin links in the student book signpost multimedia student resources found on MindTap.

Nelson MindTap for students:

- **Watch** video tutorials featuring expert teacher advice to unpack new concepts and develop your understanding.
- **Revise** using learning checks, worksheets and skillsheets to practise your skills and build your confidence.
- **Navigate** your own path, accessing the content and support as you need it.

Nelson MindTap for teachers*:

- Tailor content to different learning needs – assign directly to the student, or the whole class.
- Monitor progress using the MindTap assessment tools.
- Integrate content and assessment directly within your school's LMS for ease of access.
- Access topic tests, teaching plans and worked solutions to each exercise set.

Video playlists

Worksheets

Skillsheets

*Complimentary access to these resources is only available to teachers who use this book as part of a class set, book hire or booklist. Contact your Cengage Education Consultant for information about access and conditions.

Nelson WAmaths 11–12 series

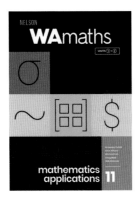

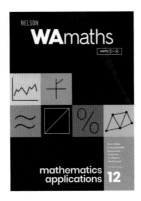

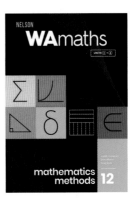

Additional credits

*modified

Chapter 1
Exercise 1.2
Worked example 4 © VCAA FM2003 1CQ6
Question 9 ©VCAA FM2009 1CQ8
Question 10 © VCAA FM2017 1CQ6
Question 12 © VCAA FM2012 1CQ6
Question 13 © VCAA FM2003 1CQ6*
Question 14 © VCAA FM2018 2CQ1a–e
Exercise 1.3
Question 1 © VCAA FM2020 1CQ6
Question 2 © VCAA FM2019 1CQ8
Question 7a © VCAA FM2016 1CQ12*
Question 7b © VCAA FM2016S 1CQ12*
Exercise 1.4
Question 9 © VCAA FM2011 1CQ11*
Question 13 © VCAA FM2013 1CQ8
Question 14 © VCAA FM2002 1CQ10*
Question 15 © VCAA FM2003 2CQ1
Cumulative examination: Calculator-assumed
Question 1 © VCAA FM2008 2CQ2

Chapter 2
Exercise 2.1
Question 5 © VCAA FM2018 1CQ10*
Question 6 © VCAA FM2019 2CQ4
Question 7 © VCAA FM2015 2CQ4
Question 8 © VCAA FM2008 2CQ4
Exercise 2.2
Question 1 © VCAA FM2009 1CQ9
Question 2 © VCAA FM2009 1CQ10
Question 5 © VCAA FM2002 1CQ12*
Question 6 © VCAA FM2019 1CQ9*
Question 7 © VCAA FM2019 1CQ10*
Question 8 © VCAA FM2011 1CQ8*
Question 9 © VCAA FM2007 1CQ7
Question 10 © VCAA FM2018 1CQ9*
Question 11 © VCAA FM2005 1CQ9*
Question 12 © VCAA FM2020 2CQ4*
Question 13 © VCAA FM2015 2CQ3
Question 14 © VCAA FM2014 2CQ4
Exercise 2.3
Question 1 © VCAA FM2004 1CQ8*
Question 2 © VCAA FM2004 1CQ9
Question 8 © VCAA FM2015 2CQ5b
Question 9 © VCAA FM2010 2CQ2*
Exercise 2.4
Question 1 © VCAA FM2008 1CQ8
Question 2 © VCAA FM2008 1CQ9*
Question 7 © VCAA FM2013 1CQ11
Question 8 © VCAA FM2018 1CQ7*
Question 9ab © VCAA FM2010 1CQ7*
Question 10 © VCAA FM2019 2CQ5
Question 11 © VCAA FM2016 2CQ3
Question 12 © VCAA FM2017 2CQ3
Question 13 © VCAA FM2018N 2CQ5
Cumulative examination: Calculator-assumed
Question 1 © VCAA FM2016S 2CQ1

Chapter 3
Cumulative examination: Calculator-assumed
Question 1 © VCAA FM2018N 2CQ4
Question 2 © VCAA FM2018 2CQ2 *

Chapter 4
Exercise 4.1
Question 5 © VCAA FM2017 1NQ1*
Question 6a © VCAA FM2019 1NQ1
Question 6b © VCAA FM2019N 1NQ1
Question 7 © VCAA FM2018N 1NQ1
Question 8a © VCAA FM2008 1NQ2*
Question 8b © VCAA FM2010 1NQ1*
Question 9 © VCAA FM2020 2NQ1*
Question 10 © VCAA FM2016 2NQ1*
Question 11b © VCAA FM2011 1NQ4*
Question 11c © VCAA FM2011 1NQ3*
Exercise 4.2
Question 1 © VCAA FM2003 1NQ4
Question 2 © VCAA FM2013 1NQ6
Question 8a © VCAA FM2013 1NQ2*
Question 8b © VCAA FM2009 1NQ1*
Question 9a © VCAA FM2020 1NQ1*
Question 9b © VCAA FM2019N 1NQ2*

Question 10a © VCAA FM2019N 1NQ6*
Question 10b © VCAA FM2018 1NQ6*
Exercise 4.3
Question 1 © VCAA FM2016 1NQ5
Question 5 © VCAA FM2017N 1NQ1*
Question 6 © VCAA FM2016S 1NQ3*
Question 7 © VCAA FM2013 2NQ1*
Exercise 4.4
Worked example 15 © VCAA FM2016S 1NQ1*
Question 2 © VCAA FM2003 1NQ8
Question 7 © VCAA FM2020 1NQ8*
Question 10 © VCAA FM2017N 1NQ4*
Question 11 © VCAA FM2018N 1NQ5
Question 12 © VCAA FM2010 2NQ1*
Question 15 © VCAA FM2017 1NQ3*
Cumulative examination: Calculator-assumed
Question 2 © VCAA FM2009 2CQ3*
Question 3 © VCAA FM2019N 2NQ1*

Chapter 5
Exercise 5.2
Worked example 5 © VCAA FM2019 2NQ1*
Question 1 © VCAA FM2018 1NQ4
Question 2 © VCAA FM2004 1NQ1*
Question 5 © VCAA FM2016 2NQ2*
Question 6 © VCAA FM2017 1NQ6*
Question 11 © VCAA FM2018 2NQ4*
Exercise 5.3
Worked example 7 © VCAA FM2017N 2NQ2*
Question 1 © VCAA FM2005 1NQ2
Question 2 © VCAA FM2019 1NQ2
Question 4 © VCAA FM2021 2NQ1*
Question 5 © VCAA FM2020 1NQ2 and
© VCAA FM2008 1NQ3*
Question 10 © VCAA FM2015 2NQ2
Cumulative examination: Calculator-free
Question 2a © VCAA FM2014 1NQ6
Question 2b © VCAA FM2021 1NQ5
Question 5 © VCAA FM2019N 1NQ7
Cumulative examination: Calculator-assumed
Question 2 © VCAA FM2017 2NQ1*
Question 3 © VCAA FM2011 2MQ1*
Question 4 © VCAA FM2021 2NQ2

Chapter 6
Exercise 6.2
Question 2 © VCAA FM2018 1CQ15*
Cumulative examination: Calculator-free
Question 3 © VCAA FM2017N 1NQ3*
Cumulative examination: Calculator-assumed
Question 3 © VCAA FM2021 1NQ4

Chapter 7
Exercise 7.1
Question 4 © VCAA FM2018N 1CQ21*
Question 5 © VCAA FM2016S 1CQ19
Exercise 7.2
Question 8b © VCAA FM2019N 1CQ18
Question 10 © VCAA FM2019 2CQ7a–c
Question 11 © VCAA FM2018 2CQ5
Question 12 © VCAA FM2017N 2CQ5di–ii
Exercise 7.3
Question 8a © VCAA FM2004 1BRMQ2
Question 8b © VCAA FM2010 1BRMQ5
Question 9 © VCAA FM2019 1CQ24
Question 10 © VCAA FM2008 1BRMQ3
Question 11 © VCAA FM2018N 2CQ8bc
Exercise 7.4
Question 8a © VCAA FM2017 1CQ19
Question 8b © VCAA FM2017 1CQ20
Question 12 © VCAA FM2020 2CQ8
Exercise 7.5
Question 8a © VCAA FM2019 1CQ23
Question 8b © VCAA FM2018 1CQ22
Question 9 © VCAA FM2012 1BRMQ3*
Question 10 © VCAA FM2020 2CQ11
Exercise 7.6
Question 1 © VCAA FM2008 1BRMQ8
Question 2 © VCAA FM2002 1BRMQ4
Question 6 © VCAA FM2017 1CQ24
Question 7 © VCAA FM2009 1BRMQ9
Question 9 © VCAA FM2007 2BRMQ2
Question 10 © VCAA FM2018N 2CQ9

Question 11 © VCAA FM2011 2BRMQ4
Question 12 © VCAA FM2016 2CQ7
Exercise 7.7
Question 1 © VCAA FM2018N 1CQ24
Question 2 © VCAA FM2011 1BRMQ9
Question 7a © VCAA FM2019N 1CQ19
Question 7b © VCAA FM2018 1CQ18
Question 8a–c © VCAA FM2016S 1CQ21–23
Question 10 © VCAA FM2019N 2CQ8
Exercise 7.8
Question 2 © VCAA FM2017N 1CQ22
Question 5a © VCAA FM2018N 1CQ22
Question 5b © VCAA FM2017N 1CQ20
Question 5c © VCAA FM2011 1BRMQ5
Question 6 © VCAA FM2014 1BRMQ2ab
Question 8 © VCAA FM2012 1BRMQ4
Exercise 7.9
Question 2 © VCAA FM2008 1BRMQ2
Question 7ab © VCAA FM2018 1CQ17–18
Question 8a © VCAA FM2018N 1CQ23
Question 8b © VCAA FM2016 1CQ23
Question 9 © VCAA FM2018 1CQ24
Question 11 © VCAA FM2017N 2CQ7
Question 12 © VCAA FM2017 2CQ7
Question 13 © VCAA FM2019N 2CQ7
Cumulative examination: Calculator-free
Question 1 © VCAA FM2017N 1NQ2*
Cumulative examination: Calculator-assumed
Question 1 © VCAA FM2006 2CQ2
Question 3 © VCAA FM2016S 2NQ1*

Chapter 8
Exercise 8.1
Question 6 © VCAA FM2018 1NQ1
Question 7 © VCAA FM2013 1NQ1
Question 11 © VCAA FM2020 1NQ5*
Question 12 © VCAA FM2016 1NQ4*
Question 13 © VCAA FM2015 2NQ1
Question 14 © VCAA FM2017 2NQ3
Exercise 8.2
Question 6 © VCAA FM2019 1NQ3
Question 7 © VCAA FM2006 1NQ6
Question 8 © VCAA FM2016 1NQ2
Question 9 © VCAA FM2017 1NQ8
Question 10 © VCAA FM2005 1NQ6
Question 11 © VCAA FM2010 1NQ7
Question 12 © VCAA FM2012 1NQ7
Question 17 © VCAA FM2013 2NQ3
Question 18 © VCAA FM2007 2NQ3
Question 19 © VCAA FM2017N 2NQ3
Exercise 8.3
Question 13 © VCAA FM2013 1NQ4
Question 14 © VCAA FM2006 2NQ1
Question 15 © VCAA FM2012 2NQ3
Question 16 © VCAA FM2019 2NQ2
Question 17 © VCAA FM2008 2NQ3
Question 18 © VCAA FM2017 2NQ2
Cumulative examination: Calculator-assumed
Question 1 © VCAA FM2020 2CQ6
Question 3 © VCAA FM2018N 2NQ2
Question 7 © VCAA FM2012 2NQ1b

Chapter 9
Exercise 9.1
Question 3 © VCAA FM2006 1NQ5
Question 4 © VCAA FM2016 1NQ6
Question 5 © VCAA FM2009 1NQ5
Question 6 © VCAA FM2019 1NQ7
Question 10 © VCAA FM2013 2NQ2a
Exercise 9.2
Question 13 © VCAA FM2016 2NQ3
Question 14 © VCAA FM2017 2NQ4
Question 17 © VCAA FM2019 2NQ3
Question 18 © VCAA FM2017N 2NQ1
Question 19 © VCAA FM2018N 2NQ4
Cumulative examination: Calculator-free
Question 2 © VCAA FM2018N 1NQ3 *
Question 4 © VCAA FM2020 1NQ6*
Question 5 © VCAA FM2021 2NQ4ab
Cumulative examination: Calculator-assumed
Question 5 © VCAA FM2018 2NQ1
Question 7 © VCAA FM2018 2NQ3

BIVARIATE DATA INTRODUCTION

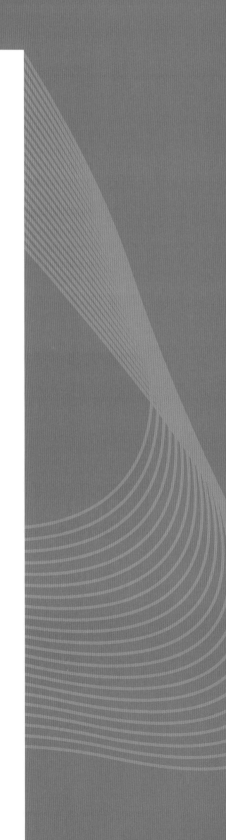

Statistical investigation process

The statistical investigation process is a cyclical process that begins with the need to solve a real-world problem and aims to reflect the way statisticians work. One description of the statistical investigation process in terms of four steps is as follows.

Step 1   **Clarify** the problem and **formulate** one or more questions that can be answered with data.

Step 2   **Design** and implement a plan to **collect** or obtain appropriate data.

Step 3   **Select** and apply appropriate graphical or numerical techniques to **analyse** the data.

Step 4   **Interpret** the results of this analysis and relate the interpretation to the original question; **communicate** findings in a systematic and concise manner.

Mathematics Applications ATAR Course Year 12 syllabus p. 27 © SCSA

① ⭤ Explanatory and response variables

Data associated with two related variables is called **bivariate data**.

Here are some examples of questions that involve analysing the association between two variables.

- Does human activity explain global warming?
- Is there a relationship between vaping and lung cancer?
- Does the number of books in a home predict academic success?
- How likely is it that the amount of time spent on social media affects ATAR results?
- Is there an association between the amount of chocolate eaten and happiness?

Identifying explanatory and response variables

It's important to decide which one of the variables is the **explanatory variable** and which is the **response variable**. An explanatory variable is a variable that we use to predict or explain the changes observed in another variable, the response variable.

Exam hack

When deciding whether a variable is explanatory or response, look for the words 'explain changes' or 'predict'. If they don't appear, think about which of the two variables is the most likely to affect the other variable.

Video playlist
Explanatory and response variables

WORKED EXAMPLE 1	Identifying explanatory and response variables

For each of the following

   i identify the two variables

   ii identify the explanatory variable, giving a reason for your answer.

Steps	Working
a An investigation is done to see whether the size of the crowd at an AFL match can be predicted by the team's position on the AFL ladder.	
i Consider the sort of data that will be collected.	*crowd size* and *position on ladder*
ii Which variable is 'explaining' or 'predicting' the other? If the wording isn't used, which variable is most likely to affect the other?	The aim of the investigation is to predict crowd size from the team's position on the AFL ladder, so *position on ladder* is the explanatory variable.
b Researchers wish to investigate the association between the age of a person and the amount of time they sleep.	
i Consider the sort of data that will be collected.	*age* and *time spent sleeping*
ii Which variable is 'explaining' or 'predicting' the other? If the wording isn't used, which variable is most likely to affect the other?	A person's age is likely to affect the time they spend sleeping. However, the time a person spends sleeping will not affect their age. So, *age* is the explanatory variable.
c A statistical analysis is done to establish whether there is a connection between the cost of avocados and the season.	
i Consider the sort of data that will be collected.	*cost of avocados* and *season*
ii Which variable is 'explaining' or 'predicting' the other? If the wording isn't used, which variable is most likely to affect the other?	The season is likely to affect the cost of avocados. However, the cost of avocados will not affect the season. So, *season* is the explanatory variable.

d A study is undertaken to establish whether a person's alcohol consumption (measured by the average number of standard alcoholic drinks per week) can explain depression levels (mild, moderate or severe).

i Consider the sort of data that will be collected.	*alcohol consumption* and *depression levels* (mild, moderate or severe)
ii Which variable is 'explaining' or 'predicting' the other? If the wording isn't used, which variable is most likely to affect the other?	The aim of the study is to explain depression levels from alcohol consumption, so *alcohol consumption* is the explanatory variable in this study. Note: it is also possible that depression levels affect alcohol consumption, so in another study, *depression levels* could be the explanatory variable.

EXERCISE 1.1 Explanatory and response variables

ANSWERS p. 424

Mastery

1 WORKED EXAMPLE 1   For each of the following

   i identify the two variables

   ii identify the explanatory variable, giving a reason for your answer.

a A researcher investigates the association between the average daily screen time of Year 12 students and their ATAR scores.

b Research is undertaken to see whether stress levels (scale of 0 to 5, where 0 is none and 5 is extremely high) can predict headache levels (1 = mild, 2 = moderate, 3 = severe).

c A study is done to establish whether there is an association between the time taken to complete a Fun Run and the age of a person.

d A statistical analysis is undertaken to investigate whether there is a relationship between a person's gender and whether they are left- or right-hand dominant.

e An experiment is conducted to see whether driving response times (in milliseconds) can be explained by levels of sleep deprivation (1 = low, 2 = medium, 3 = high).

f A researcher wants to determine whether there is a connection between Year 12 students' English study scores and the number of television sets in their homes.

Calculator-free

2 (2 marks) The temperatures (°C) at 8 am and 3 pm were measured each day in December 2021 to see whether it was possible to predict the temperature at 3 pm from the temperature at 8 am. What is the explanatory variable?

3 (2 marks) A consumer group conducts research to determine whether the price of a laptop is affected by its weight. They weigh 50 laptops and note each of their prices on a spreadsheet.

   a State the explanatory variable. (1 mark)

   b State the response variable. (1 mark)

1.2 Associations between two categorical variables

Two variables are said to be associated when they are linked in some way.

We say two categorical variables are

- associated, if the explanatory variable categories give considerably *different* results
- *not* associated, if the explanatory variable categories give *similar* results.

To find whether two categorical variables are associated or not, we compare percentages in

- percentaged two-way frequency tables
- divided column graphs.

Two-way frequency tables

A **two-way frequency table** is an effective way to display two categorical variables. Here is an example of a two-way frequency table. It looks at the association between the variables *gender* and *exercise habits*.

It's possible to have more than two categories for each variable.

Explanatory variable

Response variable	Gender		
Exercise habits	Male	Female	Total
Exercise	3028	1084	(4112) Row totals
No exercise	1532	946	(2478)
Total	(4560)	(2030)	6590

Number of males who exercise

Column totals

Video playlist
Associations between two categorical variables

Worksheets
Two-way tables 1

Two-way tables 2

Percentage tables

> 🔓 **Exam hack**
>
> The variable appearing in the columns of a two-way frequency table is usually, *but not always*, the explanatory variable.

WORKED EXAMPLE 2 | Creating two-way frequency tables

Sixty people were interviewed about their biscuit preference. Of the 35 women, 20 said they preferred chocolate biscuits, whereas 19 of the men preferred biscuits without chocolate.

a Name the explanatory variable and the response variable.

b Construct a two-way frequency table to show this information, using *gender* for the columns.

Steps	Working
a Name the explanatory and the response variables.	The explanatory variable is *gender*. The response variable is *biscuit preference*.

b 1 Create a table using *gender* as the explanatory variable for the columns and *biscuit preference* as the response variable for the rows.

Biscuit preference	Gender		Total
	Female	Male	
Chocolate			
Without chocolate			
Total			

2 Fill in the information from the question.
- 60 people
- 35 women
- 20 women preferred chocolate biscuits
- 19 men preferred biscuits without chocolate

Biscuit preference	Gender		Total
	Female	Male	
Chocolate	20		
Without chocolate		19	
Total	35		60

3 Complete the table using column and row totals.

20 + 15 = 35
35 + 25 = 60
15 + 19 = 34
6 + 19 = 25
20 + 6 = 26

Biscuit preference	Gender		Total
	Female	Male	
Chocolate	20	6	26
Without chocolate	15	19	34
Total	35	25	60

Percentaged two-way frequency tables

To find whether there are differences between categories, and therefore determine if an association between the variables exists, convert the two-way frequency table to a **percentaged two-way frequency table**. There are a number of ways to calculate the percentages in a two-way frequency table, depending on the association we are finding, but we usually calculate each data value as a percentage of the explanatory variable.

Percentaging gender in the two-way frequency table of gender and exercise habits gives the table on the right. The percentages are rounded to one decimal place.

From this percentaged two-way frequency table, we can see that there is a *difference* in exercise habits of males and females: 66.4% of males exercise whereas only 53.4% of females exercise, which is a difference of over 10%. This suggests that there is an association between gender and exercise habits.

If the male and female percentages had been approximately equal, we would have said there is no association, which would mean knowing a person's gender doesn't suggest anything about their exercise habits.

Two-way frequency table

Exercise habits	Gender		Total
	Male	Female	
Exercise	3028	1084	4112
No exercise	1532	946	2478
Total	4560	2030	6590

Percentaged two-way frequency table

Exercise habits	Gender	
	Male	Female
Exercise	66.4%	53.4%
No exercise	33.6%	46.6%
Total	100.0%	100.0%

$$\frac{\text{data value}}{\text{column total}} \times 100\% = \frac{3028}{4560} \times 100\%$$

9780170476959

Calculating column percentages for percentaged two-way frequency tables

a Convert the two-way frequency table into a percentaged two-way frequency table by percentaging the explanatory variable. Round to the nearest percentage.

b What does the table suggest about the association between a person's gender and biscuit preference? Give reasons by referring to percentages.

	Gender		
Biscuit preference	Female	Male	Total
Chocolate	28	32	60
Without chocolate	10	26	36
Total	38	58	96

Steps

Working

a 1 The explanatory variable forms the columns, so redraw the two-way table using only the column totals.

	Gender	
Biscuit preference	Female	Male
Chocolate	28	32
Without chocolate	10	26
Total	38	58

2 Calculate the required percentages.

Females preferring chocolate: $\dfrac{28}{38} \times 100 \approx 74\%$

Females preferring without chocolate: $\dfrac{10}{38} \times 100 \approx 26\%$

Males preferring chocolate: $\dfrac{32}{58} \times 100 \approx 55\%$

Males preferring without chocolate: $\dfrac{26}{58} \times 100 \approx 45\%$

3 Write the percentages in the two-way table.

	Gender	
Biscuit preference	Female	Male
Chocolate	74%	55%
Without chocolate	26%	45%
Total	100%	100%

b An association means the explanatory variable categories give considerably different results. Refer to percentages in your answer.

The table suggests there is an association between gender and chocolate biscuit preferences. 74% of females prefer biscuits with chocolate compared to only 55% of males. The difference in biscuit preference between males and females is nearly 20%, indicating that females have a greater preference for chocolate biscuits than males.

The level of water usage of 250 houses was rated in
a survey as low, medium or high, and the size of the
houses as small, standard or large. The results of
the survey are displayed in the table.

Size of house	Level of water usage		
	Low	Medium	High
Small	15	22	15
Standard	13	71	47
Large	8	11	46

a Convert the two-way frequency table into a percentaged
two-way frequency table by percentaging the
explanatory variable. Round to the nearest percentage.

b What does the table suggest about the association between the size of the house and level of water usage?
Give reasons by referring to percentages.

Steps	Working

a 1 Name the explanatory and the
response variables.

The explanatory variable is the *size of the house.*

The response variable is *level of water usage.*

2 The explanatory variable forms the
rows, so redraw the two-way table
using only the row totals.

Size of house	Level of water usage			Total
	Low	Medium	High	
Small	15	22	15	52
Standard	13	71	47	131
Large	8	11	46	65

3 Calculate the required percentages.

Small house with low water usage: $\dfrac{15}{52} \times 100 \approx 29\%$

Small house with medium water usage: $\dfrac{22}{52} \times 100 \approx 42\%$

Small house with high water usage: $\dfrac{15}{52} \times 100 \approx 29\%$

Standard house with low water usage: $\dfrac{13}{131} \times 100 \approx 10\%$

Standard house with medium water usage: $\dfrac{71}{131} \times 100 \approx 54\%$

Standard house with high water usage: $\dfrac{47}{131} \times 100 \approx 36\%$

Large house with low water usage: $\dfrac{8}{65} \times 100 \approx 12\%$

Large house with medium water usage: $\dfrac{11}{65} \times 100 \approx 17\%$

Large house with high water usage: $\dfrac{46}{65} \times 100 \approx 71\%$

4 Write the percentages in the
two-way table.

Size of house	Level of water usage			Total
	Low	Medium	High	
Small	29%	42%	29%	100%
Standard	10%	54%	36%	100%
Large	12%	17%	71%	100%

b An association means the explanatory
variable categories give considerably
different results. Refer to percentages
in your answer.

The table suggests there is an association between the size of
the house and water usage. 29% of small houses have a high
level of water usage compared to 36% of medium houses and
71% of large houses. The difference in percentages between
the rows suggests there is an association between house size
and water usage.

Exam hack

Don't use the words 'evidence' or 'proof' when discussing associations. Results are never clear-cut when it comes to associations. Also, don't use the word 'significant' because this has a specific meaning in statistics. Instead, use words such as 'indicate', 'suggest', 'considerably higher', 'similar' and 'noticeably different'.

Divided column graph

It is often easier to see patterns in data when it is displayed as a chart rather than a table. Information from a percentaged two-way frequency table can be displayed as a **divided column graph** where each cell in the table corresponds to a segment in the bar chart.

For example, the percentaged two-way frequency table of gender and exercise habits on page 6 can be shown as the divided column graph on the right.

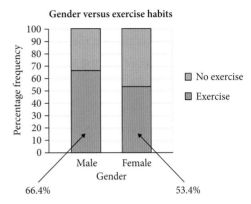

Divided column graphs are particularly useful when we are dealing with data from across several populations.

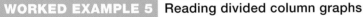

WORKED EXAMPLE 5 | Reading divided column graphs

The divided column graph shows the results of a survey of preferred car colour of males and females. Discuss whether this suggests there is an association between car colour and gender by comparing percentages.

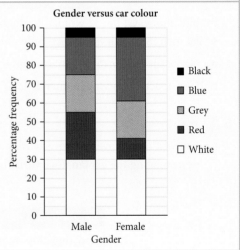

Steps	Working
Look at how many of the segments are similar and how many are different.	There are considerable differences in the percentages of males and females who prefer red cars (male 25% and female 11%) and blue cars (male 20% and female 34%), which by itself would suggest that there may be an association between car colour and gender.
	However, the three other colours had very similar percentages for males and females: white (30%), grey (20%) and black (5%). This suggests that there may be an association between car colour and gender but only for certain colours.

Recap

1 A study is done to investigate the relationship between the monthly electricity cost of a household and the number of people in the household. What is the response variable?

2 A study is done to understand the association between the breed of a dog and its life span (months). Which of the following statements is **not** true?

A One of the variables is numerical and the other is categorical.

B The explanatory variable is *breed*.

C The response variable is *life span*.

D The breed of the dog may affect its life span.

E The lifespan of the dog may affect its breed.

Mastery

3 State whether it is possible to set up a two-way frequency table for each of these pairs of variables.

a *height* and *weight*

b *stress level* (low, medium, high) and *gender* (male, female, other)

c *hair colour* and *eye colour*

d *attitude to school subject* (scale of 1 to 5 where 1 is hate and 5 is love) and *favourite reality TV series*

e *favourite colour* and *age*

4 ⚙ WORKED EXAMPLE 2   Seventy people were asked about whether or not they owned more than one mobile phone. Of the 32 women in the survey, 12 owned more than one mobile phone, whereas 15 of the men in the survey owned only one.

a Name the explanatory variable and the response variable.

b Construct a two-way frequency table to show this information, using *gender* for the columns.

5 ⚙ WORKED EXAMPLE 3

a Convert the following two-way frequency table into a percentaged two-way frequency table by percentaging the explanatory variable. Round to the nearest percentage.

Sporting club membership	Gender		Total
	Male	Female	
Sporting club member	24	38	62
Not a sporting club member	12	17	29
Total	36	55	91

b What does the table suggest about the association between a person's gender and sporting club membership? Give reasons by referring to percentages.

6 ⚙ WORKED EXAMPLE 4

a Convert the two-way frequency table into a percentaged two-way frequency table by percentaging the explanatory variable. Round to the nearest percentage.

b What does the table suggest about the association between a person's age and their income? Give reasons by referring to percentages.

Age	Income level ($000s)			Total
	40–60	60–80	100+	
17–26	23	12	4	39
27–36	15	18	16	49
37+	12	20	26	58
Total	50	50	46	146

7 `WORKED EXAMPLE 5` The divided column graph shows the results of a survey about the *type of residence* and the *number of laptops* in it. Discuss whether this suggests there is an association between the type of residence and the number of laptops by comparing percentages.

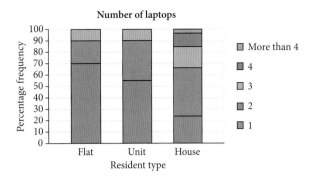

Calculator-free

8 `© SCSA` `MA2017 Q5` (9 marks) A group of university students was asked the question 'Does full attendance at school lead to an improved examination result?' The results are summarised below.

	Agree	Disagree	Undecided
Male under 20 years	8	22	6
Female under 20 years	6	20	8
Male 20 to 25 years	26	7	3
Female 20 to 25 years	30	9	5
Male over 25 years	24	3	2
Female over 25 years	18	2	1

a Copy and complete the two-way table below. (2 marks)

	Agree	Disagree	Undecided
Under 20	14		
20–25			
Over 25			3

b State the explanatory variable for these data. (1 mark)

c The incomplete table below shows row percentages.

	Percentages		
	Agree	Disagree	Undecided
Under 20		60	
20–25		20	
Over 25	84		

   i Show how the value of 20% was calculated. (2 marks)

   ii Copy and complete the table. (2 marks)

d Use the data to determine one association between the variables. Describe the association and explain your reasoning. (2 marks)

9 (2 marks) An animal study was conducted to investigate the association between *exposure to danger* during sleep (high, medium, low) and *chance of attack* (above average, average, below average). The results are summarised in the divided column graph.

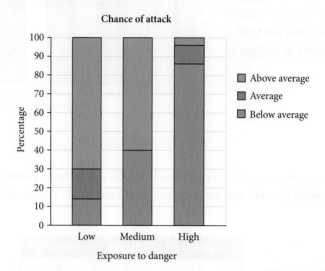

a Determine the percentage of animals whose *exposure to danger* during sleep is high, and whose *chance of attack* is below average. (1 mark)

b Determine the percentage of animals whose *exposure to danger* during sleep is low, and whose *chance of attack* is average. (1 mark)

10 (2 marks) A study was conducted to investigate the association between the *number of moths* caught in a moth trap (less than 250, 250–500, more than 500) and the *trap type* (sugar, scent, light). The results are summarised in the divided column graph.

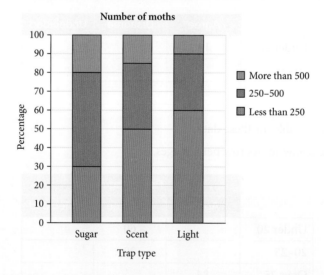

a There were 300 sugar traps. How many sugar traps caught less than 250 moths? (1 mark)

b Why does the data that exceeds 500 moths in the divided column graph support the contention that there is an association between the *number of moths* caught in a moth trap and the *trap type*? (1 mark)

9780170476959

11 © SCSA MA2016 Q6 (7 marks) Before a fitness campaign at a high school started, 50 students were chosen at random from each year group and asked the following questions:

1.2

> Question 1: Which one of the following modes of transport do you use to travel to and from school?
>
> Category A: walking/cycling
>
> Category B: public transport
>
> Category C: private car
>
> Question 2: Which year group are you in?

The campaign organisers wished to determine whether age group affected the students' likelihood of walking/cycling to and from school.

The results of the survey are shown in the table below.

	Category A	Category B	Category C	Total
Year 7	19		20	50
Year 8	12	17	21	50
Year 9	13	14	23	50
Year 10	11		21	
Year 11	10	15	25	50
Year 12	8	17	25	50
Total	73		135	300

a Copy the above table and complete the missing entries. (2 marks)

b Compare the percentages of students in Year 7 and Year 12 who use Category A as a mode of transport and comment on your results. (2 marks)

The data given in the table for part a have been displayed as a divided column graph below.

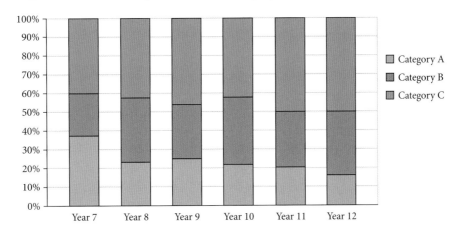

c Using the graph above or another method, comment on

   i the association between 'Year group' and 'Category A'. (1 mark)

   ii the association between 'Year group' and 'Category C'. (1 mark)

   iii the association between 'Category A' and 'Category B and C combined'. (1 mark)

12 (2 marks) The table below shows the percentage of households with and without a computer at home for the years 2017, 2019 and 2021.

	Year		
	2017	2019	2021
Households with a computer	66.4%	77.7%	84.5%
Households without a computer	33.6%	22.3%	15.5%

In the year 2019, a total of 5 170 000 households were surveyed.

a What was the number (to the nearest thousand) of households **without** a computer at home in 2019? (1 mark)

b If 6 210 000 households were surveyed in 2023 and 745 200 households did not have a computer at home, what percentage of households did have a computer at home? (1 mark)

13 (2 marks) The amount of electricity of 250 houses was rated in a survey as low, medium or high, and the size of the houses as small, standard or large. The results of the survey are displayed in the table below.

	Size of house		
Level of electricity usage	Small	Standard	Large
Low	15	14	9
Medium	22	71	11
High	15	47	46

a Name the explanatory variable. (1 mark)

b Determine the percentage of standard-sized houses rated as having a high level of electricity usage. (1 mark)

14 (4 marks) The data in Table 1 on page 15 relates to the impact of traffic congestion in 2019 on travel times in 23 cities in the United Kingdom (UK).

The four variables in this data set are:

- *city* – name of city
- *congestion level* – traffic congestion level (high, medium, low)
- *size* – size of city (large, small)
- *increase in travel time* – increase in travel time due to traffic congestion (minutes per day).

▶

Table 1

Table 1

City	Congestion level	Size	Increase in travel time (minutes per day)
Belfast	high	small	52
Edinburgh	high	small	43
London	high	large	40
Manchester	high	large	44
Brighton and Hove	high	small	35
Bournemouth	high	small	36
Sheffield	medium	small	36
Hull	medium	small	40
Bristol	medium	small	39
Newcastle-Sunderland	medium	large	34
Leicester	medium	small	36
Liverpool	medium	large	29
Swansea	low	small	30
Glasgow	low	large	34
Cardiff	low	small	31
Nottingham	low	small	31
Birmingham-Wolverhampton	low	large	29
Leeds-Bradford	low	large	31
Portsmouth	low	small	27
Southampton	low	small	30
Reading	low	small	31
Coventry	low	small	30
Stoke-on-Trent	low	small	29

Data: TomTom International BV, www.tomtom.com/en_gb/trafficindex

a Name the large UK cities with a medium level of traffic congestion. (1 mark)

b Use the data in Table 1 to copy and complete the following two-way frequency table, Table 2. (2 marks)

Table 2

Congestion level	City size	
	Small	Large
high	4	
medium		
low		
Total	16	

c What percentage of the small cities have a high level of traffic congestion? (1 mark)

(1.3) Associations between two numerical variables

Scatterplots

A **scatterplot** is used to display the results when we investigate the association between two numerical variables. A scatterplot is constructed by plotting points onto a Cartesian plane where the horizontal or x-axis is used for the explanatory variable and the vertical or y-axis is used for the response variable.

WORKED EXAMPLE 6	Interpreting scatterplots

A study was conducted on the length of time (in minutes) people of various ages spent at an exhibition at the Perth Royal Show and a scatterplot was plotted of the data.

Time spent at exhibition

(Scatterplot: Time (min) on vertical axis from 0 to 20, Age (years) on horizontal axis from 0 to 20)

Steps	Working
a What is the explanatory variable?	
The explanatory variable appears on the x-axis.	*age* (years)
b What is the response variable?	
The response variable appears on the y-axis.	*time* (min)
c How many people were in the study?	
Count the number of dots.	12 people
d What does the red dot represent?	
Read from both axes.	A 20-year-old who was at the exhibition for 15 minutes.
e How many teenagers (of ages 13 to 19) were in the study?	
Count the number of points for the ages 13 to 19.	4
f What was the longest time spent at the exhibition?	
Read from the y-axis.	20 minutes

 Exam hack

To remember which axis is used for which variable, use the fact that 'explanatory' starts with 'ex' for x-axis.

9780170476959

Scatterplots and association

There are three ways that scatterplots can be used to describe the association between two numerical variables.

1 Direction

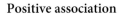

Positive association

Negative association

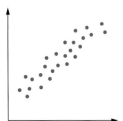 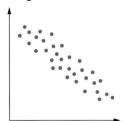

The direction of a positive association *rises* from left to right. This indicates that the y (response) variable tends to *increase* as the x (explanatory) variable increases.

The direction of a negative association *falls* from left to right. This indicates that the y (response) variable tends to *decrease* as the x (explanatory) variable increases.

Worksheets
Scatterplots and associations

A page of scatterplots

Body measurements

Relationships between variables

2 Form

Linear

Non-linear

No association

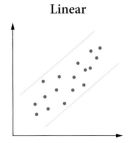  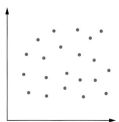

Data in general follows a straight line pattern.

Data does not occur in a straight line pattern but does follow a curved pattern.

Data is randomly scattered and shows no pattern at all.

3 Strength

Strong association

Moderate association

Weak association

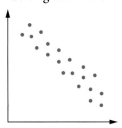  

Data points are all reasonably close together.

Data points are more spread out than a strong association.

Data points are widely spread out.

For each of the following scatterplots
  i describe the association between the two variables in terms of direction, form and strength
  ii explain what this means in terms of the variables.

a

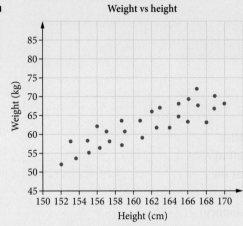

Weight vs height

b

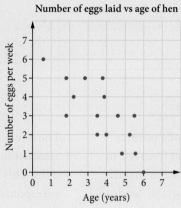

Number of eggs laid vs age of hen

c

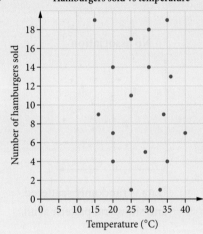

Hamburgers sold vs temperature

Steps		Working
a i Is the data sloping up or down? Does the data follow a linear pattern? How spread out are the data points?		positive, linear and strong
ii Refer to the variables.		Weight increases as height increases.
b i Is the data sloping up or down? Does the data follow a linear pattern? How spread out are the data points?		negative, linear and moderate
ii Refer to the variables.		The number of eggs a hen lays tends to decrease as the hen gets older.
c i Is the data sloping up or down? Does the data follow a linear pattern? How spread out are the data points?		no association
ii Refer to the variables.		There appears to be no association between the temperature and the number of hamburgers sold.

9780170476959

USING CAS 1 | Constructing scatterplots

The number of hot drinks sold at a café each day and the day's maximum temperature were recorded over a period of two weeks.

Temperature	35	33	27	22	18	29	38	36	24	25	29	34	21	19
No. of drinks sold	25	28	29	54	76	48	18	39	42	61	36	49	68	53

Construct a scatterplot for this information and describe the association shown in the scatterplot.

Decide first which is the explanatory variable and which is the response variable. Temperature is likely to affect the number of drinks sold, so *temperature* is the explanatory variable (*x*) and the *number of drinks sold* is the response variable (*y*).

ClassPad

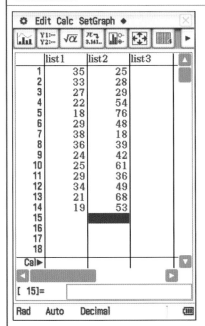 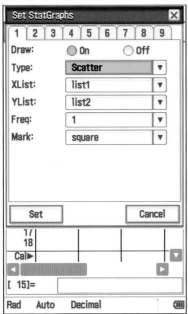 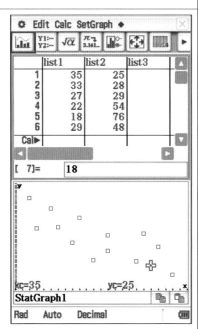

1 Tap **Menu** and open the **Statistics** application.

2 Clear all lists and enter the data from the table, as shown.

3 Tap **SetGraph > Settings** to confirm the default values shown above.

4 Tap **Set**.

5 Tap **Graph**.

6 The scatterplot will be displayed in the lower window.

7 Tap **Analysis > Trace** then use the arrow keys to display the data point values.

TI-Nspire

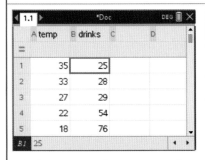 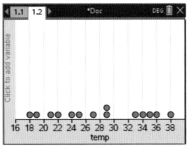 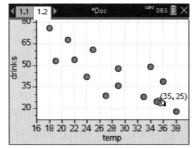

1 Start a new document and add a **Lists & Spreadsheet** page.

2 Label the columns and enter the data from the table, as shown above.

3 Insert a **Data & Statistics** page.

4 Click on the horizontal axis and select **temp**.

5 Click on the vertical axis and select **drinks**.

6 The scatterplot will be displayed.

7 Click on any data point to display the pair of values.

The association between the day's maximum temperature and the number of hot drinks sold at a café per day can be described as negative, linear and moderate.

Recap

1 A divided column graph would be an appropriate graphical tool to display the association between *month of the year* (January, February, March etc.) and the

  A *monthly average rainfall* (in millimetres).

  B *monthly mean temperature* (in degrees Celsius).

  C *annual median wind speed* (in kilometres per hour).

  D *monthly average rainfall* (below average, average, above average).

  E *annual average temperature* (in degrees Celsius).

2 Percy conducted a survey of people in his workplace. He constructed a two-way frequency table involving two variables.

One of the variables was *attitude towards shorter working days* (for, against).

The other variable could have been

  A *age* (in years).

  B *sex* (male, female).

  C *height* (to the nearest centimetre).

  D *income* (to the nearest thousand dollars).

  E *time* spent travelling to work (in minutes).

Mastery

3 WORKED EXAMPLE 6   A study was conducted on how long (in minutes) people of various ages spent at a café and a scatterplot was plotted of the data.

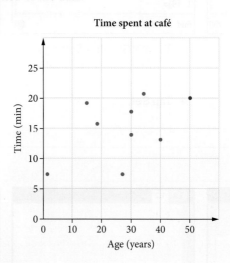

Time spent at café

  a What is the explanatory variable?

  b What is the response variable?

  c How many people were in the study?

  d What does the red dot represent?

  e How many people in their 30s were in the study?

  f Why could we reasonably conclude that the two people who stayed for the shortest time were together?

4 ⊘ WORKED EXAMPLE 7   For each of the following scatterplots

i describe the association between the two variables in terms of direction, form and strength

ii explain what this means in terms of the variables.

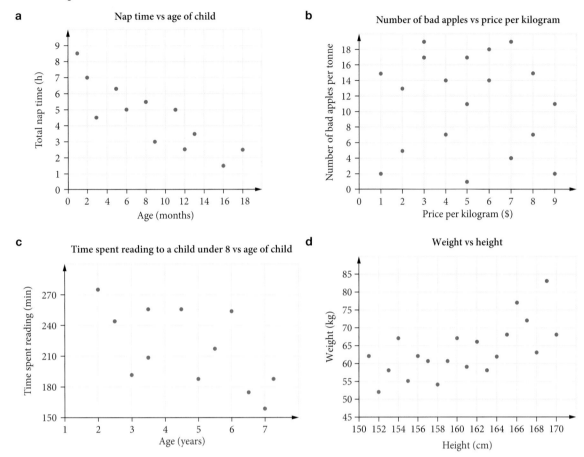

a Nap time vs age of child

b Number of bad apples vs price per kilogram

c Time spent reading to a child under 8 vs age of child

d Weight vs height

5 For each pair of variables, state whether they would generally have a positive association, a negative association or no association.

a *height* and *shoe size* of a person

b *salary level* and *lung capacity* of an employee

c *price* of a brand of car and *number of cars* of that brand sold

d *amount of time spent studying* for an exam and the *exam score*

e *number of cigarettes smoked* and *incidence of lung cancer*

f *number of police* on roads and *number of speeding cars*

6 ⊘ Using CAS 1   The hearing sensitivities (highest audible frequency measured in kilohertz) of different people from the same extended family were tested and the results are shown below.

Age	5	12	15	21	46	50	62	70	75
Frequency (kHz)	30	25	23	22	19	18	16	17	15

Use CAS to construct a scatterplot for the information and describe the association shown in the scatterplot.

7 (2 marks) What can be concluded from each of the following statements?

a There is a strong positive association between a country's Human Development Index and its carbon dioxide emissions. (1 mark)

b A large study of secondary-school male students shows that there is a negative association between the time spent playing sport each week and the time spent playing computer games. (1 mark)

8 (3 marks) Eleven people of different ages ran around an oval for 30 minutes. The number of laps they ran were recorded and the results are displayed in the scatterplot.

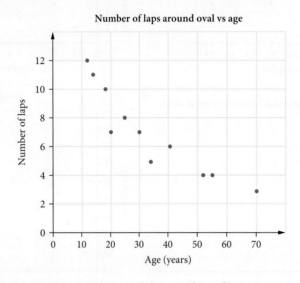

Number of laps around oval vs age

Describe the association between the age and the number of laps in terms of

a direction (1 mark)

b form (1 mark)

c strength. (1 mark)

Calculator-assumed

9 (4 marks) People at a convenience store were asked how many chocolate bars they ate on average in a month.

Age (years)	8	16	23	87	46	38	24	76	11	33	19	26	31	59	65	50
Number of chocolate bars	4	13	15	4	13	11	11	6	12	9	16	8	12	6	3	8

a Use CAS to construct a scatterplot for the information. (1 mark)

b Describe the association shown in the scatterplot. (3 marks)

9780170476959

1.4 Correlation and causation

The correlation coefficient

A scatterplot can only give us an indication of the association between two variables. To get an accurate measure, we calculate the **correlation coefficient**, r, which is a number from -1 to 1 that measures the strength and direction of *linear* associations. We will be using CAS to calculate r.

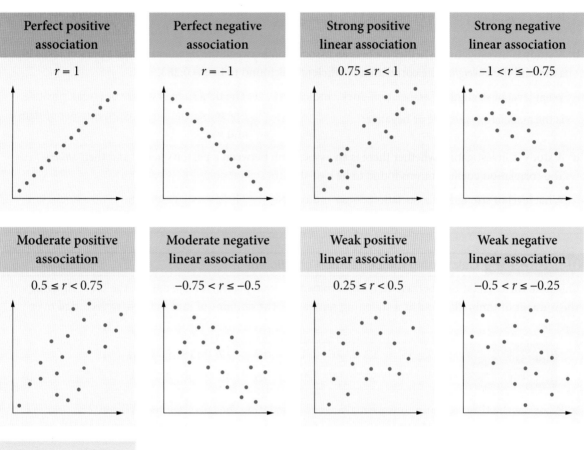

Video playlist
Correlation and causation

Perfect positive association	Perfect negative association	Strong positive linear association	Strong negative linear association
$r = 1$	$r = -1$	$0.75 \le r < 1$	$-1 < r \le -0.75$

Moderate positive association	Moderate negative linear association	Weak positive linear association	Weak negative linear association
$0.5 \le r < 0.75$	$-0.75 < r \le -0.5$	$0.25 \le r < 0.5$	$-0.5 < r \le -0.25$

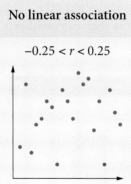

No linear association

$-0.25 < r < 0.25$

Correlation coefficient assumptions

If we are using the correlation coefficient to measure the association of two variables, we are generally making three assumptions.

1. The variables are both numerical.
2. The association is linear.
3. There are no outliers.

Interpret the correlation coefficient values and write a sentence beginning with 'The data suggests…' to describe the association for each of the following studies.

Steps	Working
a A study is investigating whether there is an association between *temperature* and the *number of heaters sold*. The correlation coefficient was found to be $r = -0.923$.	
What level of strength of association does r indicate? Is the association positive or negative?	The data suggests there is a strong negative linear association between *temperature* and the *number of heaters sold*.
b A study is investigating whether there is an association between the *number of steps* a person takes in a day and their *weight*. The correlation coefficient was found to be $r = 0.283$.	
What level of strength of association does r indicate? Is the association positive or negative?	The data suggests there is a weak positive linear association between *number of steps* and *weight*.
c A study is investigating whether there is an association between a person's *height* and their *salary*. The correlation coefficient was found to be $r = -0.031$.	
What level of strength of association does r indicate? Is the association positive or negative?	The data suggests there is no association between *height* and *salary*.

USING CAS 2 | Calculating the correlation coefficient

Calculate the correlation coefficient, correct to two decimal places, for the data in the table showing the number of umbrellas sold at a shopping centre and the amount of rain in the area recorded over 8 weeks.

Umbrellas sold	10	35	12	26	27	19	17	18
Rainfall (mm)	60	110	56	58	105	75	48	90

a What are the three assumptions we make when calculating the correlation coefficient?

b Discuss if there is an association between the number of umbrellas sold and the amount of rainfall.

Decide first which variable is the explanatory (x) and which is the response (y). Rainfall affects the number of umbrellas sold, so *rainfall* is the explanatory variable and *umbrellas sold* is the response variable.

ClassPad

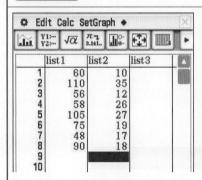

1 Tap **Menu** and open the **Statistics** application.

2 Clear all lists and enter the data from the table, as shown.

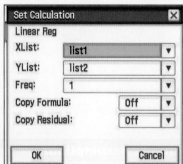

3 Tap **Calc > Regression > Linear Reg**.

4 Leave the **XList:** and **YList:** default settings of **list1** and **list2**, as shown above.

5 Tap **OK**.

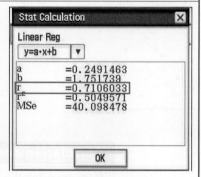

6 Tap on the **Linear Reg** dropdown menu to ensure the field is set to **y=ax+b** by tapping on the dropdown menu.

7 The r value will be displayed.

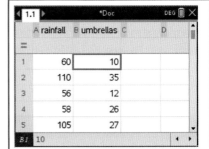
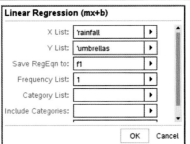

1 Start a new document and add a **Lists & Spreadsheet** page.

2 Label the columns and enter the data from the table, as shown above.

3 Press **menu > Statistics > Stat Calculations > Linear Regression (mx+b)**.

4 In the **X List:** field, select **rainfall**.

5 In the **Y List:** field, select **umbrellas**.

6 Select **OK**.

7 The linear regression labels and values will be displayed in columns **C** and **D**.

8 Scroll down to view the *r* value.

a The three assumptions are that both variables are numerical, the association is linear and there are no outliers.

b An *r* value of approximately 0.71 indicates that there is a moderate, positive, linear association between rainfall and the numbers of umbrellas sold.

🔓 Exam hack

The calculation of *r* gives the same value regardless of which variable you take to be the explanatory variable and which one to be the response variable.

Cause and effect

If two variables have a correlation or association, it doesn't necessarily mean that one *causes* the other. For example, a store recorded the following monthly sales of air conditioners and sunglasses over a year.

	Jan	Feb	Mar	Apr	May	Jun	Jul	Aug	Sep	Oct	Nov	Dec
Number of air conditioners sold	74	65	50	25	21	20	11	12	70	71	82	91
Number of sunglasses sold	310	243	180	91	63	62	35	37	223	243	266	292

These sales numbers give a correlation of $r = 0.98$ and the following scatterplot.

Number of air conditioners vs sunglasses sold

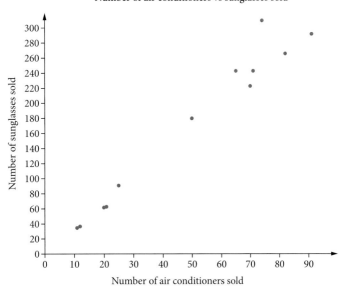

If two variables have a high correlation or association, it doesn't necessarily mean that one *causes* the other. Often when two variables have a high correlation, the cause is a third variable, which is contributing to the changes in both of them.

The correlation coefficient and scatterplot both indicate that there is an extremely strong positive correlation between the variables *number of air conditioners sold* and *number of sunglasses sold*. However, clearly neither one of these variables is *causing* the other. What is most likely happening is that another variable (*outdoor temperature*) is contributing to the changes in both variables.

Explanation for association

There are four possible explanations for an association between two variables.

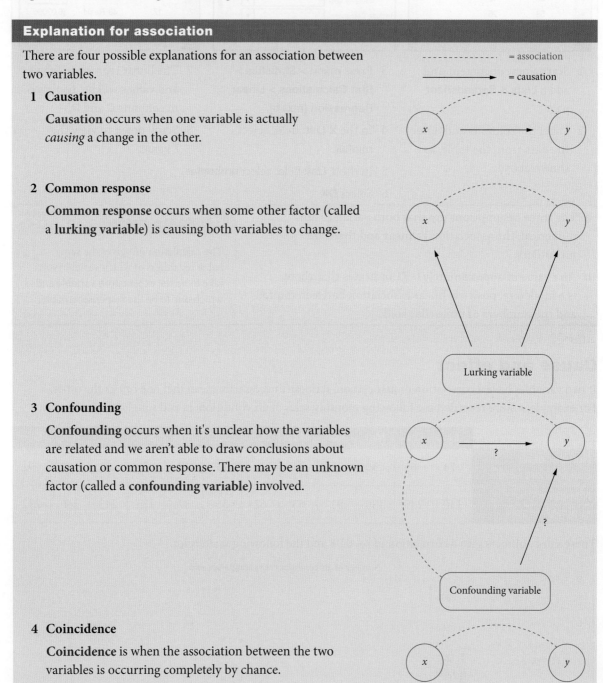

1 **Causation**

   Causation occurs when one variable is actually *causing* a change in the other.

2 **Common response**

   Common response occurs when some other factor (called a **lurking variable**) is causing both variables to change.

3 **Confounding**

   Confounding occurs when it's unclear how the variables are related and we aren't able to draw conclusions about causation or common response. There may be an unknown factor (called a **confounding variable**) involved.

4 **Coincidence**

   Coincidence is when the association between the two variables is occurring completely by chance.

Correlation measures association, not cause. It's much harder to show that one variable *causes* changes in another variable.

For each of the following correlations between pairs of variables, suggest another variable that could be the underlying cause of the correlation between the two.

Steps	Working
a A positive correlation between the *number of cars sold in Victoria* and the *number of pizza slices eaten in Victoria*.	
Which variable might be causing changes in both?	*Population* changes could be the underlying cause of the correlation between the two.
b A negative correlation between *umbrellas sold* ($) and *ice creams sold* ($).	
Which variable might be causing changes in both?	*Outdoor temperature* changes could be the underlying cause of the correlation between the two.
c A positive correlation between the *number of firefighters called out to fight a blaze* and the *level of fire damage* ($).	
Which variable might be causing changes in both?	The *size of the fire* could be the underlying cause of the correlation between the two.
d A positive correlation between *amount spent on private tutor hours per household* and *amount spent on streaming service subscriptions per household* ($).	
Which variable might be causing changes in both?	*Household income* could be the underlying cause of the correlation between the two.

1.4

WACE QUESTION ANALYSIS

© SCSA | MA2021 Q4 | Calculator-free (9 marks)

Video
WACE
question
analysis:
Bivariate data
introduction

A public opinion survey was conducted on the statement 'our overwhelming dependence on computers is a good thing', with partial results being shown in the table below.

| Age | Opinion | | | Total |
	Agree	Disagree	Undecided	
20–39 years	40	28		80
40–59 years	38		20	100
60–79 years	20		18	
Total				230

a Copy and complete the table above. (3 marks)

b Identify the response variable. (1 mark)

c Use the template below to construct a percentaged two-way frequency table showing **either** column or row percentages as appropriate, to investigate if there is an association between age and opinion. (4 marks)

| Age | Opinion | | | Total |
	Agree	Disagree	Undecided	
20–39 years				
40–59 years				
60–79 years				

d State an association that can be observed from the percentaged two-way frequency table. (1 mark)

Reading the question

- Key concepts you need to be clear on are the explanatory and response variables and determining an association from the percentaged two-way frequency table.
- Ensure you can calculate row or column percentages without a calculator.
- Highlight when you are asked for a percentage.

Thinking about the question

- Focus on key words/phrases in the context of the question.
- Which variable is affecting the other? This is the explanatory variable.
- In order to determine if an association exists, do you need to compare across the rows or down the columns?

Worked solution (✓ = 1 mark)

a The rows and columns need to add to the totals. Use addition and subtraction as required.

| | Opinion | | | |
Age	Agree	Disagree	Undecided	Total
20–39 years	40	28	$80 - 40 - 28 = \mathbf{12}$	80
40–59 years	38	$100 - 38 - 20 = \mathbf{42}$	20	100
60–79 years	20	$50 - 20 - 18 = \mathbf{12}$	18	$230 - 100 - 80 = \mathbf{50}$
Total	$40 + 38 + 20 = \mathbf{98}$	$28 + 42 + 12 = \mathbf{82}$	$12 + 20 + 18 = \mathbf{50}$	230

determines at least 3 correct entries ✓
determines at least 5 correct entries ✓
determines correctly all entries ✓

b The response variable is **opinion**. ✓

 Exam hack

Double check your answers by ensuring both rows and columns add up to the correct totals.

 Exam hack

When deciding whether a variable is explanatory or response, think about which of the two variables is the most likely to affect the other variable.

c You will need to complete the table by calculating row percentages. To calculate a row percentage, divide the value by the row total and multiple by 100.

| | Opinion | | | |
Age	Agree	Disagree	Undecided	Total
20–39 years	$\frac{40}{80} \times 100 = \mathbf{50}$	35	15	100
40–59 years	38	42	20	100
60–79 years	40	24	36	100

determines row percentages are required ✓
determines at least 3 correct entries ✓
determines at least 6 correct entries ✓
determines correctly all entries ✓

d As age increases, the percentage undecided increases. ✓

 Exam hack

Practise doing calculations without a calculator, particularly multiplying fractions or decimals and converting fractions to percentages.

 Exam hack

If the question asks, 'State an association …', you need to identify the association in terms of the context of the question.

9780170476959

Recap

Use the following information to answer Questions 1 and 2.

Consider this scatterplot, which shows the relationship between variables *M* and *H*.

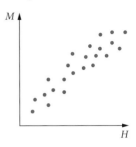

1 Which statement best describes the association between *M* and *H*?

  A no association

  B non-linear association

  C negative, linear and weak

  D positive, linear and weak

  E positive, linear and strong

2 Which of the following sentences is **correct**?

  A It can be concluded that *M* should decrease as *H* increases.

  B It has been proven that *M* will always increase as *H* increases.

  C It can be concluded that *M* is likely to increase as *H* increases.

  D There is clear evidence to support that *H* will increase as *M* increases.

  E It can be concluded that *H* is likely to increase as *M* decreases.

Mastery

3 [WORKED EXAMPLE 8] Interpret the correlation coefficient values and write a sentence beginning with 'The data suggests…' to describe the association for each of the following studies.

  a A study is investigating whether there is an association between the *amount of exercise* a person does and their *height*. The correlation coefficient was found to be $r = 0.294$.

  b A study is investigating whether there is an association between *weight* and the *number of hours a person spends sitting down*. The correlation coefficient was found to be $r = 0.528$.

  c A study is investigating whether there is an association between the *percentage of good peaches* in a container and their *number of weeks in storage*. The correlation coefficient was found to be $r = -0.910$.

  d A study is investigating whether there is an association between the number of *sales* at a local store and the *daily temperature* over a year. The correlation coefficient was found to be $r = 0.724$.

  e A study is investigating whether there is an association between the *number of pets* owned and the average *temperature* over a year. The correlation coefficient was found to be $r = -0.062$.

  f A study is investigating whether there is an association between the *number of home-cooked meals* per week and *income*. The correlation coefficient was found to be $r = -0.463$.

4 Using CAS 2 At a local cinema, people were asked their *yearly income* and the *number of times they had been to a cinema in the past year.*

Income ($000s per year)	63	125	32	19	136	49	102	67	82	25	91	42
Number of times attending a cinema in the past year	4	12	1	8	16	13	13	6	7	6	12	10

   a Calculate the correlation coefficient, correct to two decimal places, for the data in the table.

   b What are the three assumptions we make when calculating the correlation coefficient?

   c Discuss whether there is an association between *income* and the *number of times attending a cinema in the past year.*

5 Match the following scatterplots with their correlation coefficients:

    0.2   −0.9   −0.4   −0.5   0.8   0.6

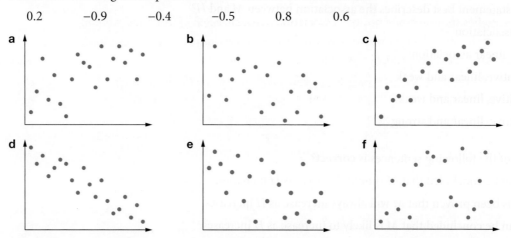

6 WORKED EXAMPLE 9 For each of the following correlations between pairs of variables, suggest another variable that could be the underlying cause of the correlation between the two.

   a A negative correlation between the *number of people who drown* and the *number of ski jackets sold.*

   b A positive correlation between the *number of weddings in Victoria* and the *total consumption of apples in Victoria* (kg).

   c A positive correlation between the *amount of money spent on eating out in a household* ($) and the *number of laptops in a household.*

   d A positive correlation between *spelling ability in children* and *foot length.*

7 Sketch a scatterplot that would have a correlation coefficient *r* closest to 0.921.

8 A study shows there is an association between secondary school results and university results. Discuss in 50 words or less what sort of association this is likely to be.

Calculator-free

9 (2 marks) For a group of 15-year-old students who regularly played computer games, the correlation between the *time spent playing computer games* and *fitness level* was found to be $r = -0.56$.

   a Is the correlation positive or negative?   (1 mark)

   b What can be concluded from this information?   (1 mark)

10 (2 marks) A worldwide study has shown that there is a strong positive correlation between the *number of cars a person owned in their lifetime* and the *age they live to.*

   a Suggest another variable that could be the underlying cause of the correlation between the two.   (1 mark)

   b Would this be considered a lurking or confounding variable?   (1 mark)

11 (7 marks) Two researchers want to find out whether taking longer strides increases running speed.

   a What are the two variables involved? (2 marks)

   The first researcher decides to look at video footage of Olympic 100 m runners, count the number of strides it takes them to complete the race, calculate the average stride length and compare this with their recorded speed.

   b What is the explanatory variable? (1 mark)

   The second researcher decides to randomly select 50 people, video and time them running 100 m races on a number of occasions under different conditions over a year, then count strides to calculate their average stride length and compare this with their speed.

   c What is the response variable? (1 mark)

   d Both studies show a strong positive association between *length of stride* and *running speed*. Which one can we reasonably say has shown that taking longer strides causes an increase in running speed? Give reasons for your answer. (2 marks)

   e List an unknown factor that could have influenced the first researchers' study. (1 mark)

12 (3 marks) Centuries ago, the people of the Hebrides, a chain of islands north of Scotland where head lice were common, were convinced that the head lice cured people who were sick with fever. They had noticed that while healthy people nearly always had head lice, sick people didn't have any.

   a What is the causal link that the people of the Hebrides were making? (1 mark)

   b What do you think was really happening to cause this association? (1 mark)

   c Explain the mistake the people of the Hebrides were making in terms of explanatory and response variables. (1 mark)

Calculator-assumed

13 (3 marks) The table shows the hourly rate of pay earned by 10 employees in a company in 2000 and in 2020.

Employee	Hourly rate of pay ($) 2000	Hourly rate of pay ($) 2020
Ben	9.53	17.02
Lani	9.15	16.71
Freya	8.88	15.10
Jill	8.60	15.93
David	7.67	14.40
Hong	7.96	13.32
Stuart	6.42	15.40
Mei Lien	11.86	19.79
Tim	14.64	23.38
Simon	15.31	25.11

   a If we wish to predict the hourly rate for an employee in 2020 from the rate in 2000, what is the response variable? (1 mark)

   b Calculate the correlation coefficient, *r*, correct to two decimal places, for the data in the table. (1 mark)

   c What level of strength of association does *r* indicate? (1 mark)

14 (3 marks) The following data was recorded from measurements made on 12 men.

Age (years)	Mass (kg)	Waist (cm)
26	84	84
29	72	74
32	67	89
32	59	75
34	97	106
37	112	114
39	67	80
40	91	101
41	98	101
43	89	94
45	117	126
51	62	82

a State the explanatory variable. (1 mark)

b Calculate the correlation coefficient, r, for mass against waist measurement correct to two decimal places, for the data in the table. (1 mark)

c What level of strength of association does r indicate? (1 mark)

15 (4 marks) The table shows the number of telephone calls made on a given day by a sample of 12 people working in a large company. Also given is the cost of each person's calls for the day.

Person	Number of calls	Cost (dollars)
A	33	4.54
B	15	1.00
C	22	5.96
D	27	4.47
E	52	8.87
F	34	8.50
G	55	11.09
H	47	8.51
I	11	3.98
J	18	2.42
K	36	11.30
L	27	7.48

a Determine the value of the correlation coefficient, r. Write your answer correct to four decimal places. (1 mark)

b What level of strength of association does r indicate? (1 mark)

c Copy and complete the following:

The value of the correlation coefficient measures the strength and direction of the _____ association between call cost and number of calls. (1 mark)

d If we wish to predict the cost of calls from the number of calls made, what is the response variable? (1 mark)

9780170476959

16 © SCSA MA2016 Q8a–c (5 marks) An experiment was conducted to determine whether there was any relationship between the maximum tidal current, in centimetres per second, and the tidal range, in metres, at a particular marine location. (The tidal range is the difference between the height of high tide and the height of low tide.) Readings were taken over a period of 12 days and the results are shown in the following table.

(1.4)

Tidal range	2.0	2.4	3.0	3.1	3.4	3.7	3.8	3.9	4.0	4.5	4.6	4.9
Maximum tidal current	15.2	22.0	25.2	33.0	33.1	34.2	51.0	42.3	45.0	50.7	61.0	59.2

a State the explanatory variable. (1 mark)

b Copy the scatterplot below and complete by plotting the last four data points and labelling the horizontal axis and the vertical axis clearly. (2 marks)

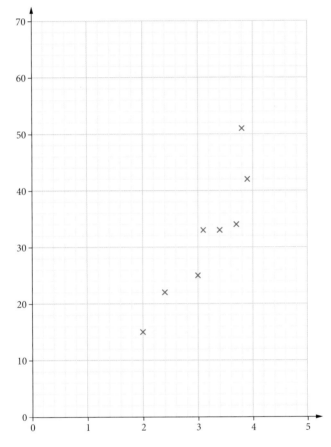

c Calculate the correlation coefficient for the data, and comment briefly on your answer with reference to the appearance of the scatterplot in part **b**. (2 marks)

Explanatory and response variables

- An **explanatory variable** is a variable that we use to predict or explain the changes observed in another variable, which is called a **response variable**.

- If the words 'explain changes' or 'predict' don't appear in the question, decide which of the two variables is the most likely to affect the other.

Associations between variables

Two variables are said to be associated when they are linked in some way.

To find whether two categorical variables are associated or not, we compare percentages in:

- percentaged two-way frequency tables

- divided column graphs.

To find whether two numerical variables are associated or not, we can use:

- scatterplots

- correlation coefficient.

Two-way tables

- **Two-way frequency tables** can be used to look at the association between two categorical variables.

- **Percentaged two-way frequency tables** gives us more information. We usually percentage the explanatory variables as either a row percentage or a column percentage.

- Information from a percentaged two-way frequency table can be displayed as a parallel divided column graph, where each cell in the table corresponds to a segment in the bar chart.

Scatterplots

There are three ways **scatterplots** can be used to describe the association between two numerical variables.

1 **Direction:** positive or negative

2 **Form:** linear, non-linear or no association

3 **Strength:** strong association, moderate association or weak association

The correlation coefficient

- The **correlation coefficient**, r, is a number from -1 to 1 that measures the strength and direction of *linear* associations.

- The calculation gives the same value regardless of which variable we take to be the explanatory variable and which to be the response variable.

It's possible to estimate the value of r from the shape of the scatterplot using these guidelines:

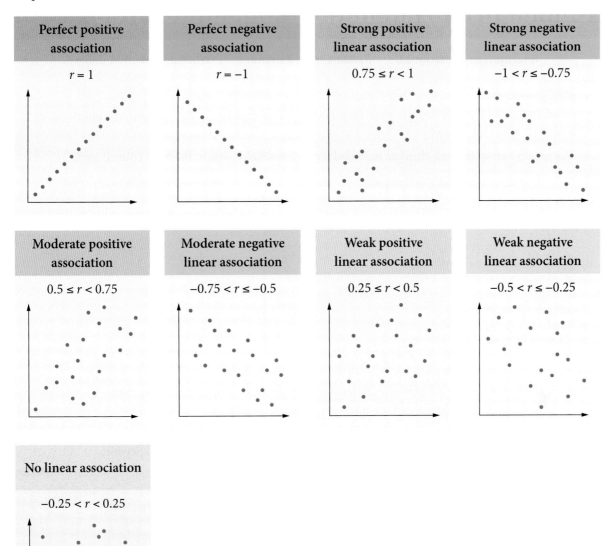

Perfect positive association	Perfect negative association	Strong positive linear association	Strong negative linear association
$r = 1$	$r = -1$	$0.75 \le r < 1$	$-1 < r \le -0.75$

Moderate positive association	Moderate negative linear association	Weak positive linear association	Weak negative linear association
$0.5 \le r < 0.75$	$-0.75 < r \le -0.5$	$0.25 \le r < 0.5$	$-0.5 < r \le -0.25$

No linear association
$-0.25 < r < 0.25$

- If we are using the correlation coefficient to measure the association of two variables, we are generally making three assumptions.
  1 The variables are both numerical.
  2 The association is linear.
  3 There are no outliers.

Causation

- Just because two variables have a high correlation or strong association, doesn't necessarily mean that one *causes* the other. There may be lurking or confounding variables present or it may be a coincidence.

Cumulative examination: Calculator-free

Total number of marks: 14   Reading time: 2 minutes   Working time: 14 minutes

1 © SCSA MA2018 Q3ad (3 marks) A local council wanted to determine whether the number of people/household, P, affected the amount of household rubbish (R, kg/day) produced and/or water (W, 100s L/day) used.

The graphs below illustrate the data recorded for 15 households within the local council area.

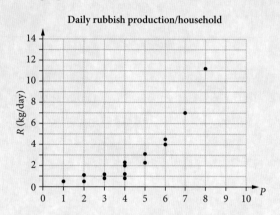

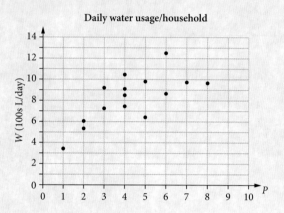

a Describe the association between the number of people/household, P, and the daily rubbish production/household, R, in terms of strength and form.   (2 marks)

b The council argued that increasing the number of people/household causes the daily water usage to increase. Provide a non-causal explanation for the association between these two variables.   (1 mark)

2 © SCSA MA2020 Q1 (5 marks) The owner of a bicycle shop recorded the type of repairs he made to bicycles with different purchase prices.

Type of repair	Purchase price			Total
	Less than $500	From $500 to $1000	Greater than $1000	
Wheels and tyres	36	6	18	60
Gears and brakes	20	12	8	40
Frame and suspension	15	2	3	20

a Identify the explanatory variable for the table above.   (1 mark)

The percentages in each row of the following table show the proportion of bicycles with different purchase prices requiring that type of repair.

b Copy and complete the table.   (2 marks)

Type of repair	Purchase price			Total
	Less than $500	From $500 to $1000	Greater than $1000	
Wheels and tyres		10		100
Gears and brakes		30	20	100
Frame and suspension	75			100

c Using the information from the table in part **b**, describe **one** association between these variables.   (2 marks)

3 ©SCSA | MA2017 Q12acde | (6 marks) The Bureau of Meteorology recorded data taken from several weather stations. The scatterplot below shows the height, h (m), of each weather station above sea level and the mean minimum temperature, t (°C), recorded at that station for the month of April.

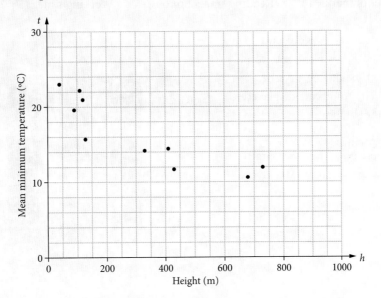

The following table provides information for three more weather stations for the month of April.

Height of weather station above sea level, h (m)	250	60	930
Mean minimum temperature, t (°C)	13.1	26.2	10.6

a Copy the scatterplot above and plot this additional information. (2 marks)

b The correlation coefficient (r) was determined for the collected data. Which value of r is most likely to be the result from the list below. (1 mark)

$r = -1.2$

$r = -0.8$

$r = -0.2$

$r = 0.5$

$r = 0.9$

c Identify whether the nature of the relationship between the height of a weather station above sea level, h (m), and the mean minimum temperature, t (°C), is linear or non-linear. (1 mark)

d A spokesperson for the Bureau of Meteorology summarised the information from parts **a–c**, saying 'It is evident that raising the height of a weather station above sea level causes the mean minimum temperature to drop'. Is this statement correct? Justify your decision. (2 marks)

Cumulative examination: Calculator-assumed

Total number of marks: 15   Reading time: 2 minutes   Working time: 15 minutes

1 (2 marks) In a large survey, Years 6, 8 and 10 girls were asked what they did (walked, sat, stood, ran) for most of the time during a typical school lunch time. The results are displayed in the divided column graph.

Does the divided column graph support the opinion that, for these girls, the lunch time activity (walked, sat or stood, ran) undertaken is associated with year level? Justify your answer by quoting appropriate percentages.

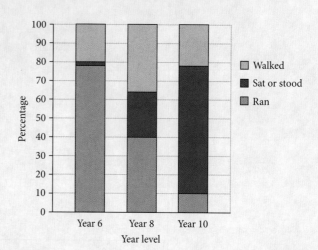

2 © SCSA   MA2018 Q10   (8 marks)

Sales of new motor vehicles in Australia, December 2016

		Passenger vehicles	Sports utility vehicles	Other vehicles	Total
Western Australia	Number of vehicles	3093	3087	1846	8026
	Percentage of vehicles	38.5%	38.5%	23.0%	100.0%
Victoria	Number of vehicles	11985	9575	*A*	27065
	Percentage of vehicles	*B*	35.4%	20.3%	100.0%
Australian Capital Territory	Number of vehicles	875	585	212	1672
	Percentage of vehicles	52.3%	35.0%	12.7%	100.0%
New South Wales	Number of vehicles	15005	11788	6839	33632
	Percentage of vehicles	44.6%	35.0%	20.3%	100.0%
Queensland	Number of vehicles	7668	6822	4762	19252
	Percentage of vehicles	39.8%	35.4%	24.7%	100.0%
South Australia	Number of vehicles	2720	2197	1547	6464
	Percentage of vehicles	42.1%	34.0%	23.9%	100.0%
Tasmania	Number of vehicles	795	648	475	1918
	Percentage of vehicles	41.4%	33.8%	24.8%	100.0%
Northern Territory	Number of vehicles	244	234	256	734
	Percentage of vehicles	33.2%	31.9%	34.9%	100.0%
Australia (Total)	Number of vehicles	42385	34936	21442	98763
	Percentage of vehicles	42.9%	35.4%	21.7%	100.0%

Note: As percentages have been rounded correctly to one decimal place, totals of percentages may not add to exactly 100%.

Use the information in the table on the previous page to answer the following questions.

a Determine the values of A and B for the Victorian data. (2 marks)

b Compare the percentage of the total new vehicle sales in Western Australia with those in South Australia. (3 marks)

c Describe the association between the number of sales of new passenger vehicles and new sports utility vehicles in Australia. (1 mark)

d Compare and comment on the percentage sales of vehicles in the Northern Territory with those in other states/territories. (2 marks)

3 © SCSA   MA2020 Q4ab MODIFIED   (5 marks) The table shows data comparing the gestation period (in days) with the birth weight (in grams) for ten Tasmanian possums.

Gestation period in days (d)	36	37	37	38	38	42	43	44	44	45
Birth weight in grams (g)	42	43	44	43	44	41	42	43	41	42

a Copy the graph below and plot the last five data points on the axes. (2 marks)

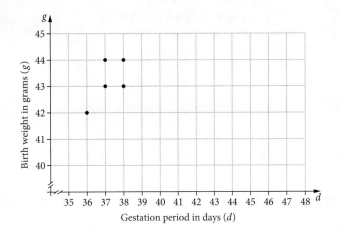

b Determine the correlation coefficient for the data to three decimal places. (1 mark)

c Describe what this correlation suggests about the general pattern of association between gestation period and birth weight. (2 marks)

CHAPTER

2

BIVARIATE DATA ANALYSIS

Syllabus coverage

TOPIC 3.1: BIVARIATE DATA ANALYSIS

Fitting a linear model to numerical data

3.1.8   identify the response variable and the explanatory variable for primary and secondary data

3.1.9   use a scatterplot to identify the nature of the relationship between variables

3.1.10   model a linear relationship by fitting a least-squares line to the data

3.1.11   use a residual plot to assess the appropriateness of fitting a linear model to the data

3.1.12   interpret the intercept and slope of the fitted line

3.1.13   use the coefficient of determination to assess the strength of a linear association in terms of the explained variation

3.1.14   use the equation of a fitted line to make predictions

3.1.15   distinguish between interpolation and extrapolation when using the fitted line to make predictions, recognising the potential dangers of extrapolation

3.1.16   write up the results of the above analysis in a systematic and concise manner

Mathematics Applications ATAR Course Year 12 syllabus p. 9 © SCSA

Video playlists (5):

2.1   The least-squares line

2.2   The coefficient of determination

2.3   Making predictions

2.4   Residual analysis

WACE question analysis Bivariate data analysis

Worksheets (7):

2.1   Lines of fit • Least-squares regression line • Height vs shoe size

2.2   Coefficient of determination

2.3   Interpolation and extrapolation

2.4   Predictions and residuals • Residual plots

Nelson MindTap

To access resources above, visit
cengage.com.au/nelsonmindtap

2.1 The least-squares line

Least-squares line

A **line of best fit** is a straight line that best shows the association between two numerical variables on a scatterplot and helps us to make predictions.

The **least-squares line** can be determined by finding the line that minimises the sum of the squares of the vertical distances (the dotted lines in the diagram) between the line and each data point in a scatterplot. These vertical distances are the differences between the predictions made by the line and the actual data points, so it makes sense that we would want to minimise these prediction errors. The vertical distances above the line are positive and the vertical distances below the line are negative.

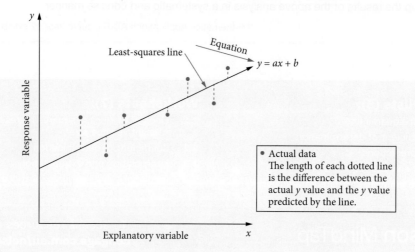

The x-axis variable is the explanatory variable and the y-axis variable is the response variable. When the equation of a least-squares line is calculated, it should be written in terms of the variable names rather than using the letters x and y.

Least-squares line

The general form of the equation for the least-squares line is

$$y = ax + b$$

where a is the **slope** or **gradient** of the line and b is the **y-intercept**.

 Exam hack

Never use rounded values in calculations. Only round for final answers.

USING CAS 1 | Finding the equation of a least-squares line

The following table shows the results of an experiment that measures the temperature of a liquid as it cools down after the source of heat is removed. Find the equation of the least-squares line, rounding the coefficients a and b to two decimal places.

Time (minutes)	5	10	15	20	25	30
Temperature (°C)	87	78	69	56	53	41

Time is the explanatory variable and *temperature* is the response variable.

ClassPad

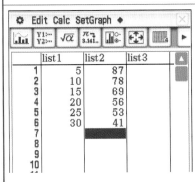  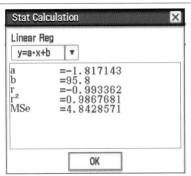

1 Open the **Statistics** application.

2 Clear all lists and enter the data from the table as shown.

3 Tap **Calc** > **Regression** > **Linear Reg**.

4 Leave the **XList:** and **YList:** default settings of **list1** and **list2** as shown above.

5 Tap **OK**.

6 Ensure the **Linear Reg** field is set to **y=ax+b** from the dropdown menu.

7 Rounding to two decimal places, $a = -1.82$ and $b = 95.8$.

TI-Nspire

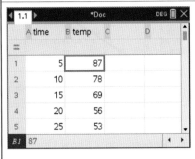 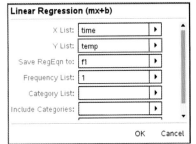 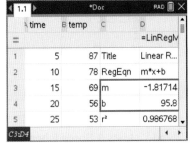

1 Start a new document and add a **Lists & Spreadsheet** page.

2 Label the columns and enter the data from the table, as shown above.

3 Press **menu** > **Statistics** > **Stat Calculations** > **Linear Regression (mx+b)**.

4 In the **X List:** field, select **time**.

5 In the **Y List:** field, select **temp**.

6 Select **OK**.

7 The linear regression labels and values will be displayed in columns **C** and **D**.

8 Rounding to two decimal places, $a = -1.82$ and $b = 95.8$.

(Note: TI uses m instead of a.)

The equation of the least-squares line is *temperature* = $-1.82 \times$ *time* + 95.8.

 Exam hack

Make sure you always write the equation for the least-squares line using the variable names. You will lose a mark if you write it using x and y.

Graph the least-squares line for the data below, which measures the temperature of a liquid on a heat source as it heats up over time.

Time (minutes)	0	2	4	6	8	10
Temperature (°C)	25	42	58	73	82	92

ClassPad

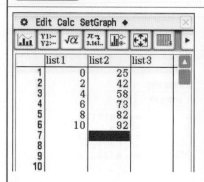

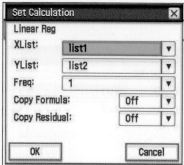

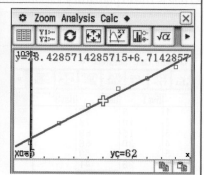

1 Open the **Statistics** application.

2 Clear all lists and enter the data from the table, as shown.

3 Tap **Calc > Regression > Linear Reg**.

4 Tap **OK**.

5 On the next screen displaying the values, tap **OK**.

6 The data points and the line of best fit will be displayed (in the lower window).

7 Tap the **Equation** tool to display the equation of the least-squares line.

TI-Nspire

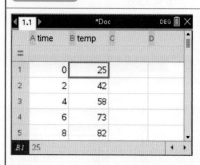

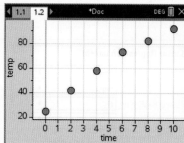

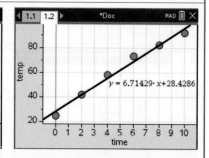

1 Start a new document and add a **Lists & Spreadsheet** page.

2 Label the columns and enter the data from the table, as shown above.

3 Insert a **Data & Statistics** page.

4 For the horizontal axis, select **time**.

5 For the vertical axis, select **temp**.

6 Press **menu > Analyze > Regression > Show Linear (mx+b)**.

7 The least-squares line will be displayed.

Interpreting the least-squares line

Let's look again at the equation of the least-squares line:

$$temperature = -1.82 \times time + 95.8$$

where the *temperature* (°C) of a liquid was measured as it cools down over *time* (minutes) after the source of heat was removed.

The y-intercept is 95.8. This tells us what the *temperature* was when we started measuring (i.e. when *time* = 0). So we know the initial temperature was 95.8°C when the source of heat was removed.

The slope is –1.82, which tells us the temperature on average decreases by 1.82°C for every 1 minute increase in time.

Interpreting the least-squares line

We interpret the equation of a least-squares line, $y = ax + b$, by saying:

- the slope is a. This means the **y variable** on average increases/decreases by **a units** for every 1 **unit** increase in the **x variable**.
- the y-intercept is b. This means the **y variable** is **b units** when the **x variable** is zero **units**.

Use the word 'increases' when a is positive and 'decreases' when a is negative.

Replace the words in **bold** with the appropriate variable names or units of measure for each variable.

x is the explanatory variable and y is the response variable.

WORKED EXAMPLE 1 | Drawing the least-squares line on a scatterplot

A study of the association between the age and price of a particular wine has resulted in the following least-squares line fitted to a scatterplot.

Age (years)	0.5	1.9	2.3	2.8	3.3	4	6.1	7
Price ($)	51	19	41	58	68	50	40	68

a Plot the points on a scatterplot.

b Determine the equation of the least-squares line using CAS, rounding a and b to three decimal places.

c Draw the least-squares line on the scatterplot by first calculating two points on the graph.

Steps	Working
a The explanatory variable is *age* and will be plotted on the x-axis, while *price* is the response variable and will be plotted on the y-axis.	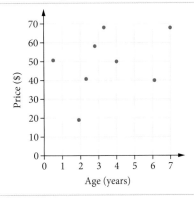
b 1 Calculate the a and b values of the equation of the least-squares line using CAS.	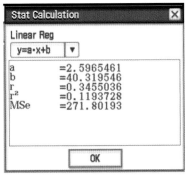
2 Write the equation of the least-squares line.	The least-squares line equation is *price* = 2.597 × *age* + 40.319.

c 1 Substitute two values into the equation in place of *age* and solve for *price*, using CAS if necessary.

$age = 1$

$price = 2.597 \times 1 + 40.319$
$= 42.9$

$age = 7$

$price = 2.597 \times 7 + 40.319$
$= 58.5$

2 On the scatterplot constructed in part **a**, draw the least-squares line by showing the two calculated points on the graph.

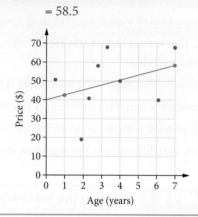

Exam hack

When finding the equation of the least-squares line, decide which is the explanatory (*x*) variable and which is the response (*y*) variable. **Don't** assume that the first variable listed is always the explanatory variable.

WORKED EXAMPLE 2 | **Interpreting the least-squares line equation**

For the following least-squares line equation

$hand\ span = 0.19 \times height - 11$

where hand span and height are measured in centimetres,

a identify and interpret the slope

b identify and interpret the *y*-intercept and comment on your result.

Steps	Working
a The slope is the value that height is multiplied by in the equation. Interpret the slope in terms of the variables and units.	The slope is +0.19. This means that on average hand span increases by 0.19 cm for every 1 cm increase in height.
b Identify and interpret the *y*-intercept and comment.	The *y*-intercept is −11. This means that hand span is −11 cm when height is 0 cm. A person of zero height can't have a −11 cm hand span. This least-squares line only applies from a certain minimum height.

9780170476959

Mastery

1 [Using CAS 1] [Using CAS 2]   The following table shows the results of an investigation by a home loan company into the association between the interest rate and the number of loan applications they had in eight consecutive years.

Interest rate (p.a.)	8.3	9.7	10.4	9.5	8.1	9.1	10.8	10.0
Number of applications	55	46	29	36	47	45	26	32

   a   Find the equation of the least-squares line.

   b   Graph the least-squares line for the data using CAS.

2 [WORKED EXAMPLE 1]   The following table shows the heights of 8 students vs their femur length.

Height (cm)	Femur length (cm)
178	50.2
173	48.4
165	45.1
164	44.6
168	45
165	42.6
155	39.9
155	38

   a   Plot the points on a scatterplot (assuming that *height* is the explanatory variable).

   b   Determine the equation of the least-squares line using CAS, rounding *a* and *b* to two decimal places.

   c   Draw the least-squares line on the scatterplot by first calculating two points on the graph.

3 [WORKED EXAMPLE 2]   A study of the association between the age (years) and price ($) of a particular car model has resulted in a least-squares line equation of

$$price = -3890 \times age + 40\,000.$$

   a   Identify and interpret the slope.

   b   Identify and interpret the *y*-intercept.

Calculator-free

4 (2 marks)   For the following least-squares line equation

$$weight = 0.7 \times height - 44$$

where height is measured in centimetres and weight is measured in kilograms

   a   identify and interpret the slope.   (1 mark)

   b   identify and interpret the *y*-intercept and comment on your result.   (1 mark)

5 (2 marks)   In a study of the association between a person's *height*, in centimetres, and *body surface area*, in square metres, the following least-squares line equation was obtained.

$$body\ surface\ area = 0.019 \times height - 1.1$$

   a   Identify and interpret the slope.   (1 mark)

   b   Identify and interpret the *y*-intercept and comment on your result.   (1 mark)

6 (3 marks) The relative humidity (%) at 9 am and 3 pm on 14 days in November 2017 is shown in the table.

A least-squares line is to be fitted to the data with the aim of predicting the relative humidity at 3 pm (*humidity 3 pm*) from the relative humidity at 9 am (*humidity 9 am*).

Relative humidity (%)	
9 am	3 pm
100	87
99	75
95	67
63	57
81	57
94	74
96	71
81	62
73	53
53	54
57	36
77	39
51	30
41	32

Data: Australian Government, Bureau of Meteorology, www.bom.gov.au

a Name the explanatory variable. (1 mark)

b Determine the values of the intercept and the slope of this least-squares line. Copy the equation below and write your answers in the appropriate boxes. Round your answer to three decimal places. (1 mark)

humidity 3 pm = [　　　　　] × *humidity 9 am* + [　　　　　]

c Determine the value of the correlation coefficient for this data set. Round your answer to three decimal places. (1 mark)

7 (2 marks) The table shows male life expectancy (*male*) and female life expectancy (*female*) for a number of countries in 2023. The scatterplot has been constructed from these data.

Life expectancy (in years) in 2023	
Male	Female
80	85
60	62
73	80
70	71
70	78
78	83
77	80
65	69
74	77
70	78
75	81
58	60
80	86
69	73
79	84
72	81
78	85
72	79
77	81
71	80

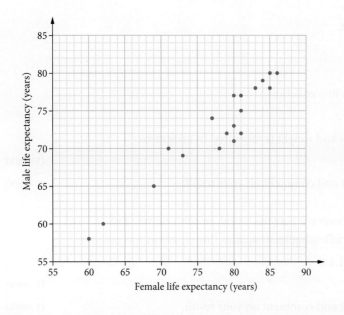

a Use the scatterplot to describe the association between *male* life expectancy and *female* life expectancy in terms of strength, direction and form. (1 mark)

b Determine the equation of a least-squares line that can be used to predict *male* life expectancy from *female* life expectancy for the year 2023. Copy and complete the equation for the least-squares line by writing the intercept and slope in the boxes. Write these values correct to two decimal places.

male = ☐ × *female* + ☐ (1 mark)

8 (4 marks) The arm spans (in cm) and heights (in cm) for a group of 13 boys have been measured. The results are displayed in the table.

The aim is to find an equation that allows arm span to be predicted from height.

Arm span (cm)	Height (cm)
152	152
153	155
174	168
141	149
170	172
165	168
163	163
155	157
165	165
152	150
143	146
156	153
174	174

a What will be the explanatory variable in the equation? (1 mark)

b Assuming a linear association, determine the equation of the least-squares line that enables *arm span* to be predicted from *height*. Write this equation in terms of the variables *arm span* and *height*, rounded to one decimal place. (2 marks)

c Using the equation that you have determined in part **b**, interpret the slope of the least-squares line in terms of the variables *height* and *arm span*. (1 mark)

(2.2) The coefficient of determination

Video playlist
The coefficient
of determination

Worksheet
Coefficient of
determination

Calculating the coefficient of determination

Now that we have used the least-squares line to model sets of data, we should ask ourselves, 'How well does our line of best fit actually represent our set of data?' To answer this question, we use the **coefficient of determination** (r^2).

The coefficient of determination

The coefficient of determination (r^2)

- can be calculated by squaring the correlation coefficient (r)
- is a value between 0 and 1
- is a measure of how useful a line of best fit is as a linear model for a particular set of data (0 means it is a totally useless measure and 1 means it is a perfect measure).

The higher the coefficient of determination

- the stronger the association between the variables
- the better the line of best fit is as a model for the data.

To calculate the coefficient of determination from a set of data, calculate the linear regression data on CAS. The r^2 value is one of the values that appears on the screen along with the values of a and b for the least-squares line equation.

USING CAS 3 | Finding the coefficient of determination

The following table shows the results of an experiment that measures the temperature of a liquid as it cools down after the source of heat is removed. Find the coefficient of determination.

Time (minutes)	5	10	15	20	25	30
Temperature (°C)	87	78	69	56	53	41

Time is the explanatory variable and *temperature* is the response variable.

ClassPad

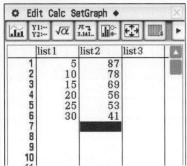  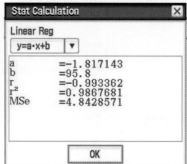

1 Open the **Statistics** application.

2 Clear all lists and enter the data from the table, as shown.

3 Tap **Calc > Regression > Linear Reg**.

4 Leave the **XList:** and **YList:** default settings of **list1** and **list2** as shown above.

5 Tap **OK**.

6 Ensure the **Linear Reg** field is set to **y=ax+b** from the dropdown menu.

TI-Nspire

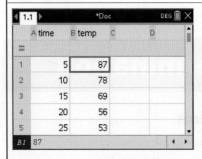 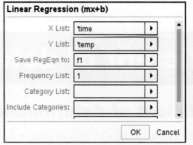 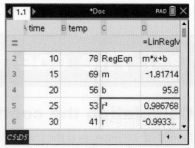

1 Start a new document and add a **Lists & Spreadsheet** page.

2 Label the columns and enter the data from the table, as shown above.

3 Press **menu > Statistics > Stat Calculations > Linear Regression (mx+b)**.

4 In the **X List:** field, select **time**.

5 In the **Y List:** field, select **temp**.

6 Select **OK**.

7 The linear regression labels and values will be displayed in columns **C** and **D**.

Rounding to two decimal places, the coefficient of determination r^2 is approximately **0.99**.

Interpreting the coefficient of determination

The coefficient of determination is usually given as a decimal between 0 and 1, but we can convert it to a percentage between 0 and 100%. This value tells us the percentage of the variation in the response variable that is explained by the explanatory variable.

For example, if $r^2 = 0.85$ for a data set comparing *height* (the explanatory variable) and *hand span* (the response variable), then we can say that 85% of the variation in hand span can be explained by the variation in height. Alternatively, we can say that 15% of the variation in the hand span is *not* explained by the variation in height.

> ### Interpreting the coefficient of determination
>
> When interpreting the coefficient of determination, the following general sentence can be used:
>
> $r^2 \times 100\%$ of the variation in the **response variable** can be explained
> by the variation in the **explanatory variable**.
>
> where the **bold text** is replaced with the appropriate percentage and variable names of the set of data being investigated.
>
> A coefficient of determination of 70% and above is considered high. A high coefficient of determination indicates that the line of best fit is an appropriate model for the data.

Exam hack

The coefficient of determination measures the predictive power of the association and is affected by outliers. Removing an outlier increases the coefficient of determination. See the example below.

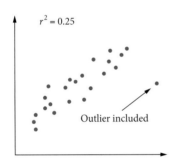

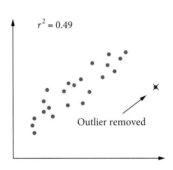

WORKED EXAMPLE 3 | Interpreting the coefficient of determination

Data was collected to investigate the association between the minimum daily temperature and the maximum daily temperature and is displayed in the table below.

Minimum temperature (°C)	12	15	13.5	14.1	10.2	11.7	13.2	11.3
Maximum temperature (°C)	23.4	25.6	23.1	25.3	20	21.1	22.6	21

a Assuming that *minimum temperature* is the explanatory variable, calculate the coefficient of determination, correct to three decimal places.

b Interpret the coefficient of determination.

c What percentage of variation in the *maximum temperature* is **not** explained by the variation in the *minimum temperature*? Round your answer to the nearest whole number.

d What is the least-squares line equation that models these data? Round *a* and *b* to one decimal place.

e Do you think that the model is appropriate? Justify your answer.

Steps	Working
a We are told the *minimum temperature* is the explanatory variable. If we weren't told, we would have to determine this. Use CAS to find r^2. The *a* and *b* values for the equation of the least-squares line will also be calculated. **ClassPad** 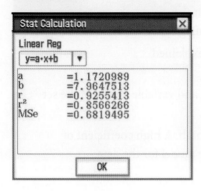 **TI-Nspire** 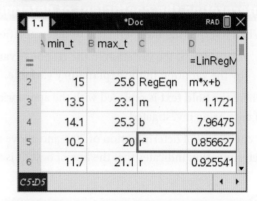 $r^2 = 0.857$	
b Calculate $r^2 \times 100\%$, then interpret the result using the general sentence.	85.7% of the variation in the *maximum temperature* can be explained by the variation in the *minimum temperature*.
c Subtract the coefficient of determination percentage from 100% and round to the nearest whole number.	$100 - 85.7 = 14.3$ 14% of variation in the *maximum temperature* is not explained by the variation in the *minimum temperature*.
d 1 Read the *a* and *b* values for the equation of the least-squares line from the screen and write the equation using these values.	$y = 1.1721x + 7.964\,75$
2 Round *a* and *b* to one decimal place.	$y = 1.2x + 8.0$
3 Replace *x* and *y* in the equation with the correct variable names.	*maximum temperature* = 1.2 × *minimum temperature* + 8.0
e Determine the appropriateness of the model, using r^2 to support your decision.	Yes, this is an appropriate model because of the high r^2 value of 85.7%.

Positive or negative association

The coefficient of determination is a squared number, so it will always be positive. It doesn't tell us about the direction (positive or negative) of the association. For example, if

$r = 0.54$,   then $r^2 = (0.54)^2$ $= 0.2916$

and also if

$r = -0.54$,   then $r^2 = (-0.54)^2$ $= 0.2916$

So if we know, for example, that $r^2 = 0.2916$ then the correlation coefficient could be either

$r = \sqrt{0.2916} = 0.54$

or

$r = -\sqrt{0.2916} = -0.54$

 Exam hack

Your calculator will most likely only give the positive square root, so remember when finding r from r^2, you need to use the extra information in the question to work out if r is positive or negative.

Finding the correlation coefficient from the coefficient of determination

If we know the coefficient of determination r^2 and want to find the correlation coefficient r, we calculate the square root *and* decide whether it is positive or negative, by looking at either

- the equation of the least-squares line to see if the slope a is positive or negative, or
- the scatterplot to see if the association is positive or negative.

WORKED EXAMPLE 4	Finding the correlation coefficient from the coefficient of determination

Find the value of the correlation coefficient, correct to two decimal places, for each of the following.

Steps	Working
a The coefficient of determination for the data displayed in the scatterplot is 0.78.	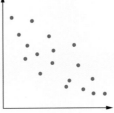
1 Calculate the square root of the coefficient of determination, rounded to the given decimal places. Remember that this value could be positive or negative.	$r = \pm\sqrt{0.78}$ $\approx \pm 0.88$
2 Is the association shown on the scatterplot positive or negative?	The association is negative.
3 The association allows us to determine if r is positive or negative.	$r = -0.88$
b The least-squares line that enables the percentage mark on a test to be determined from the number of hours spent studying is *% mark* = 9.5 × *study hours* + 15 The coefficient of determination for these data is 0.92.	
1 Calculate the square root of the coefficient of determination rounded to the given decimal places. Remember that this value could be positive or negative.	$r = \pm\sqrt{0.92}$ $\approx \pm 0.96$
2 Find the slope of the least-squares line and check if it is positive or negative.	The slope of the least-squares line is positive 9.5.
3 The sign of the slope allows us to determine if r is positive or negative.	$r = 0.96$

Recap

Use the following information to answer Questions 1 and 2.

The table lists the average life span (in years) and average sleeping time (in hours/day) of 12 animal species.

Species	Life span (years)	Sleeping time (hours/day)
baboon	27	10
cow	30	4
goat	20	4
guinea pig	8	8
horse	46	3
mouse	3	13
pig	27	8
rabbit	18	8
rat	5	13
red fox	10	10
rhesus monkey	29	10
sheep	20	4

1 Using *sleeping time* as the explanatory variable, a least-squares line is fitted to the data. The equation of the least-squares line is closest to

  A *life span* = −2.36 × *sleeping time* + 38.9

  B *life span* = −0.185 × *sleeping time* + 11.7

  C *life span* = −11.7 × *sleeping time* − 0.185

  D *sleeping time* = −0.185 × *life span* + 11.7

  E *sleeping time* = −2.36 × *life span* + 38.9

2 The value of the correlation coefficient for *life span* and *sleeping time* is closest to

  A −0.6603   **B** −0.4360   **C** −0.1901   **D** 0.4360   **E** 0.6603

Mastery

3 WORKED EXAMPLE 3   Using CAS 3   The shoe sizes and heights of some students were measured and the results are shown below.

Shoe size	8	7.5	11	6	7	7.5	7	7.5	10	8	8
Height (cm)	169	167	182	162	163	157	168	171	180	171	168

  a Assuming that *height* is the explanatory variable, calculate the coefficient of determination, correct to three decimal places.

  b Interpret the coefficient of determination.

  c What percentage of variation in the *height* is **not** explained by the variation in the *shoe size*? Round your answer to the nearest whole number.

  d What is the least-squares line equation that models these data?

  e Do you think the model is appropriate? Justify your answer.

4 ▷ **WORKED EXAMPLE 4** Find the value of the correlation coefficient, correct to two decimal places, for each of the following.

a The coefficient of determination for the data displayed in the scatterplot is 0.91.

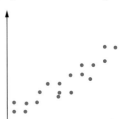

b The coefficient of determination for the data displayed in the scatterplot is 0.62.

c The least-squares line that enables the distance travelled to be determined from the time spent travelling is *distance travelled* = 1.5 × *time* + 31 and the coefficient of determination for these data is 0.88.

d The least-squares line that enables a person's score on a fitness test to be determined by their weight is *fitness test score* = −2.3 × *weight* + 142 and the coefficient of determination for these data is 0.31.

Calculator-free

5 (3 marks) A person's weight is known to be positively associated with their height.
To investigate this association for 12 men, a scatterplot is constructed as shown.

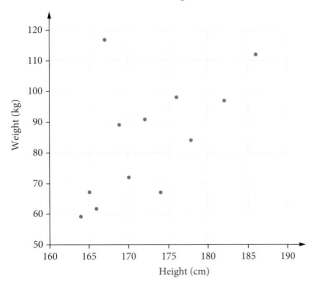

a Describe the association in terms of strength, direction and form. (1 mark)

b Identify the outlier. (1 mark)

c When a least-squares line is used to model these data, the coefficient of determination is found to be 0.3146. If the outlier is removed from the data, and a least-squares line is refitted to the data of the remaining 11 men, what will happen to the value of the coefficient of determination? (1 mark)

6 (2 marks) A least-squares line is used to model the relationship between the monthly *average temperature* and *latitude* recorded at seven different weather stations.
The equation of the least-squares line is found to be

   average temperature = −0.877 447 × *latitude* + 42.9842

a Round the numbers in this equation to three decimal places. (1 mark)

b Identify and interpret the slope. (1 mark)

7 (2 marks) The coefficient of determination for Question **6** was calculated to be 0.893 743.

    a Determine the value of the correlation coefficient, rounded to three decimal places.   (1 mark)

    b Do you think that the model is appropriate? Justify your answer.   (1 mark)

8 (3 marks) When blood pressure is measured, both the systolic (or maximum) pressure and the diastolic (or minimum) pressure are recorded. The table displays the blood pressure readings, in mmHg, that result from fifteen successive measurements of the same person's blood pressure.

Reading number	Blood pressure	
	Diastolic	Systolic
1	73	121
2	75	126
3	73	141
4	73	125
5	67	122
6	74	126
7	70	129
8	72	130
9	69	125
10	65	121
11	66	118
12	77	134
13	70	125
14	64	127
15	69	119

    a What is the explanatory variable?   (1 mark)

    b Determine the correlation coefficient, rounded to three decimal places.   (1 mark)

    c From the fifteen blood pressure measurements for this person, what is the percentage of the variation in systolic blood pressure that is explained by the variation in diastolic blood pressure?   (1 mark)

9 (2 marks) The lengths and diameters (in mm) of a sample of jellyfish were recorded and displayed in a scatterplot. The least-squares line for these data is shown.

The equation of the least-squares line is

    length = 0.87 × *diameter* + 3.5

and the correlation coefficient is *r* = 0.9034.

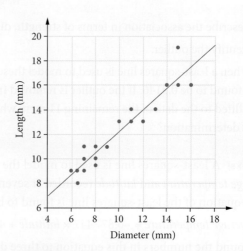

    a Determine the coefficient of determination as a percentage.   (1 mark)

    b What percentage of variation in the *length* is not explained by the variation in the *diameter*? Round your answer to the nearest whole number.   (1 mark)

10 (2 marks) The scatterplot displays the *resting pulse rate*, in beats per minute, and the *time spent exercising*, in hours per week, of 16 students. A least-squares line has been fitted to the data.

The coefficient of determination is 0.9336.

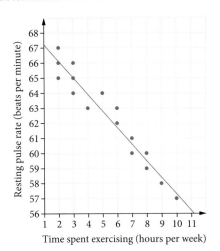

a Determine the correlation coefficient, *r*, to three decimal places. (1 mark)

b What percentage of variation in the *resting pulse rate* is explained by the variation in the *time spent exercising*? Round your answer to the nearest whole number. (1 mark)

11 (2 marks) Eighteen students sat for a 15-question multiple-choice test. In the scatterplot, the number of errors made by each student on the test is plotted against the time they reported studying for the test. A least-squares line has been determined for these data and is also displayed on the scatterplot.

The equation for the least-squares line is

number of errors = −0.120 × *study time* + 8.8

and the coefficient of determination is 0.7744.

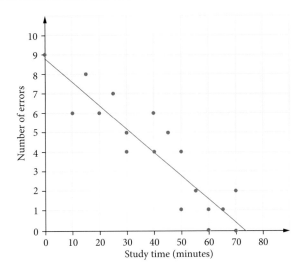

a Identify and interpret the slope. (1 mark)

b Determine the value of the correlation coefficient, *r*, for these data, correct to two decimal places. (1 mark)

12 (3 marks) The *age*, in years, *body density*, in kilograms per litre, and *weight*, in kilograms, of a sample of 12 men aged 23–25 years are shown in the table.

Age (years)	Body density (kg/L)	Weight (kg)
23	1.07	70.1
23	1.07	90.4
23	1.08	73.2
23	1.08	85.0
24	1.03	84.3
24	1.05	95.6
24	1.07	71.7
24	1.06	95.0
25	1.07	80.2
25	1.09	87.4
25	1.02	94.9
25	1.09	65.3

> ### 🔓 Exam hack
>
> When you are asked to round your answer to a certain number of decimal places, don't round anything until the very last step.

a A least-squares line is to be fitted to the data with the aim of predicting *body density* from *weight*.

   i Name the explanatory variable for this least-squares line. (1 mark)

   ii Determine the slope of this least-squares line. Round your answer to three decimal places. (1 mark)

b What percentage of the variation in *body density* can be explained by the variation in *weight*? Round your answer to the nearest percentage. (1 mark)

13 (3 marks) The scatterplot plots male life expectancy (*male*) against female life expectancy (*female*) in 1950 for a number of countries. A least-squares line has been fitted to the scatterplot as shown.

The slope of this least-squares line is 0.88.

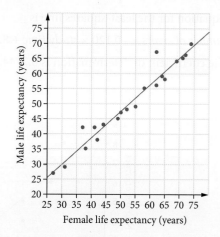

a Interpret the slope in terms of the variables *male* life expectancy and *female* life expectancy. (1 mark)

b The equation of this least-squares line is

   male = 0.88 × *female* + 3.6

In a particular country in 1950, *female* life expectancy was 35 years. Use the equation to predict *male* life expectancy for that country. (1 mark)

c The coefficient of determination is 0.95. Interpret the coefficient of determination in terms of male life expectancy and female life expectancy. (1 mark)

9780170476959

14 (2 marks) The scatterplot shows the *population density*, in people per square kilometre, and the *area*, in square kilometres, of 38 inner suburbs of a city.

For this scatterplot, $r^2 = 0.141$.

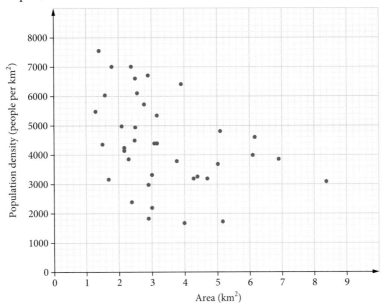

  a Describe the association between the variables *population density* and *area* for these suburbs in terms of strength, direction and form. (1 mark)

  b Determine the value of the correlation coefficient, *r*, for these data, correct to two decimal places. (1 mark)

15 © SCSA MA2021 Q9a–f (10 marks) A real estate agent is analysing data on the sale of houses over the last six months. The table shows the average sale price of houses, in thousands of dollars ($'000), and their distance from the ocean, to the nearest kilometre.

Distance from the ocean (km)	1	2	3	4	5	6	7	8	9
Average sale price ($'000)	1758	1909	2265	1934	1228	1641	751	967	676

  a State the explanatory variable. (1 mark)

  b Copy the scatterplot below, and then plot the last two data points from the table. (1 mark)

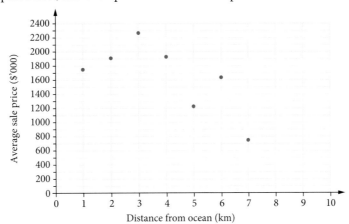

  c Determine the equation of the least-squares line for this data. (1 mark)

  d Interpret the slope of the least-squares line from part **c** in the context of this question. (2 marks)

  e **i** State the value of the correlation coefficient for these data. (1 mark)

    ii What does the correlation coefficient measure? (1 mark)

    iii Describe the association between the variables in terms of direction and strength. (2 marks)

  f What percentage of the variation in average sale price can be explained by the variation in the distance from the ocean? (1 mark)

Video playlist
Making
predictions

Worksheet
Interpolation
and
extrapolation

(2.3) Making predictions

Interpolation and extrapolation

It is possible to make predictions directly from a least-squares line graph. For example, it is possible to predict from the graph below that an interest rate of 10% will have between 30 and 31 applicants; see the red cross on the graph. However, this method is not accurate as it's difficult to determine by eye whether the prediction is closer to 30 or 31.

A better way to predict values is to use the equation of the least-squares line by substituting the given values into the equation and then solving. The equation can be used to predict *within* the original data range, which is called **interpolation**, as well as *outside* the original data range, which is called **extrapolation**.

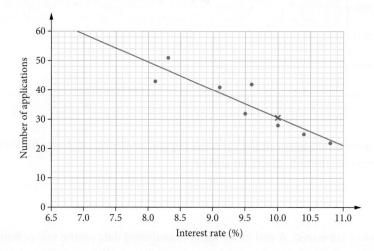

Predictions based on extrapolation are not as reliable as those based on interpolation because we can't be certain that the equation applies to values outside the range of data values we have.

WORKED EXAMPLE 5 | Making predictions from the line of best fit

Data was collected from people aged between 7 and 19 years of age, and a least-squares line was found to have the equation

$height$ (cm) = 5 × age (years) + 90

a Predict the height of a 15-year-old. Does this involve interpolation or extrapolation?

b Predict the height of a 24-year-old. Does this involve interpolation or extrapolation?

c Which of the predictions in parts **a** and **b** is more reliable? Justify your answer.

d Predict the age of a person of height 145 cm.

Steps	Working
a Substitute the value into the equation in place of *age* and solve for *height*.	$height = 5 \times age + 90$ $= 5 \times 15 + 90$ $= 165$ cm
Was the value used within or outside the original data range?	15 is within the original data range of 7–19 years, so this involves interpolation.
b Substitute the value into the equation in place of *age* and solve for *height*.	$height = 5 \times age + 90$ $= 5 \times 24 + 90$ $= 210$ cm
Was the value used within or outside the original data range?	24 is outside of the original data range of 7–19 years, so this involves extrapolation.

c Decide which of the two is the more reliable prediction and justify your decision.	The prediction for the 15-year-old involves interpolation, so it is more reliable than the prediction for a 24-year-old, which involves extrapolation.
d Substitute the value into the equation in place of *height* and solve for *age*, using CAS if necessary.	$145 = 5 \times age + 90$ $5 \times age = 145 - 90$ (subtracting 90 from both sides) $5 \times age = 55$ $age = 55 \div 5$ (dividing both sides by 5) $age = 11$ years

ClassPad

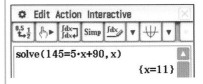

1 Tap **Interactive** > **Advanced** > **solve**.

2 In the dialogue box, enter the equation in the **Equation:** field.

3 In the **Variable field**, keep the default variable **x**.

4 Tap **OK**.

TI-Nspire

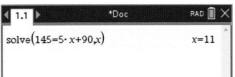

1 Press **menu** > **Algebra** > **Solve**.

2 Enter the equation followed by **,x**, where *x* represents the variable *age*.

3 Press **enter**.

Therefore, a person with a height of 145 cm is predicted to be 11 years old.

EXERCISE 2.3 Making predictions

ANSWERS p. 429

Recap

Use the following information to answer Questions 1 and 2.

The average rainfall and temperature range at several different locations in the South Pacific region are displayed in the scatterplot.

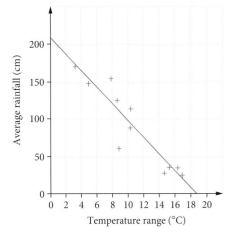

1 Describe the association between the variables *temperature* and *average rainfall* for these locations in terms of strength, direction and form.

2 The value of the correlation coefficient, *r*, for the data, is $r = -0.9260$. The value of the coefficient of determination is

  A -0.8575 **B** -0.9260 **C** 0.8575 **D** 0.9260 **E** 0.9623

3 The least-squares line with equation

number of laps = −0.16 × *age* + 10.87

models the association between the number of laps run around an oval in 30 minutes and the age of the runner (in years). Use the model to predict, correct to two decimal places, the number of laps run in 30 minutes by a runner aged

a 12

b 34

c 55

> 🔓 **Exam hack**
>
> When stating the predicted value, make sure that you include the units of measure as well.

4 The least-squares line with the equation given below models data relating the outside temperature (°C) to the amount of gas consumed (kWh) by a household over a three-month period.

gas consumed = −125 × *outside temperature* + 2212

Use the model to predict the outside temperature, to the nearest degree, when the gas consumed is

a 2000 kWh

b 1200 kWh

c 0 kWh

5 WORKED EXAMPLE 5 Data was collected from people aged between 5 and 15 years of age, and a least-squares line was found to have the equation

height (cm) = 4 × *age* (years) + 95

a Predict the height of an 11-year-old. Does this involve interpolation or extrapolation?

b Predict the height of a 22-year-old. Does this involve interpolation or extrapolation?

c Which of the predictions in parts **a** and **b** is more reliable? Justify your answer.

d Predict the age of a person of height 151 cm.

Calculator-free

6 (5 marks) The weight (in grams) of boxes of chocolates containing between 15 and 50 chocolates were recorded. The least-squares line for the data was found to have the equation

weight = 5 × *number of chocolates* + 20

a Predict the weight of a box containing 25 chocolates. Does this involve interpolation or extrapolation? (1 mark)

b Predict the weight of a box containing 5 chocolates. Does this involve interpolation or extrapolation? (1 mark)

c Predict the weight of a box containing 70 chocolates. Does this involve interpolation or extrapolation? (1 mark)

d Which of these predictions is the most reliable? Justify your answer. (1 mark)

e Predict the number of chocolates that would be in a box weighing 125 grams. (1 mark)

7 (7 marks) A least-squares line modelling the height (in cm) of 20 children aged between 5 and 15 years was found to have the equation

$height = 2.59 \times age + 99.97$

a What is the predicted height for a 14-year-old? (1 mark)

b What is the predicted age of a child who is 120.69 cm? (1 mark)

c State whether the following are true or false. (5 marks)

    i Predicting the height of a child aged 9 years is an example of interpolation.

    ii For every one year of increase in age, the model predicts a child's height increases by 2.59 cm.

    iii Predicting the height of a 6-year-old will be more reliable than predicting the height of a 12-year-old.

    iv A child's height will increase as their age increases.

    v Predicting the height of a 10-year-old is more likely to be reliable than predicting the height of an 18-year-old.

8 (3 marks) In 1975, the life expectancies in Australia and the UK were very similar. From 1975, the gap between the life expectancies in the two countries increased, with people in Australia having a longer life expectancy than people in the UK. To investigate the difference in life expectancies, least-squares lines were fitted to the data for both Australia and the UK for the period 1975 to 2010. The results are shown.

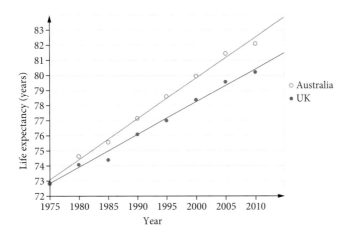

The equations of the least-squares lines are as follows:

Australia: $life\ expectancy = 0.2657 \times year - 451.7$

UK: $life\ expectancy = 0.2143 \times year - 350.4$

a Use these equations to predict the difference between the life expectancies of people living in Australia and people living in the UK in 2030. Give your answer correct to the nearest year. (2 marks)

b Explain why this prediction may be of limited reliability. (1 mark)

9 (4 marks) In the scatterplot, average annual *female income*, in dollars, is plotted against average annual *male income*, in dollars, for 16 countries. A least-squares line is fitted to the data.

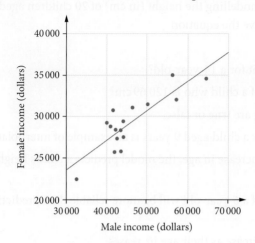

The equation of the least-squares line for predicting female income from male income is

female income = 0.35 × *male income* + 13 000

a What is the explanatory variable? (1 mark)

b Copy and then complete the following statement by filling in the missing information.

From the least-squares line equation, it can be concluded that, for these countries, on average, female income increases by $ _____ for each $1000 increase in male income. (1 mark)

c **i** Use the least-squares line equation to predict the average annual female income (in dollars) in a country where the average annual male income is $15 000. (1 mark)

ii The prediction made in part **c i** is not likely to be reliable. Explain why. (1 mark)

 Exam hack

For 'explain' questions, refer only to the information in the question. Don't offer your own personal views.

10 © SCSA MA2021 Q9g–i (5 marks) A real estate agent is analysing data on the sale of houses over the last six months. The table shows the average sale price of houses, in thousands of dollars ($000s), and their distance from the ocean, to the nearest kilometre.

Distance from the ocean (km)	1	2	3	4	5	6	7	8	9
Average sale price ($'000)	1758	1909	2265	1934	1228	1641	751	967	676

a In 6 months' time, a homebuyer will have saved enough money for a deposit on a house. He would like to live about four kilometres from the ocean. The equation of the least-squares line is $p = -174.58d + 2331.7$.

i Use the equation of the least-squares line to predict the average sale price of houses four kilometres from the ocean. (1 mark)

ii Explain why your prediction is different from the average sale price given in the table. (1 mark)

b Give a reason why extrapolation in the context of this question would not make sense. (1 mark)

c The real estate agent was talking to some potential buyers and was heard to make the statement, 'Having property closer to the ocean causes higher selling prices'. Comment on this statement. (2 marks)

9780170476959

2.4 Residual analysis

Residual values

As we have seen, it is important to know whether the association between the variables is linear or not. We have used the following to help determine whether the association is linear:

- the form of the scatterplot
- fitting a least-squares line
- seeing how close the correlation coefficient, r, is to 1 or −1
- the value of the coefficient of determination, r^2.

However, one of the best ways to test the assumption that the association is linear is to calculate the residual values.

A **residual** is the vertical distance between each data point and the least-squares line.

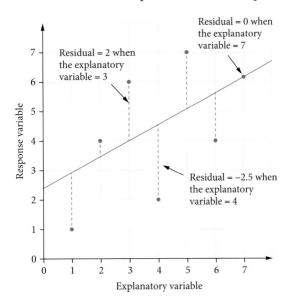

Video playlist
Residual
analysis

Worksheet
Predictions
and residuals

Residual values and actual values

residual value = actual value y – predicted value $\hat{y}$

The actual value can be read from a scatterplot or a table.

The predicted value needs to be calculated from the least-squares line.

Actual values that lie

- *above* the least-squares line will have a positive residual value
- *below* the least-squares line will have a negative residual value
- *on* the least-squares line will have a residual value of zero.

The further the actual value is from the least-squares line, the larger the residual value and the less accurate the prediction.

WORKED EXAMPLE 6 | Finding residual values from the least-squares line

Find the residual values for each of the following.

Steps	Working

a A study was conducted to establish the association between the age of a child (years) and the cost of their favourite toy ($). Find the residual values for the following directly from the graph.

   i Elke who is $1\frac{1}{2}$ years old.

   ii Ethan who is 6 years old.

   iii Cara, the 9-year-old with the cheapest favourite toy.

   iv Carson, the 9-year-old with the most expensive favourite toy.

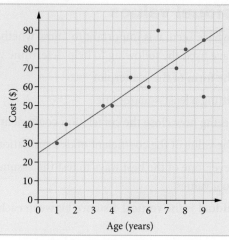

Read the vertical distance from the point to the least-squares line from the graph. The value is negative if it is below the line and zero if it is on the line. Include the response variable's units.

  i residual value for Elke = $5.00

  ii residual value for Ethan = −$5.00

  iii residual value for Cara = −$30.00

  iv residual value for Carson = $0.00

b For a study conducted to establish the association between the age of children (years) and their height (cm), the least-squares line equation was found to be

   $height = 2.5 \times age + 100$

Find the residual values for the following, correct to one decimal place.

  i Martha who is 12 years old and 142 cm tall.

  ii Mo who is 11 years old and 118 cm tall.

1 Calculate the predicted height by substituting the age into the least-squares line equation.

  i Martha's predicted *height*
    $= 2.5 \times age + 100$
    $= 2.5 \times 12 + 100$
    $= 130\,cm$

  ii Mo's predicted *height*
    $= 2.5 \times age + 100$
    $= 2.5 \times 11 + 100$
    $= 127.5\,cm$

2 Find the residual value using the formula:

residual value = actual value y − predicted value $\hat{y}$

Include the response variable's units.

  i residual value $= y - \hat{y}$
    $= 142 - 130$
    $= 12\,cm$

  ii residual value $= y - \hat{y}$
    $= 118 - 127.5$
    $= -9.5\,cm$

9780170476959

Create a table of residual values for the set of data on the right, stating all answers correct to one decimal place.

x	1	2	3	4	5	6	7
y	1	4	6	2	7	4	5

ClassPad

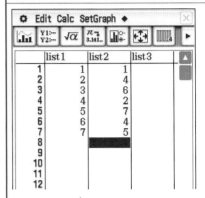

1 Open the **Statistics** application.

2 Clear all lists and enter the data from the table, as shown.

3 Tap **Calc** > **Regression** > **Linear Reg**.

4 Leave the **XList:** and **YList:** default settings of **list1** and **list2**.

5 Change the **Copy Residual:** field from **Off** to **list3**.

6 Tap **OK**.

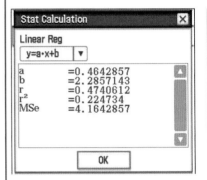

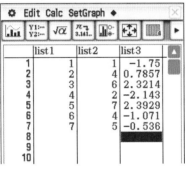

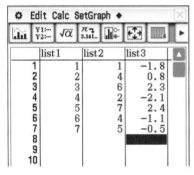

7 The linear regression results will appear as before.

8 Tap **OK**.

9 Tap **Menu** and open the **Statistics** application again.

10 Tap in the upper window and select **Resize** to expand it.

11 Scroll left and the residual values will be displayed in **list3**.

12 Change the **Basic Format** > **Number Format** to **Fix 1** to display the residual values to one decimal place.

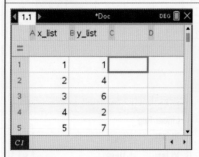

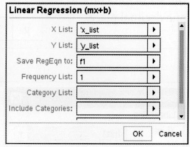

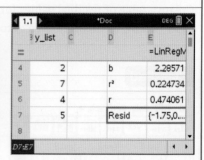

1 Start a new document and add a **Lists & Spreadsheet** page.

2 Label the columns and enter the data from the table as shown above (avoid using x and y for list names).

3 Place the cursor in column **C**, which will be used later for the residual values.

4 Press **menu > Statistics > Stat Calculations > Linear Regression (mx+b)**.

5 In the **X List:** field select **x_list**.

6 In the **Y List:** field select **y_list**.

7 Press **OK**.

8 Scroll down column **E** to view the residual values, which are now in a list. We will now link this list to column **C**.

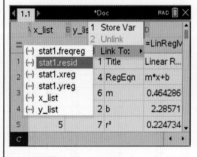

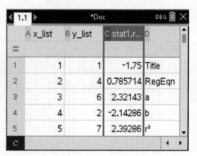

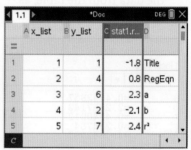

9 Place the cursor in the column **C** heading.

10 Press **var**.

11 Select **Link To:**.

12 From the dropdown menu, select **stat1.resid**.

13 The residual values will be displayed in column **C**.

14 Change the **Document Settings > Display Digits** to **Fix 1** to display the residual values to one decimal place.

Residual plots

Once all of the residual values have been found, a **residual plot** can be constructed. A residual plot is like a scatterplot with the explanatory variable on the *x*-axis and the residual values on the *y*-axis. Residual values are negative as well as positive so the residual plot will always have both positive and negative *y*-axes.

 Exam hack

Think of the zero line in a residual plot as the line of best fit in a scatterplot made horizontal.

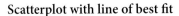

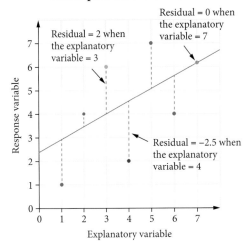

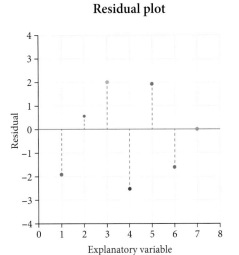

Most residual plots will fall into one of the following three types.

Type 1: Randomly scattered

The residual values are randomly scattered above and below the *x*-axis. This lack of a clear pattern suggests that the association between the explanatory and response variables is *linear*.

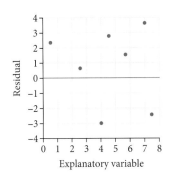

Exam hack

For residual plots, a random scattering indicates that there is a *linear association* between explanatory and response variables. For a scatterplot, a random scattering indicates there is *no association* between the explanatory and response variables. Don't get the two confused.

Type 2: Hill

The residual values show a hill shape. This suggests that the association between the explanatory and response variables is *non-linear*.

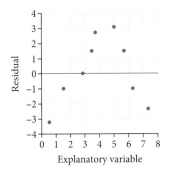

Type 3: Valley

The residual values show a 'u' or 'v' valley shape. This also suggests that the association between the explanatory and response variables is *non-linear*.

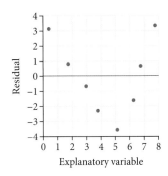

Worksheet
Residual plots

Graph a residual plot for the data on the right and use it to decide whether the data being investigated is linear or non-linear.

x	1	2	3	4	5	6	7
y	1	4	6	2	7	4	5

ClassPad

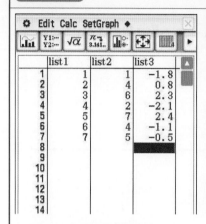 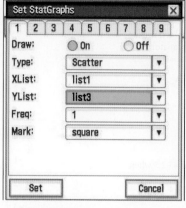 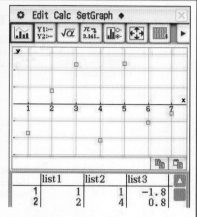

1 Open the **Statistics** application and enter the data (or use the previous data from Using CAS 4).

2 Insert the residual values into **list3**.

3 Tap **SetGraph > Setting**.

4 Keep the default settings except change the **YList:** field to **list3**.

5 Tap **Set**.

6 Tap **Graph**.

7 The graph of the residual values will appear in the lower window (note: the windows have been swapped).

TI-Nspire

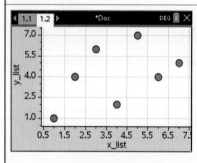 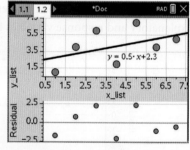

1 Insert a **Lists & Spreadsheet** application and enter the data into **x_list** and **y_list** (or use the previous data from Using CAS 4).

2 Insert a **Data & Statistics** application.

3 For the horizontal axis, select **x_list**.

4 For the vertical axis, select **y_list**.

5 Press **menu > Analyze > Regression > Show Linear (mx+b)** (note the mode setting is **Fix 1**).

6 Press **menu > Analyze > Residuals > Show Residual Plot**.

7 The residual plot will be displayed below the scatterplot.

The data is probably linear because the residual values appear randomly scattered above and below the x-axis.

MA2018 Q15 Calculator-assumed (12 marks)

2.4

Ali is researching mobile phone carriers and has found several plans with monthly contracts. The table below shows the data allowance, GB (d) and the monthly cost \$ (C), of ten plans that he is considering.

Data allowance GB (d)	10	2.5	0.5	15	5	1	6	6	25	10
Monthly cost \$ (C)	70	50	35	135	50	55	95	38	195	80

The graphs below show a scatterplot and a residual plot for the information in the table, with two points missing on both graphs.

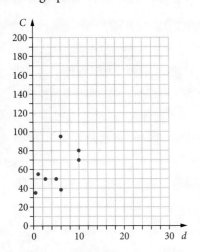

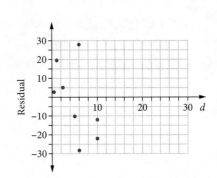

a Copy the scatterplot and then on it plot the two missing points. (2 marks)

b **i** Determine the equation of the least-squares line for the information in the table and state the correlation coefficient. (2 marks)

  ii Describe the linear association between Data allowance and Monthly cost. (2 marks)

  iii Approximately how much does the cost change for every additional GB of data allowance? (1 mark)

  iv What percentage of the variation in monthly cost can be explained by the variation in the data allowance? (1 mark)

c **i** Copy the residual plot, calculate the two missing residuals and include them on the residual plot. (2 marks)

  ii What feature of the residual plot indicates that a linear model would be appropriate for the data? (1 mark)

d Predict the monthly cost of a plan with a data allowance of 20 GB. (1 mark)

Video
WACE question analysis: Bivariate data analysis

Reading the question

- Key concepts you need to be clear on are the equation of the least-squares line, calculating residual values and plotting points on a residual.
- Understand the difference between correlation coefficient and correlation of determination.
- Take note of the 2-mark questions.

Thinking about the question

- Some 2-mark questions require a single answer with two steps, while others require two separate answers.
- The formulas for the residual value and the least-squares line are on the formula sheet.
- A reference to 'what percentage of the variation ...' means that the coefficient of determination needs to be calculated.

Worked solution (✓ = 1 mark)

a Identify the missing points on the scatterplot and plot them correctly.

(15, 135) ✓ and (25, 195) ✓

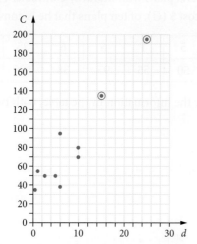
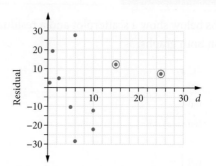

b i Using CAS, determine the equation of the least-squares line and correlation coefficient.

$C = 6.30d + 29.25$ ✓ and $r = 0.932$ ✓

> 🔒 **Exam hack**
>
> Ensure you use the variables given in the question.

ii Describe the linear association in terms of strength and direction. **Positive** ✓ and **strong** ✓ as the correlation coefficient is strong.

iii The cost changes by **$6.30** ✓ for every 1 GB change as shown in the slope that was calculated in the least-squares line equation.

iv Using the correlation of determination, the percentage of the variation in monthly cost can be explained by the variation in the data allowance. Approximately **87%** ✓

c i residual value = actual value − predicted value

Determine the predicted value using CAS. When $d = 15$, $C = 123.79$.

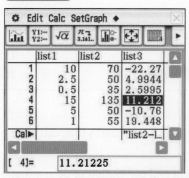

The residual value for 15 is 135 − 123.79 = 11.21.

The predicted value for $d = 25$ is $C = 186.81$ using CAS.

The residual value for 25 is 195 − 186.81 = 8.19.

calculates correct residuals ✓

correctly plots residuals on graph ✓

ii Plots are random, i.e. no pattern evident ✓

d Determine the predicted value using CAS when $d = 20$, $C = \$155.30$ ✓

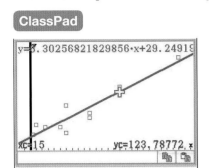

ClassPad

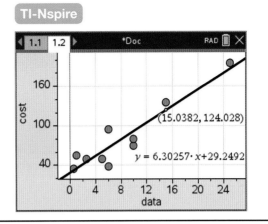

TI-Nspire

EXERCISE 2.4 Residual analysis

ANSWERS p. 429

Recap

Use the following information to answer Questions 1 and 2.

The weights (in g) and lengths (in cm) of 12 fish were recorded and plotted in the scatterplot. The least-squares line that enables the weight of these fish to be predicted from their length has also been plotted.

1 The least-squares line predicts that the weight (in grams) of a fish of length 30 cm would be closest to

  A 240

  B 252

  C 262

  D 274

  E 310

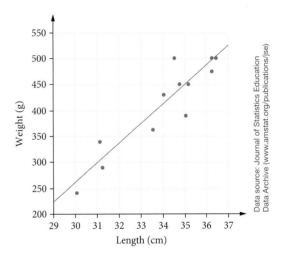

Data source: Journal of Statistics Education
Data Archive (www.amstat.org/publications/jse)

2 The least-squares line predicts that the length of a fish of weight 450 g would be closest to

  A 33 cm   **B** 34 cm   **C** 35 cm   **D** 36 cm   **E** 37 cm

Mastery

3 WORKED EXAMPLE 6

  a A study was conducted to establish the association between the time (minutes) spent studying for a test and the percentage mark for the test. Find the residual values directly from the graph for

  i Selby who studied for 25 minutes for the test

  ii Surinam who studied for 60 minutes for the test

  iii Normie who studied for over 90 minutes for the test and achieved 65%

  iv Nissal who also studied for over 90 minutes for the test and achieved less than 50%.

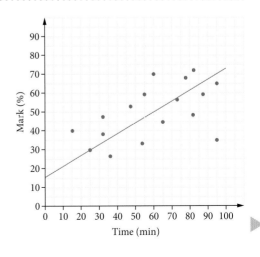

b For a study conducted to establish the association between the height of a building and the number of levels, the least-squares line equation was found to be

$$height = 0.57 \times number\ of\ levels + 5$$

Find the residual values, correct to one decimal place, using the equation for a building with

  i 12 levels that is 12.3 metres tall

  ii 5 levels that is 7.2 metres tall.

🔒 **Exam hack**

Don't lose easy marks. *Always* check whether the question has asked for a specific number of decimal places.

4 For each of the following residual plots, state whether it suggests a linear or non-linear relationship. Provide a reason for your answer.

a

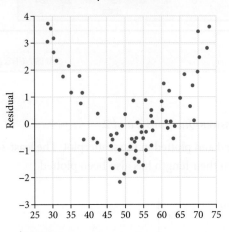

b

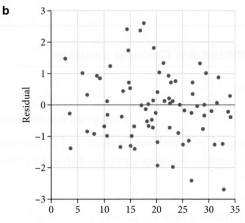

c

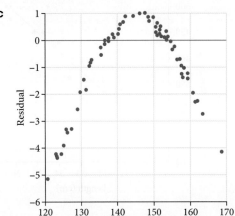

d

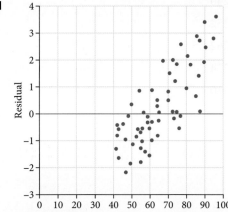

5 🔗 Using CAS 4   🔗 Using CAS 5   Create a table of residual values for the following set of data (assuming that *height* is the explanatory variable), giving all answers correct to one decimal place. Graph a residual plot and use it to decide whether the data being investigated is linear or non-linear.

Height (cm)	Femur length (cm)
178	50.2
173	48.4
165	45.1
164	44.6
168	45
165	42.6
155	39.9
155	38

6 (3 marks)

a Does the below residual plot suggest a linear association? (1 mark)

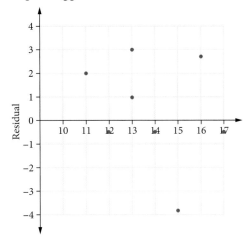

b Using the residual plot in part **a**, determine the residual value when $x = 11$. (1 mark)

c Explain why the residual plot does not match the scatterplot and least-squares line below, referring specifically to the data value at $x = 11$. (1 mark)

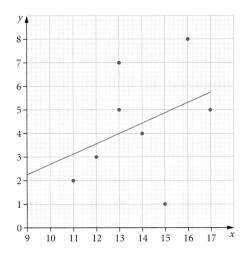

7 (2 marks)

a Graph a residual plot for the following data. (1 mark)

b Decide whether the data being investigated is linear or non-linear. (1 mark)

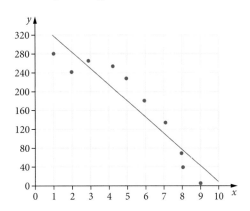

8 (2 marks) The scatterplot displays the *resting pulse rate*, in beats per minute, and the *time spent exercising*, in hours per week, of 16 students. A least-squares line has been fitted to the data.

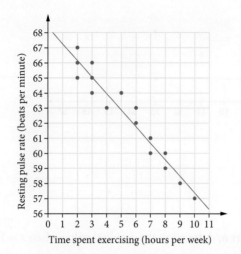

Use this least-squares line to model the association between resting pulse rate and time spent exercising.

a Graph a residual plot using CAS. (1 mark)

b Is the data being investigated linear or non-linear? (1 mark)

9 (3 marks) The *height* (in cm) and *foot length* (in cm) for each of eight Year 12 students were recorded and displayed in the scatterplot below. A least-squares line has been fitted to the data as shown.

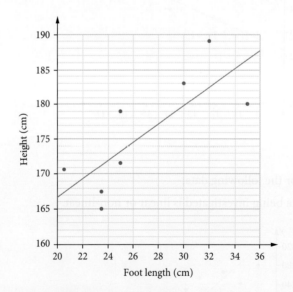

a By inspection, what is the value of the correlation coefficient (r)? (1 mark)

b Graph a residual plot using CAS. (1 mark)

c Is the data being investigated linear or non-linear? (1 mark)

10 (7 marks) The scatterplot shows the atmospheric pressure, in hectopascals (hPa), at 3 pm (*pressure 3 pm*) plotted against the atmospheric pressure, in hectopascals, at 9 am (*pressure 9 am*) for 23 days in November 2017 at a particular weather station.

A least-squares line has been fitted to the scatterplot as shown.

The equation of this line is

   pressure 3 pm = 0.8894 × *pressure 9 am* + 111.4

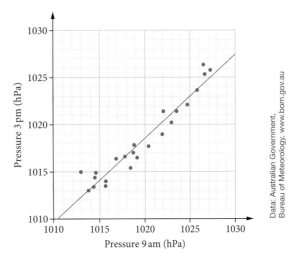

a Interpret the slope of this least-squares line in terms of the atmospheric pressure at this weather station at 9 am and at 3 pm. (1 mark)

b Use the equation of the least-squares line to predict the atmospheric pressure at 3 pm when the atmospheric pressure at 9 am is 1025 hPa.

Round your answer to the nearest whole number. (1 mark)

c Is the prediction made in part **b** an example of extrapolation or interpolation? (1 mark)

d Determine the residual when the atmospheric pressure at 9 am is 1013 hPa. Round your answer to the nearest whole number. (1 mark)

e The correlation coefficient for these data, rounded to three decimal places, is $r = 0.966$. What percentage of the variation in *pressure 3 pm* is explained by the variation in *pressure 9 am*? Round your answer to one decimal place. (1 mark)

 Exam hack

When the word 'use' appears in a question, make sure you use the method suggested even if it's possible to get the answer another way.

f The residual plot associated with the least-squares line is shown.

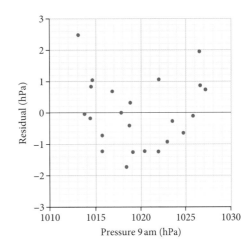

i The residual plot above can be used to test one of the assumptions about the nature of the association between the atmospheric pressure at 3 pm and the atmospheric pressure at 9 am. What is this assumption? (1 mark)

ii The residual plot above does not support this assumption. Explain why. (1 mark)

11 (8 marks) The data in the table shows a sample of actual temperatures and apparent temperatures recorded at a weather station. A scatterplot of the data is also shown. The data will be used to investigate the association between the variables *apparent temperature* and *actual temperature*.

Apparent temperature (°C)	Actual temperature (°C)
24.7	28.5
24.3	27.6
24.9	27.7
23.2	26.9
24.2	26.6
22.6	25.5
21.5	24.4
20.6	23.8
19.4	22.3
18.4	22.1
17.6	20.9
18.7	21.2
18.2	20.5

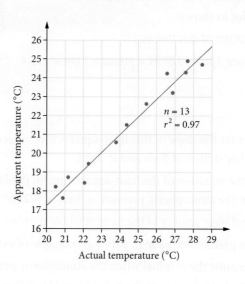

a Use the scatterplot to describe the association between *apparent temperature* and *actual temperature* in terms of strength, direction and form. (1 mark)

b **i** Determine the equation of the least-squares line that can be used to predict the *apparent temperature* from the *actual temperature*. Copy and complete the following by adding the values of the intercept and slope of this least-squares line in the appropriate boxes. Round your answers to two decimal places.

apparent temperature = ☐ × *actual temperature* + ☐ (3 marks)

ii Interpret the intercept of the least-squares line in terms of the variables *apparent temperature* and *actual temperature*. (1 mark)

c The coefficient of determination for the association between the variables *apparent temperature* and *actual temperature* is 0.97. Interpret the coefficient of determination in terms of these variables. (1 mark)

d The residual plot obtained when the least-squares line was fitted to the data is shown.

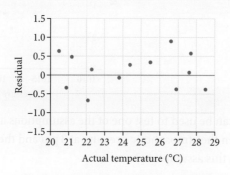

i A residual plot can be used to test an assumption about the nature of the association between two numerical variables. What is this assumption? (1 mark)

ii Does the residual plot above support this assumption? Explain your answer. (1 mark)

12 (6 marks) The *number of male moths* caught in a trap set in a forest and the *egg density* (eggs per square metre) in the forest are shown in the table.

2.4

Number of male moths	35	37	45	49	65	74	77	86	95
Egg density (eggs per square metre)	471	635	664	997	1350	1100	2010	1640	1350

a Determine the equation of the least-squares line that can be used to predict the *egg density* in the forest from the *number of male moths* caught in the trap. Copy and complete the following by writing the values of the slope and intercept of this least-squares line in the appropriate boxes as shown. Round your answers to one decimal place.

egg density = [   ] × *number of male moths* + [   ]   (2 marks)

b The *number of female moths* caught in a trap set in a forest and the *egg density* (eggs per square metre) in the forest can also be examined. A scatterplot of the data is shown.

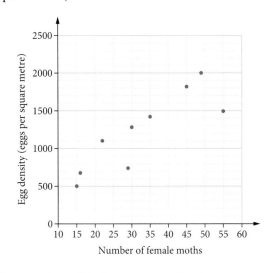

> 🔓 **Exam hack**
>
> Take a ruler into the exam and use it for these sorts of questions. Watch out for horizontal axes that don't start with 0.

The equation of the least-squares line is

   egg density = 31.3 × *number of female moths* + 191

i Copy the above scatterplot and draw a least-squares line on the graph.   (1 mark)

ii Interpret the slope of the least-squares line in terms of the variables *egg density* and *number of female moths* caught in the trap.   (1 mark)

iii The *egg density* is 1500 when the *number of female moths* caught is 55. Determine the residual value if the least-squares line is used to predict the *egg density* for this number of female moths.

> 🔓 **Exam hack**
>
> Make sure you get the sign right in residual questions. If rounding isn't asked for, give the full answer.   (1 mark)

iv The correlation coefficient is *r* = 0.862. Determine the percentage of the variation in *egg density* in the forest explained by the variation in the *number of female moths* caught in the trap. Round your answer to one decimal place.

> 🔓 **Exam hack**
>
> When asked for rounding, do not round until the very last step.   (1 mark)

▶ **13** (7 marks) The scatterplot shows the *weight*, in grams, and the *head length*, in millimetres, of 110 birds.

The equation of the least-squares line fitted to these data is

$$weight = 1.739 \times head\ length - 24.83$$

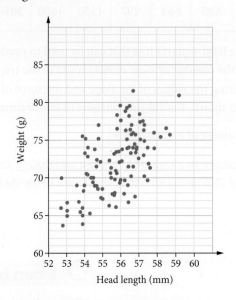

a Copy the above axes and draw this least-squares line. (1 mark)

b Use the equation to predict the *weight*, in grams, of a bird with a *head length* of 49.0 mm. Round your answer to one decimal place. (1 mark)

c Is the prediction made in part **b** an example of interpolation or extrapolation? Explain your answer briefly. (1 mark)

d When the least-squares line is used to predict the *weight* of a bird with a *head length* of 59.2 mm, the residual value is 2.78. Calculate the actual weight of this bird. Round your answer to one decimal place. (2 marks)

e The correlation coefficient, *r*, is equal to 0.5957. Given this information, what percentage of the variation in the *weight* of these birds is **not** explained by the variation in *head length*? Round your answer to one decimal place. (1 mark)

f The residual plot obtained when the least-squares line is fitted to the data set is shown.

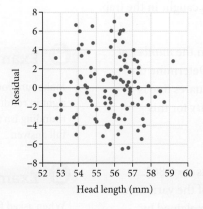

What does the residual plot indicate about the association between *head length* and *weight* for these birds? (1 mark)

 Exam hack

Always check the real-life context for your answer. If your answer looks to be impossible in real life, then it is almost certainly wrong.

14 © SCSA MA2017 Q9 (13 marks) The World Health Organization produces tables showing child growth standards. The median lengths (cm) for girls at various times during the first five years of life are shown below.

Age (months)	0	3	12	21	27	42	48	60
Median length (cm)	49.1	59.8	74.0	83.7	88.3	99.0	102.7	109.4
Predicted length (cm)	58.2	61.0	69.5	77.9	A	97.7	B	114.7
Residual	−9.1	−1.2	4.5	5.8	4.7	1.3	C	D

a **i** Determine the equation of the least-squares line for predicting the median length from a girl's age. (1 mark)

    ii Use the equation from part **a i** to determine the predicted median lengths A and B in the above table. (2 marks)

    iii What increase in median length can be expected for each additional year? (1 mark)

    iv Given that the correlation coefficient is 0.97, describe the association between age and median length in terms of its direction and strength. (2 marks)

    v What percentage of the variation in the median length can be explained by the variation in age? (1 mark)

b **i** Determine the residuals C and D in the table. (2 marks)

    ii Copy the residual plot below and complete by plotting the last four residual values. (2 marks)

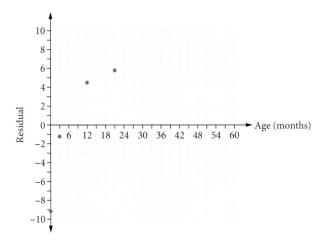

    iii Use the residual plot to assess the appropriateness of fitting a linear model to the data. (2 marks)

Line of best fit

- A **line of best fit** is a straight line that is the best approximation for a set of data.
- The equation for the **least-squares line** is $y = ax + b$ where
  - a is the **slope**. This means the y variable on average increases/decreases by a units for every 1-unit increase in the x variable
  - b is the **y-intercept**. This means the y variable is b units when the x variable is zero units.

The coefficient of determination (r^2)

- The **coefficient of determination (r^2)**:
  - can be calculated by squaring the **correlation coefficient** (r)
  - is a value between 0 and 1
  - is a measure of how useful a least-squares line is as a linear model for a particular set of data (0 means it is a totally useless measure and 1 means it's a perfect measure)
  - is usually converted to a percentage.
- $r^2 \times 100\%$ of the variation in the **response variable** can be explained by the variation in the **explanatory variable**.
- A high coefficient of determination (70% and above) indicates that the least-squares line is an appropriate model for the data.
- When calculating r from r^2, use the slope of the least-squares line on the scatterplot to see if r is positive or negative.

Making predictions

- Values can be predicted by substituting values into the equation of the least-squares line and solving.
- Predicting *within* the original data range is called **interpolation**.
- Predicting *outside* the original data range is called **extrapolation**.
- Predictions based on extrapolation are not as reliable as those based on interpolation.

Residual analysis

- A **residual** is the vertical distance between each data point and the least-squares line.
  residual value = actual value y – predicted value $\hat{y}$
  - The actual value can be read from a scatterplot or a table.
  - The predicted value needs to be calculated from the equation of the least-squares line.
- Actual values that lie
  - above the least-squares line will have a positive residual value
  - below the least-squares line will have a negative residual value
  - on the least-squares line will have a residual value of zero.
- The further the actual value is from the least-squares line, the larger the residual value.

Residual plot

- A **residual plot** is like a scatterplot with the explanatory variable on the x-axis and the residual values on the y-axis.
- When residual values
  - are randomly scattered above and below the x-axis, the association between the original two variables, x and y, is probably *linear*
  - show a hill or a valley shape, the association between the original two variables, x and y, is probably *non-linear*.

Cumulative examination: Calculator-free

Total number of marks: 16   Reading time: 2 minutes   Working time: 16 minutes

1 © SCSA MA2019 Q2 (9 marks) Katie is a hobby farmer who has been experimenting with a species of tomato plant growing under the same soil and climatic conditions. She varied the amount of water (W), in millimetres, used during each week and recorded the total number of tomatoes (T) produced by each plant. The scatterplot showing her results is drawn below.

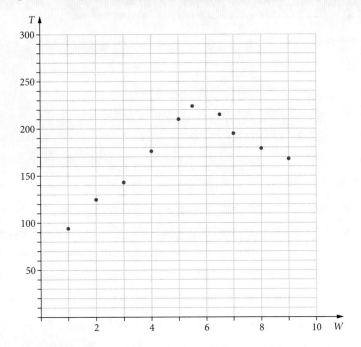

Katie determined the following summary information:

- $r_{WT} = 0.66$
- the equation of the least-squares line is $T = 10.55W + 119.11$.

a Identify the response variable. (1 mark)

b Use the equation of the least-squares line to predict the total number of tomatoes produced when 10 millimetres of water are given to a plant during each week. (2 marks)

c Copy the scatterplot above and fit the least-squares line to it. (2 marks)

Katie decided to draw a residual plot to gather more information about her results.

d **i** Copy the axes below and sketch a residual plot she would have likely drawn for the given data. Note: you do not have to calculate actual values. (2 marks)

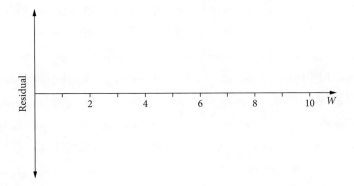

ii Use your residual plot to discuss the appropriateness of fitting a linear model to the data. (2 marks)

2 © SCSA MA2021 Q7 (7 marks) The ages in years, and salaries in thousands of dollars ($'000), of eight employees at a company are shown below. The equation of the least-squares line for these data is $y = 0.2x + 38$.

Age (x)	35	37	41	43	45	47	53	55
Salary (y)	42	44	47	50	52	51	49	45

The table below shows the predicted y values, obtained from the equation of the least-squares line, and the corresponding residuals.

x	y	Predicted y value	Residual
35	42	45.0	−3.0
37	44	45.4	−1.4
41	47	46.2	0.8
43	50	46.6	3.4
45	52	47.0	5.0
47	51	47.4	3.6
53	49	48.6	0.4
55	45	A	B

a Determine the value of *A* and *B*. (2 marks)

b Copy the graph below and plot the last two residuals. (2 marks)

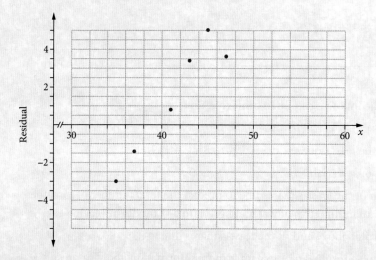

c Justify, using the residual plot in part **b**, whether the least-squares line is a good model for these data. (2 marks)

The calculated correlation coefficient for these data is 0.42.

d Describe how this supports your response in part **c**. (1 mark)

Total number of marks: 25   Reading time: 3 minutes   Working time: 25 minutes

1 (3 marks) The divided column graph shows the age distribution of people in three countries, Australia, India and Japan, for the year 2010.

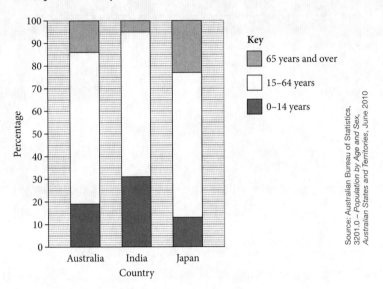

a What percentage of people in Australia were aged 0–14 years in 2010?   (1 mark)

b In 2010, the population of Japan was 128 000 000. How many people in Japan were aged 65 years and over in 2010?   (1 mark)

c From the graph, it appears that there is no association between the percentage of people in the 15–64 age group and the country in which they live. Explain why, quoting appropriate percentages to support your explanation.   (1 mark)

2 © SCSA   MA2020 Q10   (15 marks) A football club records body measurements for all of their players. Shown below are the waistline measurements (cm) and percentage body fat for eleven players.

Player	1	2	3	4	5	6	7	8	9	10	11
Waistline measurement (w)	89	100	87	96	94	83	81	83	84	97	98
Percentage body fat (p)	14	17	11	19	17	12	9	10	8	14	19

Research has shown that estimates for percentage body fat can be determined by using waistline measurements.

a Calculate the correlation coefficient r_{wp} for these data.   (1 mark)

b Determine the equation of the least-squares line for these data.   (1 mark)

c In the context of this question, interpret the slope of the line found in part **b**.   (2 marks)

d The residual plot shown below is for the first 10 players' data. Copy the residual plot, calculate the residual for player number 11 and plot this point on the graph. (2 marks)

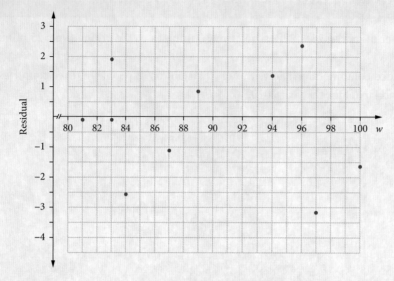

e Comment on the appropriateness of fitting a linear model to the data. Justify your answer. (2 marks)

f What percentage of the variation in the percentage body fat measurements is **unexplained** by the variation in the waistline measurements? (2 marks)

g Wayne is player number 12 and has a waistline measurement of 105 cm.

   i Determine his predicted percentage of body fat. (1 mark)

   ii Comment on the validity of the prediction and give a justification for your answer. (2 marks)

h Player number 13 has a residual of –2.6. What information does this provide about the percentage body fat for this player? (2 marks)

3 © SCSA MA2019 Q16 (7 marks) The table below records the altitude (metres above sea level), latitude (°S) and mean maximum temperature (°C) during January for eight cities in the southern hemisphere.

Altitude (A)	Latitude (L)	Mean maximum temperature (T)
15	31.95	25
20	43.53	20
24	42.88	18
314	45.03	16
8	6.18	28
154	12.05	26
37	12.46	29
8	34.60	25

Comparing altitude and the mean maximum temperature, it was determined that the least-squares line for these data was $T = -0.022A + 24.97$ and $r_{AT} = -0.50$.

a Determine the coefficient of determination for altitude and the mean maximum temperature and interpret this value. (2 marks)

b Determine the equation of the least-squares line for comparing latitude and the mean maximum temperature and state the correlation coefficient. (2 marks)

Rio de Janeiro has a latitude of 22.93° S and an altitude of 9 metres.

c Use the two least-squares lines above to predict the mean maximum temperature in January for Rio de Janeiro. Which prediction is more valid? Justify your choice. (3 marks)

CHAPTER

3

GROWTH AND DECAY IN SEQUENCES

TOPIC 3.2: GROWTH AND DECAY IN SEQUENCES

The arithmetic sequence

3.2.1   use recursion to generate an arithmetic sequence

3.2.2   display the terms of an arithmetic sequence in both tabular and graphical form and demonstrate that arithmetic sequences can be used to model linear growth and decay in discrete situations

3.2.3   deduce a rule for the nth term of a particular arithmetic sequence from the pattern of the terms in an arithmetic sequence, and use this rule to make predictions

3.2.4   use arithmetic sequences to model and analyse practical situations involving linear growth or decay

The geometric sequence

3.2.5   use recursion to generate a geometric sequence

3.2.6   display the terms of a geometric sequence in both tabular and graphical form and demonstrate that geometric sequences can be used to model exponential growth and decay in discrete situations

3.2.7   deduce a rule for the nth term of a particular geometric sequence from the pattern of the terms in the sequence, and use this rule to make predictions

3.2.8   use geometric sequences to model and analyse (numerically, or graphically only) practical problems involving geometric growth and decay

Sequences generated by first-order linear recurrence relations

3.2.9   use a general first-order linear recurrence relation to generate the terms of a sequence and to display it in both tabular and graphical form

3.2.10   generate a sequence defined by a first-order linear recurrence relation that gives long term increasing, decreasing or steady-state solutions

3.2.11   use first-order linear recurrence relations to model and analyse (numerically or graphically only) practical problems

Mathematics Applications ATAR Course Year 12 syllabus p. 10 © SCSA

Video playlists (4):

3.1   Arithmetic sequences

3.2   Geometric sequences

3.3   First-order linear recurrence relations

WACE question analysis Growth and decay in sequences

Worksheets (4):

3.1   Arithmetic sequences • Arithmetic progressions

3.2   Geometric sequences • Geometric progressions

⁂ Nelson MindTap

To access resources above, visit
cengage.com.au/nelsonmindtap

3.1 Arithmetic sequences

A **sequence** is an ordered list of numbers separated by commas; for example:

2, 4, 6, 8, 10 …

Each number is called a **term**. In this example, 2 is the 'first term', 4 is the 'second term' and 10 is the 'fifth term'.

We use the notation T_n to refer to the nth term in a sequence. The subscript n changes depending on the position. In the above example, $T_1 = 2$ and $T_4 = 8$.

In an **arithmetic sequence**, there is a common difference between consecutive terms. The starting term is a and the common difference is d.

For $a = 5$ and $d = 10$, the sequence is 5, 15, 25, 35 …

For $a = 3$ and $d = -4$, the sequence is 3, −1, −5, −9 …

Arithmetic growth occurs when the terms of a sequence increase by a constant amount, which is represented by a positive d value. **Arithmetic decay** occurs when the terms of a sequence decrease by a constant amount. This is represented by a negative d value.

Sequences form the mathematical basis for modern technologies including sound editing, digital photography compression, earthquake protection and medical imaging. In Chapter 7, sequences will be used to study compound interest investments, loans, depreciating assets and annuities.

Video playlist
Arithmetic sequences

Worksheets
Arithmetic sequences

Arithmetic progressions

Recurrence relations for arithmetic sequences

A sequence can be defined by a **recurrence relation**, also known as a **recursive rule**. A recurrence relation should have two components:

1 an equation explaining how to get the following term from any term

2 an equation identifying the initial term.

The recurrence relation for arithmetic sequences takes the form $T_{n+1} = T_n + d$ and $T_1 = a$. T_{n+1} refers to 'the term one after T_n.'

Recurrence relation for an arithmetic sequence

An arithmetic sequence is defined by a starting term, a, and a common difference, d, between consecutive terms. The recurrence relation for an arithmetic sequence is given by two statements:

$$T_{n+1} = T_n + d$$

$$T_1 = a$$

 Exam hack

Letters other than 'T' can also be used for sequences. The important thing is to use whatever letter is in the question.

For each of the following arithmetic sequences

    i write the recursive rule

    ii find the 6th term.

a 3, 9, 15, 21 …

b 2, −6, −14, −22 …

Steps	Working
a **i** **1** Identify the first term and the common difference.	$a = 3$ $d = 6$
2 Write the equation explaining how to get the $(n + 1)$th term from the nth term. Write the equation for the first term.	$T_{n+1} = T_n + 6$ $T_1 = 3$
ii Use the recursive rule to find the desired term.	$T_5 = T_4 + 6 = 21 + 6 = 27$ $T_6 = T_5 + 6 = 27 + 6 = 33$ $T_6 = 33$
b **i** **1** Identify the first term and the common difference.	$a = 2$ $d = -8$
2 Write the equation explaining how to get the $(n + 1)$th term from the nth term. Write the equation for the first term.	$T_{n+1} = T_n - 8$ $T_1 = 2$
ii Use the recursive rule to find the desired term.	$T_5 = T_4 - 8 = -22 - 8 = -30$ $T_6 = T_5 - 8 = -30 - 8 = -38$ $T_6 = -38$

The rule for the nth term for arithmetic sequences

Consider the sequence 3, 5, 7, 9, 11 … If we are asked to find the 15th term, do we need to count up in twos 14 times? No. The 15th term is given by $3 + 14 \times 2$.

The common difference in arithmetic sequences allows us to find any term without needing to recursively find every prior term. Because we already start at T_1, to find T_n, we only need to count up $n - 1$ times. Therefore, the rule for the nth term is given by

    $T_n = a + (n - 1)d$

where a is the first term, T_1, and d is the common difference. n is the position of the desired term and it remains as a variable in the equation. Remember that the expression $(n - 1)d$ is equivalent to $(n - 1) \times d$.

> **Rule for the nth term for an arithmetic sequence**
>
> The rule for the nth term of an arithmetic sequence is
>
> $T_n = a + (n - 1)d$
>
> where a is the first term, T_1, and d is the common difference.

9780170476959

3.1

For each of the following arithmetic sequences

   i identify the first term (a) and the common difference (d)

   ii use $T_n = a + (n - 1)d$ to write the rule for the nth term

   iii use the rule for the nth term to find the 11th term.

a 12, 7, 2, –3 …

b –10, –7, –4, –1 …

Steps	Working
a **i** Identify the first term, a, and the common difference, d.	$a = 12$ $d = -5$
ii **1** Substitute a and d into $T_n = a + (n - 1)d$. Remember there is a × in between $(n - 1)$ and d. Leave the n as a variable.	$T_n = 12 + (n - 1) \times -5$
2 Expand the brackets and simplify the expression. Recall that a positive number times a negative number will give a negative number.	$T_n = 12 + -5(n - 1)$ $= 12 - 5(n - 1)$ $= 12 - 5n + 5$ $= 17 - 5n$
iii To find the 11th term, substitute 11 for n.	$T_{11} = 17 - 5 \times 11$ $= 17 - 55$ $= -38$
b **i** Identify the first term, a, and the common difference, d.	$a = -10$ $d = 3$
ii **1** Substitute a and d into $T_n = a + (n - 1)d$. Remember there is a × in between $(n - 1)$ and d. Leave the n as a variable.	$T_n = -10 + (n - 1) \times 3$
2 Expand the brackets and simplify the expression. Recall that a positive number times a negative number will give a negative number.	$T_n = -10 + 3(n - 1)$ $= -10 + 3n - 3$ $= -13 + 3n$
iii To find the 11th term, substitute 11 for n.	$T_{11} = -13 + 3 \times 11$ $= -13 + 33$ $= 20$

Note that after expanding and simplifying the rule for the nth term,

- the constant value is the term 'prior' to the first term
- the coefficient of n is the difference d, which is negative in arithmetic decay.

For example, in the sequence 12, 7, 2, –3 … , the term 'prior' to the first term is 17 and the difference between terms is –5. Therefore, the rule for the nth term simplifies to: $T_n = 17 - 5n$.

 Exam hack

Practise doing calculations without a calculator. Always show working, even for 2-mark questions.

WORKED EXAMPLE 3 | Finding the value of n in an arithmetic sequence by hand

In the sequence –19, –10, –1, 8 … , in what position is the term 53?

Steps	Working
1 Find the rule for the nth term.	$\begin{aligned} T_n &= a + (n-1)d \\ &= -19 + (n-1) \times 9 \\ &= -19 + 9n - 9 \\ &= -28 + 9n \end{aligned}$
2 Substitute the desired value for T_n.	$53 = -28 + 9n$
3 Rearrange for n.	$\begin{aligned} 53 + 28 &= 9n \\ 81 &= 9n \\ n &= 9 \end{aligned}$ 53 is the 9th term.

USING CAS 1 | Finding the value of n in an arithmetic sequence

In the sequence –19, –10, –1, 8 …, what is the position of the value 188?

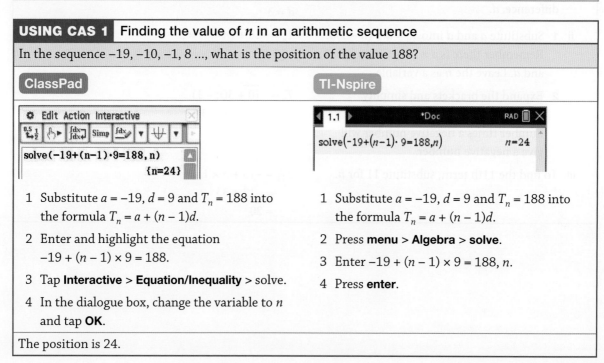

ClassPad

solve(-19+(n−1)·9=188, n)
{n=24}

1 Substitute $a = -19$, $d = 9$ and $T_n = 188$ into the formula $T_n = a + (n-1)d$.

2 Enter and highlight the equation $-19 + (n-1) \times 9 = 188$.

3 Tap **Interactive > Equation/Inequality > solve**.

4 In the dialogue box, change the variable to n and tap **OK**.

TI-Nspire

solve(-19+(n−1)·9=188,n)   n=24

1 Substitute $a = -19$, $d = 9$ and $T_n = 188$ into the formula $T_n = a + (n-1)d$.

2 Press **menu > Algebra > solve**.

3 Enter $-19 + (n-1) \times 9 = 188, n$.

4 Press **enter**.

The position is 24.

USING CAS 2 Generating the terms of an arithmetic sequence from a rule

Consider the arithmetic sequence 37, 31, 25 …

Generate the first 30 terms of the sequence, then determine the values of T_{10}, T_{20} and T_{30}.

For the arithmetic sequence 37, 31, 25 …, the first term is $a = 37$ and the common difference is $d = 31 - 37 = 25 - 31 = -6$.

ClassPad

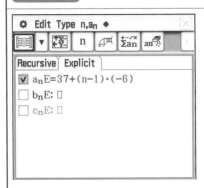

1 Open the **Sequence** application.

2 Tap on the **Explicit** tab.

3 On the first line, enter the expression as shown above and press **EXE**.

4 Tap the **Table** tool.

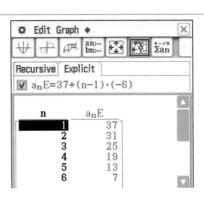

5 The arithmetic sequence will appear in the lower window.

6 Tap the **Sequence Table Input** tool.

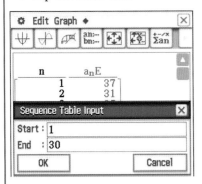

7 Change the **End:** value to **30**.

8 Tap **OK**.

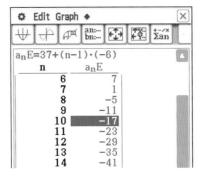

9 The table of values in the lower window will now extend to 30.

10 Tap **Resize** then scroll down to find the values at $n = 10$, 20 and 30.

$T_{10} = -17$, $T_{20} = -77$, $T_{30} = -137$

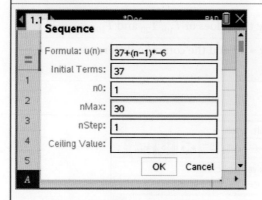

1 Add a **Lists & Spreadsheet** page.

2 Place the cursor to the cell immediately under the **A**.

3 Press **menu > Data > Generate Sequence**.

4 Complete the fields of the dialogue box as shown above. (Note the asterisk * before the **–6** is the multiplication sign ×.)

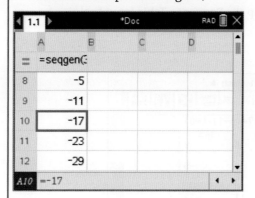

5 The arithmetic sequence values will be generated in column **A**.

6 Press the **down arrow** to scroll down to find the values at $n = 10$, 20 and 30.

$T_{10} = -17$, $T_{20} = -77$, $T_{30} = -137$

The graph of an arithmetic sequence

In an arithmetic sequence, the values of n and T_n can be written in tabular form.

Consider the sequence 5, 8, 11, 14 …, and its tabular form:

n	1	2	3	4	…
T_n	5	8	11	14	…

The paired values for n and T_n can be read as coordinates (n, T_n) and plotted on a graph. n values are shown on the horizontal axis and T_n values are shown on the vertical axis.

A positive slope indicates growth and a negative slope indicates decay. In arithmetic sequences, the form is linear, representing the constant difference between consecutive terms.

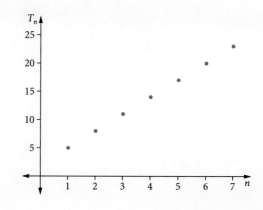

Growth and decay graphs of arithmetic sequences

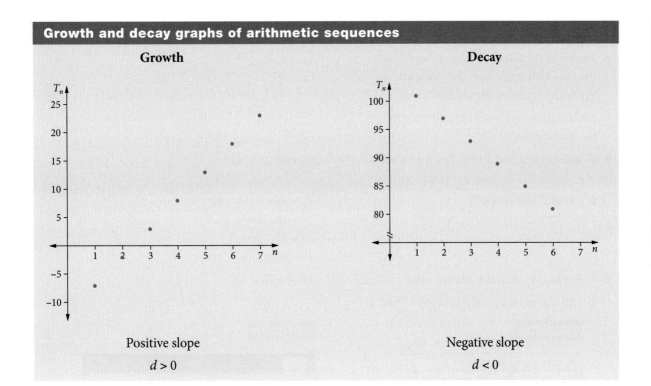

Growth

Positive slope

$d > 0$

Decay

Negative slope

$d < 0$

🔒 Exam hack

Do not connect the dots when plotting a sequence on a graph. Separate dots show that there are gaps between the terms of a sequence.

WORKED EXAMPLE 4 | **Graphing arithmetic sequences**

The graph shows the first 8 terms of an arithmetic sequence. The vertical axis provides the value of each term, Q_n, for a given n on the horizontal axis.

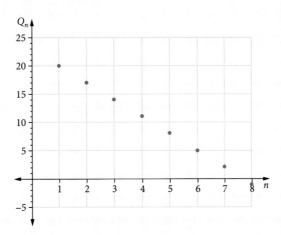

a Write the recursive rule for the arithmetic sequence.

b Write the rule for the nth term for the arithmetic sequence.

c Determine which term of the sequence is -244.

Steps	Working
a Write the equation explaining how to get the $(n + 1)$th term from the nth term. Write the equation for the first term. Make sure to use the letter provided in the question.	$Q_{n+1} = Q_n - 3$ $Q_1 = 20$
b 1 Substitute a and d into $T_n = a + (n - 1)d$. **2** Expand and simplify.	$T_n = a + (n - 1)d$ $\quad = 20 + (n - 1) \times -3$ $T_n = 20 + -3(n - 1)$ $\quad = 20 - 3(n - 1)$ $\quad = 20 - 3n + 3$ $\quad = 23 - 3n$
c 1 Replace T_n with the desired value. **2** Use the solve function on CAS to find n.	$-244 = 23 - 3n$

ClassPad

```
⚙ Edit Action Interactive        ✕
0.5 1  ◻▸  ∫dx⌐ Simp ∫dx◻ ▼ ⊞ ▼ ▸
└▶2      ∫dx◻
solve(-244=23-3·n, n)            ▲
                    {n=89}       ▣
```

TI-Nspire

```
◀ 1.1 ▶              *Doc        RAD ▯ ✕
solve(-244=23-3· n,n)            n=89
```

$n = 89$

-244 is the 89th term.

Summing the first n terms of an arithmetic sequence

The sum of the first n terms of an arithmetic sequence is equal to n multiplied by the average of the 1st term and the nth term. For example, to sum 1, 2, 3, 4, 5, 6, 7, 8, 9, take the average of 1 and 9, which is 5, and multiply it by the number of terms: $5 \times 9 = 45$.

Using arithmetic sequence notation, this is given by

$$\frac{\overset{\text{first term}}{\boxed{a}} + \overset{\text{nth term}}{\boxed{a + (n - 1)d}}}{2} \times n$$

This can be written more simply as

$$\frac{T_1 + T_n}{2} \times n$$

Summing the first n terms of an arithmetic sequence

The sum of the first n terms is given by

$$\frac{T_1 + T_n}{2} \times n$$

🔓 **Exam hack**

Adding the first n terms of an arithmetic sequence can be done by summing every term individually. However, using $\dfrac{T_1 + T_n}{2} \times n$ is much faster.

WORKED EXAMPLE 5 | Summing the first n terms of an arithmetic sequence

Jade wants to improve her fitness. The number of skips she jumps each day is modelled by the recursive rule

$$K_{n+1} = K_n + 8, K_1 = 7$$

Jade stops increasing her daily total after she goes over 300 daily skips.

a Determine a rule for the number of skips Jade does on the nth day, before she stops adding to her daily total.

b Hence, determine the number of skips Jade jumps on day 30.

c Find the total number of skips Jade has jumped after 30 days.

d What is the first day Jade does not increase her daily total?

Steps	Working
a 1 Identify the first term, a, and the common difference, d.	$a = 7$ $d = 8$
2 Substitute a and d into $K_n = a + (n - 1)d$.	$K_n = 7 + (n - 1) \times 8$
3 Simplify.	$K_n = 7 + 8(n - 1)$ $= 7 + 8n - 8$ $= -1 + 8n$
b 1 The word 'hence' requires us to use what we found in the previous part.	$K_n = -1 + 8n$
2 Substitute the value $n = 30$.	$K_{30} = -1 + 8 \times 30$ $= -1 + 240$ $= 239$
c 1 Recall that the sum of the first n terms is given by $\dfrac{K_1 + K_n}{2} \times n$. Write K_1, K_n, and n.	$K_1 = 7$ $K_{30} = 239$ $n = 30$
2 Find $\dfrac{K_1 + K_n}{2}$.	$\dfrac{K_1 + K_n}{2} = \dfrac{7 + 239}{2} = 123$
3 Multiply $\dfrac{K_1 + K_n}{2}$ by n.	$\dfrac{K_1 + K_n}{2} \times n = 123 \times 30 = 3690$ The total number of skips Jade has jumped after 30 days is 3690.

Chapter 3 | Growth and decay in sequences **97**

d 1 Use the rule for the nth term to substitute the desired value for K_n.

Jade stops adding to her daily skips after she passes 300 skips.

$$300 = -1 + 8n$$

2 Solve for n using CAS.

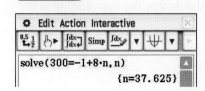

{n=37.625}

TI-Nspire

solve(300=-1+8· n,n) n=37.625

$n = 37.625$

3 Interpret the result.

On day 37, Jade has not yet reached 300 daily skips, so she increases again on day 38. On day 38, Jade passes 300 daily skips. She does not increase her daily total on day 39.

EXERCISE 3.1 Arithmetic sequences

ANSWERS p. 432

Mastery

1 For each of the following mathematical expressions, choose the correct description from the box.

a T_{11}

b T_n

c $T_{n+1} = T_n + 3$, $T_1 = 8$

d $T_n = 5 + 3n$

e 11

f $\dfrac{T_1 + T_{10}}{2} \times 10$

A rule for the nth term	A recursive rule
The 11th term	The sum of the first 10 terms of an arithmetic sequence
The nth term	The number of terms from T_3 to T_{13}, inclusive

2 WORKED EXAMPLE 1 For each of the following arithmetic sequences

   i write the recursive rule

   ii find the 6th term.

a 1, 9, 17, 25 …

b −10, −4, 2, 8 …

c 51, 49, 47, 45 …

d −7, −11, −15, −19 …

3 WORKED EXAMPLE 2 For each of the following arithmetic sequences

   i identify the first term a and the common difference d

   ii use $T_n = a + (n - 1)d$ to write the rule for the nth term

   iii use the rule for the nth term to find the 9th term.

a 3, 5, 7, 9 …

b 98, 88, 78, 68 …

c −9, −6, −3, 0 …

d 71, 64, 57, 50 …

4 WORKED EXAMPLE 3 In the sequence

a 30, 33, 36, 39 … , in what position is the value 72?

b −6, −11, −16, −21 … , in what position is the value −136?

5 Using CAS 1 In the sequence

   a 11, 22, 33, 44 … , in what position is the value 374?

   b −5, −8, −11, −14 … , in what position is the value −53?

6 Using CAS 2

   a Consider the arithmetic sequence 17, 6, −5 …
      Find T_{10}, T_{11} and T_{12}.

   b Consider the arithmetic sequence 2, 16, 30 …
      Find V_{21}, V_{22} and V_{23}.

7 WORKED EXAMPLE 4 The graph shows the first 8 terms of an arithmetic sequence. The vertical axis provides the value of each term, P_n, for a given n on the horizontal axis.

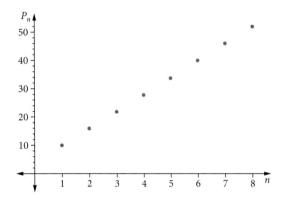

   a Write the recursive rule for the arithmetic sequence.

   b Write the rule for the nth term for the arithmetic sequence.

   c Determine which term of the sequence is 136.

8 WORKED EXAMPLE 5 In 2070, the West Coast Eagles draft promising youngster Chris Shuey. In Chris Shuey's first season, the number of disposals, D_n, he collects in round n is given by the recursive rule

$$D_{n+1} = D_n + 3, D_1 = 5$$

   a Deduce a rule for the number of disposals he gets in round n.

   b Hence, determine the number of disposals Chris Shuey obtains in round 22, the final round of the season (assume there are no bye rounds).

   c Find the total number of disposals Chris Shuey collects across the first nine rounds.

   d Find the total number of disposals Chris Shuey collects across the entire season.

Calculator-free

9 © SCSA MA2021 Q1 (5 marks) Hanai is a successful college basketball player. His coach has warned him that he will lose his scholarship if he scores 54% or below on a weekly assessment. On his first three weekly assessments, he scored 84%, 81% and 78% respectively. Assume Hanai's weekly assessments continue to follow this pattern.

   a Deduce a rule for the nth term of this sequence.   (2 marks)

   b Determine Hanai's score on his sixth weekly assessment.   (1 mark)

   c Predict when Hanai will lose his scholarship.   (2 marks)

10 ©️ SCSA | MA2017 Q1 | (8 marks) Consider the following recurrence relation

$T_{n+1} = T_n - 3$, $T_3 = 2$.

    a Copy the axes and display the first six terms of this sequence. Label the axes clearly. (3 marks)

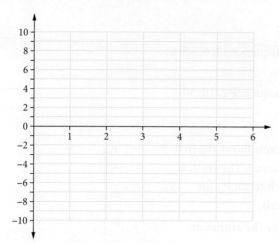

    b **i** Deduce a rule for the nth term of this sequence. (2 marks)

        ii Hence, determine the first term in the sequence that is less than −500. (3 marks)

Calculator-assumed

11 ©️ SCSA | MA2020 Q7 | (6 marks) The world's tallest man was recorded as 60 cm long at birth. He grew 28 cm in his first year, 26 cm in his second year and so on, always 2 cm less than in the previous year until he stopped growing.

    a Calculate his annual growth (in cm) in his fourth and fifth years. (1 mark)

    b Deduce the rule for his annual growth in the nth year, until he stopped growing. (2 marks)

    c In which year did he first not grow any taller? (1 mark)

    d Calculate his maximum height. (2 marks)

12 ©️ SCSA | MA2019 Q7 | (6 marks) A water tank is full. When a tap at the bottom of the tank is opened, 84 litres run out in the first minute, 78 litres in the second minute and 72 litres in the third minute. This pattern continues until the tank is empty.

    a Write a rule for the nth term of a sequence in the form $T_n = A + Bn$, which will model this situation where T_n is the amount of water that runs out in the nth minute. (2 marks)

    b How many litres run out in the seventh minute? (1 mark)

    c How many litres have run out after 8 minutes? (1 mark)

    d What is the capacity of the tank? (2 marks)

(3.2) Geometric sequences

Geometric sequences have a **common ratio**, r, between consecutive terms. Each new term is equal to its preceding term multiplied by this constant number. The letter a is used to represent the first term.

    For $a = 3$ and $r = 2$, the sequence is 3, 6, 12, 24 …

    For $a = 5$ and $r = 4$, the sequence is 5, 20, 80, 320 …

    For $a = 120$ and $r = 0.5$, the sequence is 120, 60, 30, 15 …

Geometric growth occurs when r is greater than 1. **Geometric decay** occurs when r is between 0 and 1.

Finding and interpreting r

The common ratio, r, can be found by dividing any term by its previous term. For example, in the sequence 50, 60, 72, 86.4 …, the common ratio is $\dfrac{60}{50} = 1.2$, but the same $r = 1.2$ can also be found by calculating $\dfrac{72}{60}$ or $\dfrac{86.4}{72}$.

Finding the common ratio, r

In a geometric sequence, the common ratio, r, between consecutive terms can be found by dividing any term by its previous term.

$$r = \frac{\text{any term}}{\text{previous term}} = \frac{T_2}{T_1} = \frac{T_3}{T_2} = \frac{T_4}{T_3} = \cdots$$

To interpret r as a percentage change, subtract 1 and multiply by 100.

r	$(r - 1) \times 100$	Interpret the result
1.2	20%	20% increase
0.65	−35%	35% decrease
2.5	150%	150% increase

Recurrence relations for geometric sequences

Just like an arithmetic sequence, a geometric sequence can be defined by a recurrence relation, also known as a recursive rule. The recurrence relation should have two components:

1. an equation explaining how to get the following term from any term
2. an equation identifying the initial term.

The recurrence relation for geometric sequences takes the form $T_{n+1} = rT_n$ and $T_1 = a$. T_{n+1} refers to 'the term one after T_n'. Recall that rT_n means $r \times T_n$.

Recurrence relation for a geometric sequence

A geometric sequence is defined by a starting term, a, and a common ratio, r, between consecutive terms. The recurrence relation for a geometric sequence is given by two statements:

$T_{n+1} = rT_n$

$T_1 = a$

 Exam hack

Remember to write the statement identifying the initial term. Students often forget it.

WORKED EXAMPLE 6 Finding the recursive rule for a geometric sequence

For each of the following geometric sequences

  i identify the first term (a) and the common ratio between consecutive terms (r)

  ii write a recursive rule for the sequence

  iii state the percentage change between consecutive terms.

a 50, 60, 72, 86.4 …

b 800, 720, 648, 583.2 …

Steps	Working
a **i** Identify the first term, a. Find the common ratio, r, by dividing the second term by the first term.	$a = 50$ $r = \dfrac{60}{50}$

ClassPad | **TI-Nspire**

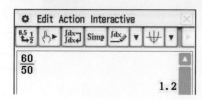 |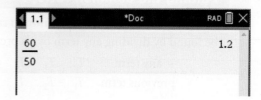

$r = 1.2$

| **ii** Write the equation explaining how to get the $(n + 1)$th term from the nth term. Write the equation for the first term. | $T_{n+1} = 1.2T_n$ $T_1 = 50$ |
| **iii** Subtract 1 from the common ratio and convert it to a percentage. Interpret it. | $1.2 - 1 = 0.2$ $0.2 \times 100 = 20$ 20% increase |

| **b** **i** Identify the first term, a. Find the common ratio, r, by dividing the second term by the first term. | $a = 800$ $r = \dfrac{720}{800}$ |

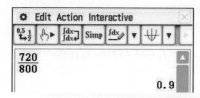 |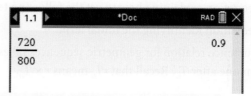

$r = 0.9$

| **ii** Write the equation explaining how to get the $(n + 1)$th term from the nth term. Write the equation for the first term. | $T_{n+1} = 0.9T_n$ $T_1 = 800$ |
| **iii** Subtract 1 from the common ratio and convert it to a percentage. Interpret it. | $0.9 - 1 = -0.1$ $-0.1 \times 100 = -10$ 10% decrease |

9780170476959

USING CAS 3 | Generating the terms of a geometric sequence from a recurrence relation

Consider the sequence generated by the recurrence relation $T_{n+1} = 1.5T_n$, $T_1 = 10$. Find T_{10}, T_{11} and T_{12}, correct to one decimal place.

ClassPad

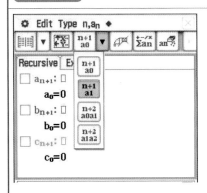

1 Open the **Sequence** application.

2 Tap on the **Recursive** tab.

3 Tap on the **n+1/a0** tool and select the **n+1/a1** option from the dropdown menu.

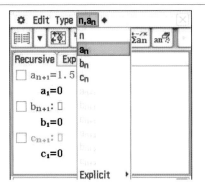

4 Enter **1.5**.

5 Tap the **n,a_n** menu and select **a_n** from the dropdown menu.

6 Press **EXE** to select the equation.

7 Enter **a_1 = 10**, then press **EXE**.

8 Tap the **Table** tool.

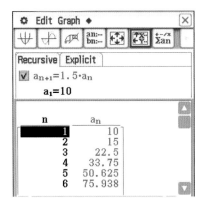

9 The recursive sequence values will appear in the lower window.

10 Tap the **Sequence Table Input** tool.

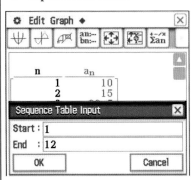

11 In the dialogue box, change the **End: value** to **12**.

12 Tap **OK**.

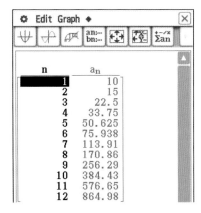

13 The table of values in the lower window will now extend to 12.

14 Tap **Resize** then scroll down to find the a_n values at $n = 10$, 11 and 12.

$T_{10} = 384.4$, $T_{11} = 576.7$, $T_{12} = 865.0$

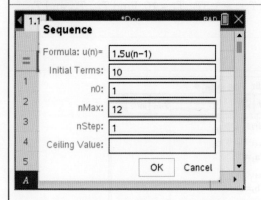

1 Open a **Lists & Spreadsheet** application.

2 Place the cursor to the cell immediately under the **A**.

3 Press **menu** > **Data** > **Generate Sequence**.

4 Complete the fields of the dialogue box as shown above, then press **enter**.

5 The recursive sequence values will be generated in column **A**.

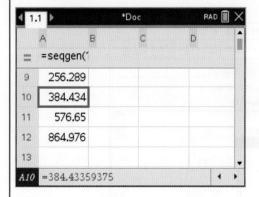

6 Press the **down arrow** to scroll down to find the values at 10, 11 and 12.

$T_{10} = 384.4$, $T_{11} = 576.7$, $T_{12} = 865.0$

The rule for the nth term for geometric sequences

Consider the sequence 3, 6, 12, 24, 48 … To find the 15th term, we could start with 3 and multiply by 2 fourteen times: $3 \times 2 \times 2 \times … \times 2$. However, if we recall that $\underbrace{2 \times 2 \times … \times 2}_{14 \text{ times}} = 2^{14}$, then the 15th term is easily calculated by 3×2^{14}.

The common ratio in geometric sequences allows us to find any term without needing to recursively find every prior term. Because we already start at T_1, to find T_n, we only need to multiply $n - 1$ times. Therefore, the rule for the nth term is given by

$T_n = ar^{n-1}$

where a is the first term, T_1, and r is the common ratio. n is the position of the desired term and it remains as a variable in the equation.

Rule for the *n*th term for a geometric sequence

The rule for the *n*th term of a geometric sequence is

$$T_n = ar^{n-1}$$

where a is the first term, T_1, and r is the common ratio.

3.2

🔒 Exam hack

Remember that the expression ar^{n-1} is equivalent to $a \times r^{n-1}$. As such, the r is often put in brackets to show that the power does not apply to the a. For example, $T_n = 25(0.8)^{n-1}$.

WORKED EXAMPLE 7 | Working with the rule for the *n*th term for a geometric sequence

For each of the following recursive rules:

   i state the first three terms

  ii identify the first term a and the common ratio between consecutive terms r

 iii use $T_n = ar^{n-1}$ to find the rule for the *n*th term

 iv use the rule for the *n*th term to find the value of T_7. Round to two decimal places where appropriate.

a $T_{n+1} = 5T_n$, $T_1 = 4$

b $T_{n+1} = \dfrac{1}{4}T_n$, $T_2 = 60$

Steps	Working
a **i** Use the recursive rule to generate the first three terms.	$T_1 = 4$ $T_2 = 5 \times 4 = 20$ $T_3 = 5 \times 20 = 100$ The first three terms are 4, 20, 100.
ii Identify the first term, a. Identify the common ratio, r.	$a = 4$ $r = 5$
iii Substitute a and r into $T_n = ar^{n-1}$. Keep the r in brackets to show that the power applies just to it.	$T_n = 4(5)^{n-1}$
iv To find the 7th term, substitute 7 for n.	$T_7 = 4(5)^{7-1}$ $ = 4(5)^6$

ClassPad

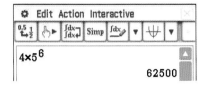

TI-Nspire

The 7th term is 62 500.

---

b **i** Use the recursive rule to generate the first three terms.

Note that the term provided is T_2.

$$T_2 = 60$$

$$T_3 = \frac{1}{4} \times 60 = 15$$

To find T_1, calculate $T_2 \div r$:

$$T_1 = T_2 \div r = 60 \div \frac{1}{4}$$

$$T_1 = 60 \times 4$$
$$= 240$$

The first three terms are 240, 60, 15.

ii Identify the first term, a. Identify the common ratio, r.

$$a = 240$$

$$r = \frac{1}{4}$$

iii Substitute a and r into $T_n = ar^{n-1}$. Keep the r in brackets to show that the power applies just to it.

$$T_n = 240\left(\frac{1}{4}\right)^{n-1}$$

iv To find the 7th term, substitute 7 for n.

$$T_7 = 240\left(\frac{1}{4}\right)^{7-1}$$

$$= 240\left(\frac{1}{4}\right)^{6}$$

ClassPad

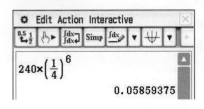

TI-Nspire

The 7th term is 0.06 to two decimal places.

WORKED EXAMPLE 8	Finding the value of n in a geometric sequence

In the geometric sequence 4, 1, 0.25 …, what is the

a rule for the nth term

b position of 0.015 625

c first term smaller than 0.001, correct to five decimal places?

Steps	Working
a 1 Identify the first term, a. Find the common ratio, r, by dividing the second term by the first term.	$a = 4$ $r = \dfrac{1}{4} = 0.25$
2 Substitute a and r into $T_n = ar^{n-1}$.	$T_n = 4(0.25)^{n-1}$
b 1 Substitute the desired value for T_n.	$0.015\,625 = 4(0.25)^{n-1}$
2 Use the solve function on CAS to find the value of n.	

ClassPad

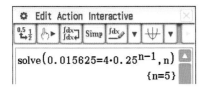

solve(0.015625=4·0.25ⁿ⁻¹,n)

{n=5}

TI-Nspire

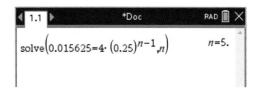

solve$\left(0.015625 = 4\cdot (0.25)^{n-1}, n\right)$ $n=5.$

$n = 5$

c 1 Substitute the desired value for T_n.	$0.001 = 4(0.25)^{n-1}$
2 Use the solve function on CAS to find the value of n.	

ClassPad

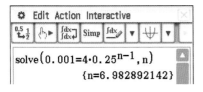

solve(0.001=4·0.25ⁿ⁻¹,n)

{n=6.982892142}

TI-Nspire

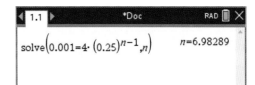

solve$\left(0.001 = 4\cdot (0.25)^{n-1}, n\right)$ $n=6.98289$

$n = 6.98$

3 Interpret the result.	0.001 occurs when $n = 6.98$. Therefore, T_6 is larger than 0.001 and T_7 is smaller than 0.001. Therefore, T_7 is the desired term.
4 Find the desired term.	$T_7 = 4(0.25)^{7-1}$ $\quad = 0.000\,98$ 0.000 98 is the first term smaller than 0.001.

a Find T_3 if $T_n = 3(4)^{n-1}$.

b Find T_5 if $T_n = 90\left(\dfrac{2}{3}\right)^{n-1}$. Express your answer as a fraction.

Steps	Working
a Substitute the desired value of *n* into the rule for the *n*th term.	$T_3 = 3(4)^{3-1}$
Simplify.	$T_3 = 3(4)^2$ $= 3 \times 4 \times 4$ $= 48$
b Substitute the desired value of *n* into the rule for the *n*th term.	$T_5 = 90\left(\dfrac{2}{3}\right)^{5-1}$
Simplify. Instead of evaluating the power directly, split it into parts and simplify with other numbers in the expression.	$T_5 = 90\left(\dfrac{2}{3}\right)^4$ $= 90 \times \dfrac{2 \times 2 \times 2 \times 2}{3 \times 3 \times 3 \times 3}$ $= 90 \times \dfrac{4 \times 4}{9 \times 9}$ $= 10 \times \dfrac{16}{9}$ $= \dfrac{160}{9}$

The graph of a geometric sequence

Consider the geometric sequence 4, 6, 9, 13.5, 20.25, 30.38 … The common ratio is 1.5. This is its tabular form:

n	1	2	3	4	5	6	...
T_n	4	6	9	13.5	20.25	30.38	...

The paired values for *n* and T_n can be read as coordinates (n, T_n) and plotted on a graph. *n* values are shown on the horizontal axis and T_n values are shown on the vertical axis.

A positive slope indicates growth and *r* larger than 1. A negative slope indicates decay and *r* between 0 and 1. The graph is a curve and not a straight line and therefore the form is non-linear.

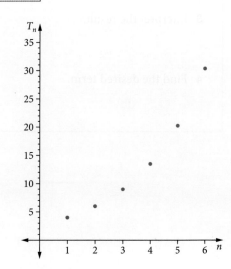

9780170476959

Growth and decay graphs of geometric sequences

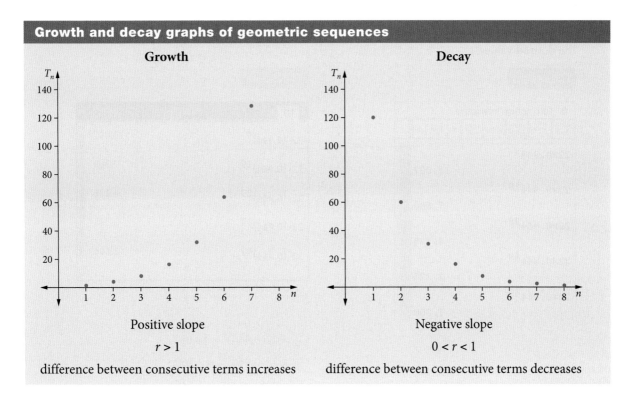

Growth	Decay
Positive slope	Negative slope
$r > 1$	$0 < r < 1$
difference between consecutive terms increases	difference between consecutive terms decreases

3.2

WORKED EXAMPLE 10 | Graphing geometric sequences

The concentration of algae in a pond is being measured. At the beginning of the first minute, the concentration is 22 units per litre. The concentration is decreasing by 5.6% each minute.

a Write a rule for C_n, the concentration of algae at the beginning of the nth minute.

b Plot the concentration of algae at the beginning of the 10th, 20th, 30th, 40th and 50th minutes. Label each point with its coordinates and give the concentration correct to one decimal place.

c Does the graph display arithmetic decay or geometric decay? Explain why.

Steps	Working
a 1 Identify the first term, a.	$a = 22$
2 Find the common ratio, r, by converting the percentage into a decimal. Subtract this from 1 due to a decreasing concentration.	$1 - \dfrac{5.6}{100} = 0.944$
3 Substitute a and r into $C_n = ar^{n-1}$.	$C_n = 22(0.944)^{n-1}$

b 1 Use the rule for the nth term to calculate the desired values.

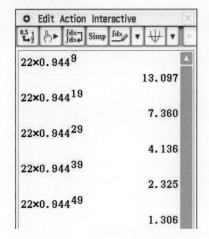

$$C_{10} = 22(0.944)^9 = 13.097$$
$$C_{20} = 22(0.944)^{19} = 7.360$$
$$C_{30} = 22(0.944)^{29} = 4.136$$
$$C_{40} = 22(0.944)^{39} = 2.325$$
$$C_{50} = 22(0.944)^{49} = 1.306$$

2 Plot a graph with n (minutes) on the horizontal axis and C_n (units/L) on the vertical axis. Round the C_n values to one decimal place.

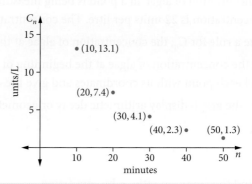

c Arithmetic growth/decay is represented by a straight line. Geometric growth/decay is represented by a curved line.

Geometric decay.

The difference between consecutive points is decreasing and the graph is not a straight line.

Alternative form of the rule for the nth term for geometric sequences

In practical applications of sequences, the value n is often used to represent time. For example, B_n could be the number of birds in a sanctuary n years after 2025; C_n could be the concentration of a chemical n minutes after a reaction; and D_n could be the dollars in a savings account n months after investing. B_1, C_1 and D_1 refer to values after one minute, one year and one month, respectively. In each scenario, it is reasonable to ask: 'How much was there at the very beginning?' We represent such values with B_0, C_0 and D_0.

The rule for the nth term is most commonly written as $T_n = ar^{n-1}$, where $a = T_1$. If, instead of a, we want to use T_0, then we need to construct the rule differently. The alternative form is derived as follows:

$$T_n = ar^{n-1} = T_1 r^{n-1}$$
$$= T_1 \frac{r^n}{r}$$
$$= \frac{T_1}{r} r^n$$
$$= T_0 r^n$$

Alternative form of the rule for the nth term for geometric sequences

For geometric sequences, there are two forms of the rule for the nth term.

Form 1:

$$T_n = ar^{n-1}, \text{ where } a = T_1$$

Form 2:

$$T_n = T_0 r^n$$

Both forms will give the same results and are valid ways of calculating T_n.

Form 1 is the most common.

 Exam hack

Form 2 might appear in practical scenarios where n represents time, and you are provided T_0, a value before time started being counted.

WORKED EXAMPLE 11 | **Writing rules for the nth term of geometric sequences**

At the beginning of 2030, a YouTube video has 5 views. The number of views grows by a constant rate each month. The views at the end of the first three months are shown in the table.

January	February	March
15	45	135

T_n is the number of views n months after the beginning of 2030.

a Write a recursive rule in the form $T_{n+1} = rT_n$, $T_0 = u$.

b Write the rule for the nth term in the form $T_n = vr^{n-1}$ by choosing the correct value for v.

c Write the rule for the nth term in the form $T_n = wr^n$ by choosing the correct value for w.

d How many views will there be at the end of September 2031? Answer to the nearest billion.

Steps	Working
a 1 Identify T_0.	T_0 is the number of views at the beginning of 2030.
	$u = 5$
Identify the common ratio, r.	$r = 3$
2 Write the equation explaining how to get the $(n + 1)$th term from the nth term. Write the equation for the beginning term T_0.	$T_{n+1} = 3T_n$
	$T_0 = 5$

b When there is $n - 1$ in the power, the rule for the nth term is written as $T_n = ar^{n-1}$, where a is T_1.	$T_1 = 15$ $T_n = 15(3)^{n-1}$
c When there is only n in the power, the rule for the nth term is written as $T_n = T_0(r)^n$.	$T_0 = 5$ $T_n = 5(3)^n$
d 1 Calculate the value of n needed to get to the required date.	The end of September 2031 is 1 year and 9 months after the beginning of 2030. $12 + 9 = 21$ months $n = 21$
2 Substitute the value of n into one of the rules for the nth term.	$T_{21} = 15(3)^{20}$ $= 52\,301\,766\,015$
3 Divide by $1\,000\,000\,000$ and round to the nearest billion.	52 billion views

---

EXERCISE 3.2 Geometric sequences

ANSWERS p. 432

Recap

1 Consider the arithmetic sequence 33, 27, 21, 15 …

   a Write a recursive rule for this sequence.

   b Write the rule for the nth term.

   c In what position is the value -39?

2 Consider the arithmetic sequence with rule for the nth term: $A_n = 2 + 7n$. What is the sum of the first 10 terms?

Mastery

3 WORKED EXAMPLE 6   For each of the following geometric sequences

   i identify the first term (a) and the common ratio between consecutive terms (r)

   ii write a recursive rule for the sequence

   iii state the percentage change between consecutive terms.

   a 200, 220, 242, 266.2 …   **b** 30, 21, 14.7, 10.29 …   **c** 3, 7.5, 18.75, 46.875 …

4 Using CAS 3

   a Consider the sequence generated by the recurrence relation $T_{n+1} = 1.75T_n$, $T_1 = 8$. Find T_{10}, T_{11} and T_{12} correct to one decimal place.

   b Consider the sequence generated by the recurrence relation $V_{n+1} = 0.75V_n$, $V_1 = 10\,000$. Find V_{20}, V_{21} and V_{22} correct to one decimal place.

5 WORKED EXAMPLE 7   For each of the following recursive rules

   i state the first three terms

   ii identify the first term (a) and the common ratio between consecutive terms (r)

   iii use $T_n = ar^{n-1}$ to find the rule for the nth term

   iv use the rule for the nth term to find the value of T_7. Round to two decimal places where appropriate.

   a $T_{n+1} = 3T_n$, $T_1 = 2$ **b** $T_{n+1} = \frac{1}{2}T_n$, $T_1 = 1200$ **c** $T_{n+1} = 1.2T_n$, $T_2 = 36$

6 [WORKED EXAMPLE 8] In the geometric sequence 2, 8, 32, 128 …, what is the

   a rule for the nth term

   b position of the value 32 768

   c first term larger than 5000?

3.2

7 [WORKED EXAMPLE 9] Find

   a T_3 if $T_n = 5(3)^{n-1}$.

   b T_4 if $T_n = 3(2)^{n-1}$.

   c T_4 if $T_n = 32\left(\dfrac{3}{4}\right)^{n-1}$. Express your answer as a fraction.

   d T_5 if $T_n = 18\left(\dfrac{2}{3}\right)^{n-1}$. Express your answer as a fraction.

8 [WORKED EXAMPLE 10] The concentration of bacteria in a sample is increasing by 7.1% each minute. After 1 minute, the concentration is 20 cfu/mL.

   a Write a rule for C_n, the concentration of bacteria in the sample after n minutes.

   b Copy the axes and plot the concentration of bacteria after 10, 20, 30, 40 and 50 minutes.

      Label each point with its coordinates and give the concentration correct to one decimal place.

   c Does the graph represent arithmetic growth or geometric growth? Why?

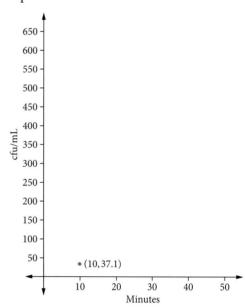

9 A fast-growing tech company generated revenue of $253 million in 2022 and $322 million in 2023. Assume that the revenue continues to grow at the same rate each year until the end of 2040.

   a Find the common ratio, to two decimal places, between the revenues generated in consecutive years.

   b Using the rounded value found in part **a**, write the rule for R_n, where R_n is the revenue, in millions of dollars, generated in year $(2021 + n)$.

   c Find the revenue generated in 2026 to the nearest million dollars.

   d In which year will the revenue exceed $2 billion?

10 `WORKED EXAMPLE 11` A population of 400 quokkas is introduced into a protected area near Albany at the beginning of 2025. Immediately, the area is invaded by a red fox. The population of quokkas declines at a rate of 5% each month. The population at the end of each month, to the nearest whole number, is shown in the table.

January 2025	February 2025	March 2025
380	361	343

Take Q_n as the quokka population in the protected area n months after the quokkas were first introduced.

a Write a recursive rule in the form $Q_{n+1} = rQ_n$, $Q_0 = u$.

b Write the rule for the nth term in the form $Q_n = vr^{n-1}$ by choosing the correct value for v.

c Write the rule for the nth term in the form $Q_n = wr^n$, by choosing the correct value for w.

d To the nearest whole number, what is the quokka population at the end of April 2026?

Calculator-free

11 (6 marks) The graph shows the first 3 terms of a geometric sequence. The vertical axis provides the value of each term, A_n, for a given n on the horizontal axis.

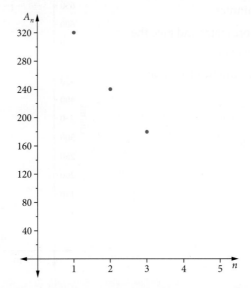

a Write the recursive rule for the geometric sequence. (3 marks)

b Write the rule for the nth term for the geometric sequence, expressing the ratio, r, as a fraction. (1 mark)

c Find the value of the 5th term, expressing your answer as a fraction. (2 marks)

12 © SCSA MA2018 Q7 (6 marks) A researcher compared the performance of various golf balls. The graph below shows the height reached above the ground by a particular golf ball after each of the first three bounces. It was initially dropped from a height of 54 cm.

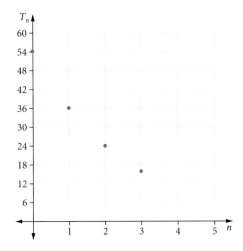

a   Write the recursive rule for this sequence.   (3 marks)

b   Write the rule for the nth term of this sequence.   (1 mark)

c   Show that the height reached by the golf ball above the ground after the fifth bounce is $\dfrac{64}{9}$ cm.   (2 marks)

Calculator-assumed

13 © SCSA MA2019 Q10 (7 marks) Ruby Ducks Coffee shops commenced operations in 1992 and had 15 stores open by the end of the year. They have been so successful over the years that the number of stores worldwide has continued to grow exponentially since then. The number of shops operating, T, at the end of 2017 was 22 579 and at the end of 2018 was 30 256.

The number of shops operating at the end of n years can be represented by the recursive rule

$T_n = 1.34T_{n-1}, T_1 = 15.$

a   Show mathematically that the common ratio is approximately 1.34.   (1 mark)

b   Write the rule for the nth term of this sequence.   (1 mark)

c   Determine the first year in which there is likely to be over 200 000 Ruby Ducks Coffee shops.   (2 marks)

Typically, each store has twelve employees working during the day across different shifts. Each employee earns, on average, $114.80 per day.

d   Calculate the total daily wages for all stores at the beginning of 2012.   (3 marks)

14 ©SCSA MA2021 Q11 (8 marks) Judith monitors the water quality in her garden pond at the same time every day. She likes to maintain the concentration of algae at between 200 and 250 units per 100 litres (L). Her measurements show that the concentration increases daily according to the recursive rule $C_{n+1} = 1.025C_n$, where $C_1 = 200$ units per 100 L (the minimum concentration).

When the concentration gets above the 250 units per 100 L limit, she treats the water to bring the concentration back to the minimum 200 units per 100 L.

a If Judith treated the water on Sunday, 6 December 2020, determine

   i the concentration on Wednesday, 9 December 2020   (2 marks)

   ii the day and date when she next treated the water.   (2 marks)

b During the first week of January 2021, Judith monitored the water and recorded the following readings.

Day	1	2	3	4	5	6	7
Concentration (C)	200	206	212.18	218.55	225.10	231.85	238.81

   i Determine the revised recursive rule.   (2 marks)

   ii If she treated the water on 10 January and went on holiday until 20 January, when she next treated the water, calculate the concentration of the water on her return, assuming the recursive rule from part **b i** is used.   (2 marks)

Video playlist
First-order
linear
recurrence
relations

(3.3) First-order linear recurrence relations

First-order linear recurrence relations

The recurrence relations for arithmetic and geometric sequences are examples of **first-order linear recurrence relations**. In this section, we will study the more general form.

'First-order' means that the only term used to recursively create T_{n+1} is its prior term, T_n; for example, $T_{n+1} = T_n + 2$ or $T_{n+1} = 0.82T_n$. The famous Fibonacci sequence, 1, 1, 2, 3, 5, 8, 13, 21 ..., where each term is the sum of the previous two terms ($T_{n+1} = T_n + T_{n-1}$), is a second-order recurrence relation and is not studied in this course.

'Linear' means that the relationship between T_{n+1} and T_n is linear. Recall that a linear equation is in the form $y = a + bx$. Replace y with T_{n+1} and x with T_n to get $T_{n+1} = a + bT_n$. Because the letter a means something else in sequences, we use c instead and arrive at the general form: $T_{n+1} = bT_n + c$. As with any recurrence relation, also state the initial term: $T_1 = a$.

> **First-order linear recurrence relations**
>
> A first-order linear recurrence relation is given by two statements:
> $$T_{n+1} = bT_n + c$$
> $$T_1 = a$$

An arithmetic sequence is produced when $b = 1$ and a geometric sequence is produced when $c = 0$. In this section, we will examine all other cases. For example, $T_{n+1} = 0.87T_n + 15$, $T_1 = 80$. In such cases, there is no common difference or common ratio between consecutive terms. Therefore, finding a rule for the nth term is mathematically complex and beyond the level of this course.

USING CAS 4	Generating sequences through recursive computation

Write the first four terms generated by this first-order linear recurrence relation

$$T_{n+1} = 4T_n - 9, \; T_1 = 2$$

ClassPad

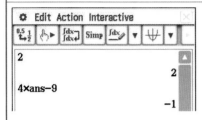
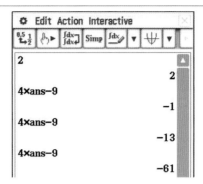

1 Input the T_1 value into the calculator and press **EXE**.

2 Enter the rule for T_{n+1}, using '**ans**' from the Soft Keyboard.

3 Press **EXE** repeatedly to generate the terms of the sequence. Because you used 'ans', you do not need to write out each line.

TI-Nspire

1 Input the T_1 value into the calculator and press **enter**.

2 Enter the rule for T_{n+1}, using '**ans**' by pressing **ctrl + (-)**.

3 Press **enter** repeatedly to generate the terms of the sequence. Because you used 'ans', you do not need to write out each line.

The first four terms are 2, −1, −13, −61.

WORKED EXAMPLE 12	Finding the constant in a first-order linear recurrence relation

Find the value of k in each of these first-order linear recurrence relations, given the first three terms of the sequence.

a 200, 180, 158 …

  $T_{n+1} = 1.1T_n + k, \; T_1 = 200$

b 10, 20, 27 …

  $T_{n+1} = 0.7T_n + k, \; T_1 = 10$

Steps	Working
a 1 Multiply T_1 by the coefficient.	$200 \times 1.1 = 220$
2 Determine the addition/subtraction needed to arrive at T_2.	To get from 220 to 180, subtract 40. $k = -40$
b 1 Multiply T_1 by the coefficient.	$10 \times 0.7 = 7$
2 Determine the addition/subtraction needed to arrive at T_2.	To get from 7 to 20, add 13. $k = 13$

 Exam hack

For these types of sequences, to find something like T_{30}, you need to generate each term recursively on the calculator. Use the shortcut techniques below.

Consider the sequence generated by the recurrence relation $M_{n+1} = 0.92M_n + 6$, $M_1 = 5$.

Generate a list of terms from M_1 to M_{50}. Record the values of M_{10}, M_{20}, M_{30}, M_{40} and M_{50}, correct to one decimal place.

ClassPad

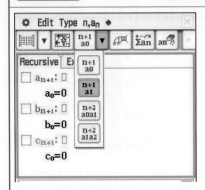

1 Open the **Sequence** application.

2 Tap on the **Recursive** tab.

3 Tap on the **n+1/a0** tool and select the **n+1/a1** option from the dropdown menu.

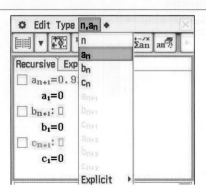

4 Enter **0.92**.

5 Tap the **n,aₙ** menu and select **aₙ** from the dropdown menu.

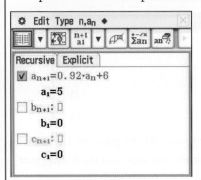

6 Enter **+6** to complete the equation, then press **EXE**.

7 Enter a_1 = **5**, then press **EXE**.

8 Tap the **Table** tool.

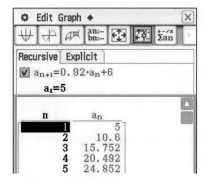

9 The recursive sequence values will appear in the lower window.

10 Tap the **Sequence Table Input** tool.

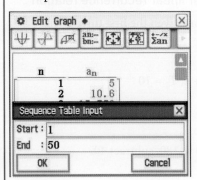

11 In the dialogue box, change the **End:** value to **50**.

12 Tap **OK**.

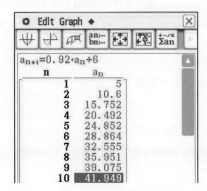

13 The table of values in the lower window will now extend to 50.

14 Tap **Resize** then scroll down to find the a_n values at $n = 10, 20, 30, 40$ and 50.

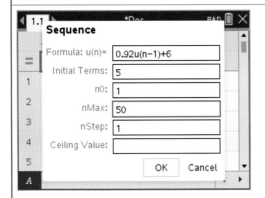

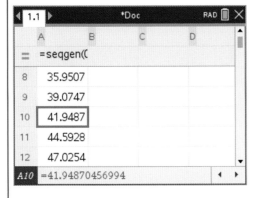

1 Open a **Lists & Spreadsheet** application.

2 Place the cursor to the cell immediately under the **A**.

3 Press **menu** > **Data** > **Generate Sequence**.

4 Complete the fields of the dialogue box as shown above, then press **enter**.

5 The recursive sequence values will be generated in column **A**.

6 Press the **down arrow** to scroll down to find the values at 10, 20, 30, 40 and 50.

Record the values of M_{10}, M_{20}, M_{30}, M_{40} and M_{50}, correct to one decimal place.

$M_{10} = 41.9$

$M_{20} = 60.6$

$M_{30} = 68.8$

$M_{40} = 72.3$

$M_{50} = 73.8$

The steady state of a first-order recurrence relation

The terms of a first-order recurrence relation reach a **steady state** if $T_{n+1} = T_n$. In other words, a steady state will have occurred if the values of the sequence no longer change.

There are three ways that the terms of a first-order recurrence relation can reach a steady state:

1 Decay to a long-term steady state

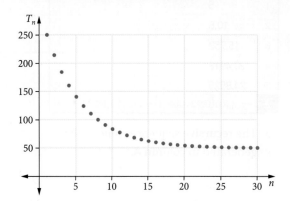

2 Grow to a long-term steady state

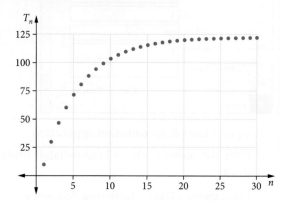

3 Start with and maintain a steady state

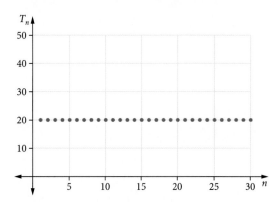

Not all first-order recurrence relations will reach a steady state. If a steady state exists, the steps to find it are:

1 take the recursive part of the recurrence relation and set both T_{n+1} and T_n to x

2 solve for x, by hand or by CAS.

For example, if the recurrence relation is $T_{n+1} = \dfrac{2}{3} T_n + 10$, $T_1 = 7$, then the equation used to find the steady state is: $x = \dfrac{2}{3} x + 10$. T_1 is not relevant. The equation can be solved by hand or by CAS, as shown in the worked examples.

> **Finding the steady state from a first-order linear recurrence relation**
>
> For $T_{n+1} = bT_n + c$, $T_1 = a$, if a steady state exists, the steps to find it are:
>
> 1 take the recursive part of the recurrence relation and set both T_{n+1} and T_n to x
>
> 2 solve for x, by hand or by CAS.

WORKED EXAMPLE 13	Finding the steady state by hand

What is the long-term steady state of the sequence defined by the following recurrence relations?

a $T_{n+1} = \dfrac{3}{4} T_n + 6$, $T_1 = 8$

b $H_{n+1} = \dfrac{1}{5} H_n + 16$, $H_1 = 10\,006$

Steps	Working
a 1 A steady state is achieved when the $(n + 1)$th term is the same as the nth term. Represent this idea with an x on both sides.	$T_{n+1} = \dfrac{3}{4} T_n + 6$ $x = \dfrac{3}{4} x + 6$
2 Move the x term from the right side of the equation to the left side. Simplify and rearrange for x.	$x - \dfrac{3}{4} x = 6$ Remember that 1 minus $\dfrac{3}{4}$ equals $\dfrac{1}{4}$. $\dfrac{x}{4} = 6$ $x = 24$ The long-term steady state is 24.
b 1 A steady state is achieved when the $(n + 1)$th term is the same as the nth term. Represent this idea with an x on both sides.	$H_{n+1} = \dfrac{1}{5} H_n + 16$ $x = \dfrac{1}{5} x + 16$
2 Move the x term from the right side of the equation to the left side. Simplify and rearrange for x.	$x - \dfrac{1}{5} x = 16$ Remember that 1 minus $\dfrac{1}{5}$ equals $\dfrac{4}{5}$. $\dfrac{4x}{5} = 16$ $\dfrac{x}{5} = 4$ $x = 20$ The long-term steady state is 20.

Finding the constant that produces a desired steady state

A desired steady state can often be achieved by changing the value of c in the recursive rule $T_{n+1} = bT_n + c$, $T_1 = a$. The steps to find the value of c to produce a desired steady state are:

1 take the recursive part of the recurrence relation and set both T_{n+1} and T_n to the desired steady state

2 solve for c, by hand or by CAS.

For example, if the recurrence relation is $T_{n+1} = \frac{2}{3}T_n + c$, $T_1 = 7$, then the equation used to find a desired

steady state of 50 is: $50 = \frac{2}{3} \times 50 + c$. T_1 is not relevant. The equation can be solved by hand or by CAS,

as shown in the worked examples.

> ## Finding the constant that produces a desired steady state
>
> For $T_{n+1} = bT_n + c$, $T_1 = a$, the steps to find the value of c that produces
> a desired steady state are:
>
> 1 take the recursive part of the recurrence relation and set both T_{n+1} and
> T_n to the desired steady state
>
> 2 solve for c, by hand or by CAS.

> ## Exam hack
>
> A shortcut for finding the constant is
>
> $c = (1 - b) \times$ steady state.
>
> But remember to show workings to get full marks.

WORKED EXAMPLE 14 Finding the constant that produces a steady state by hand

Consider the sequence generated by the first-order linear recurrence relation

$$F_{n+1} = \frac{3}{4}F_n + k, \quad F_1 = 32$$

What value of k is needed to produce a long-term steady state of 80?

Steps	Working
1 A steady state is achieved when the $(n + 1)$th term is the same as the nth term.	$F_{n+1} = \frac{3}{4}F_n + k$
Substitute the desired steady state into the equation.	$80 = \frac{3}{4} \times 80 + k$
2 Simplify.	$80 = 3 \times 20 + k$ $80 = 60 + k$
3 Rearrange for k.	$80 - 60 = k$ $k = 20$

Consider the sequence generated by the first-order linear recurrence relation

$$T_{n+1} = 0.57T_n + k, \quad T_1 = 99$$

a If k is equal to 9, what is the long-term steady state of the sequence, correct to two decimal places?

b What value of k would produce a long-term steady state of 40?

Steps	Working
a 1 A steady state is achieved when the $(n + 1)$th term is the same as the nth term. Represent this idea with an x on both sides.	$T_{n+1} = 0.57T_n + 9$ $x = 0.57x + 9$
2 Use the solve function on CAS to find the value of x.	

ClassPad

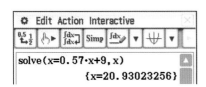

TI-Nspire

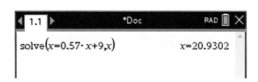

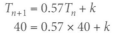

$x = 20.93$

The long-term steady state is 20.93.

b 1 A steady state is achieved when the $(n + 1)$th term is the same as the nth term. Substitute the desired steady state into the equation.	$T_{n+1} = 0.57T_n + k$ $40 = 0.57 \times 40 + k$
2 Use the solve function on CAS to find the value of k.	

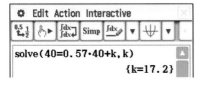

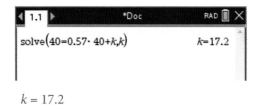

$k = 17.2$

Graphing the terms of a first-order linear recurrence relation

Terms can be generated from the recursive rule $T_{n+1} = bT_n + c$, $T_1 = a$ and plotted on a graph.
The coordinates are (n, T_n). n values are shown on the horizontal axis and T_n values are shown on the vertical axis.

The graph will always reach a steady state if b is between 0 and 1. However, unlike arithmetic and geometric sequences, growth or decay cannot be predicted by analysing b or c alone. Here are examples of graphs for different combinations of a, b and c.

$T_1 = 30$
$T_{n+1} = 1.2T_n - 5$

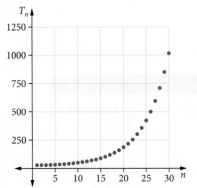

Values are increasing

$T_1 = 20$
$T_{n+1} = 1.2T_n - 5$

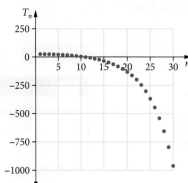

Values are decreasing

$T_1 = 25$
$T_{n+1} = 1.2T_n - 5$

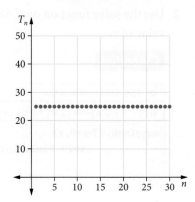

Values are a constant steady state

$T_1 = 30$
$T_{n+1} = 0.85T_n + 15$

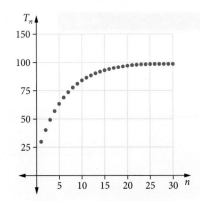

Values increase, then
level out to a steady state

$T_1 = 200$
$T_{n+1} = 0.85T_n + 15$

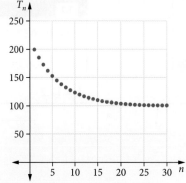

Values decrease, then
level out to a steady state

$T_1 = 100$
$T_{n+1} = 0.85T_n + 15$

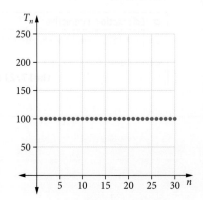

Values are a constant
steady state

🔓 Exam hack

A first-order linear recurrence relation, with $T_{n+1} = bT_n + c$, $T_1 = a$, will always reach a steady state if b is between 0 and 1.

9780170476959

Plotting the terms of a sequence

Consider the sequence generated by the recurrence relation $M_{n+1} = 0.92M_n + 6$, $M_1 = 5$.

a Find M_{10}, M_{20}, M_{30}, M_{40} and M_{50}, correct to one decimal place.

b Graph the values of M_{10}, M_{20}, M_{30}, M_{40} and M_{50}. Label the vertical axis M_n and the horizontal axis n. Label the coordinates of each point correct to one decimal place.

c Identify whether the long-term trend is increasing, is decreasing or levels out to a steady state.

Steps	Working
a Follow the steps in Using CAS 5 to find the desired terms.	These values were found in Using CAS 5. $M_{10} = 41.9$ $M_{20} = 60.6$ $M_{30} = 68.8$ $M_{40} = 72.3$ $M_{50} = 73.8$

ClassPad

b

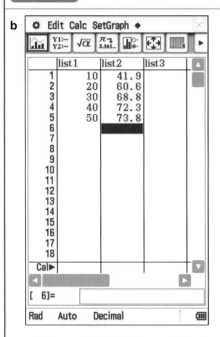 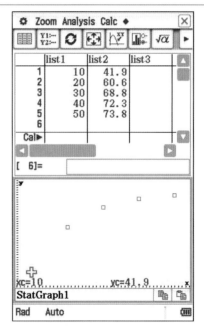

1 Open the **Statistics** application.

2 Enter the values from part **a** into **list1** and **list2**, as shown above.

3 Tap **Graph**.

4 The points will be plotted in the lower window.

5 Tap **Analysis > Trace** and press the left and right arrow keys to display the coordinates of the points (optional).

b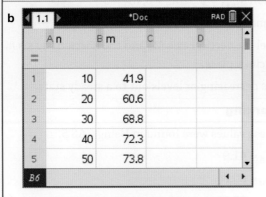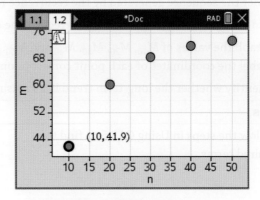

1   Add a **Lists & Spreadsheet** page.

2   Label column **A** as **n** and column **B** as **m**.

3   Enter the values into columns **A** and **B** as shown above.

4   Add a **Data & Statistics** page.

5   For the horizontal axis, select **n**.

6   For the vertical axis, select **m**.

7   The points will be plotted on the graph.

8   Press **menu** > **Analyze** > **Graph Trace** and press the left and right arrow keys to display the coordinates of the points (optional).

Plot the points and label the coordinates correct to one decimal place. Label the axes.

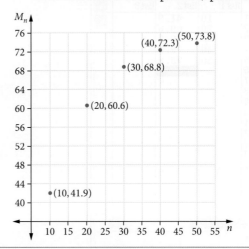

c   Observe the graph and describe the long-term trend.

In the long term, the graph levels out to a steady state.

Applying first-order linear recurrence relations

Applying first-order linear recurrence relations

The recurrence relation $T_{n+1} = bT_n + c$, $T_1 = a$ is often used to solve practical problems where

- n is a measure of time (minutes, days, weeks, months, years)
- b represents a percentage increase or decrease, which occurs during each period
- c is an addition or a subtraction, which happens after the percentage increase or decrease, each period.

Exam hack

Remember that the addition or subtraction happens after the percentage increase or decrease.

WORKED EXAMPLE 16 | Applying first-order linear recurrence relations

The number of trout in a dam decreases by 18% each year due to fishing. At the end of each year, 250 trout are added to the lake. There are 1020 trout in the dam initially. Take T_n as the number of trout in the dam at the beginning of the nth year.

a Determine a recursive rule for the number of trout in the dam.

b How many trout will there be in the long run, to the nearest whole number?

c How many trout would need to be added each year to maintain a steady state of 2000?

Some years down the track, following a natural disaster, the trout population drops to 800. The trout population then exhibits an annual natural growth rate of 7%. The new population model becomes

$$C_{n+1} = 1.07C_n + u, \quad C_1 = 800$$

d What value of u would be required to maintain a constant trout population? What does this value represent in the context of the question?

Steps	Working
a The first-order linear recurrence relation is of the form $T_{n+1} = bT_n + c$, $T_1 = a$.	
1 Identify the first term, a.	$a = 1020$
2 Find the coefficient, b, by converting the percentage change into a decimal. Subtract this from 1 due to a decreasing population.	$1 - \dfrac{18}{100} = 0.82$ $b = 0.82$
3 Find c, the number added or subtracted after the percentage change.	$c = 250$
4 Write both equations to define the recursive rule.	$T_{n+1} = 0.82T_n + 250$, $T_1 = 1020$

b 1 A steady state is achieved when the $(n + 1)$th term is the same as the nth term.

$$T_{n+1} = 0.82T_n + 250$$

Represent this idea with an x on both sides.

$$x = 0.82x + 250$$

2 Use the solve function on CAS to find the value of x.

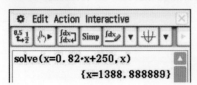

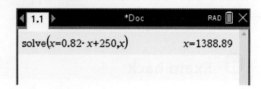

$$x = 1388.889$$

To the nearest whole number, in the long run, there will be 1389 trout in the lake.

c 1 A steady state is achieved when the $(n + 1)$th term is the same as the nth term.

$$T_{n+1} = 0.82T_n + k$$
$$2000 = 0.82 \times 2000 + k$$

Substitute the desired steady state into the equation.

2 Use the solve function on CAS to find the value of k.

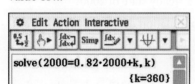

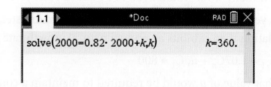

$$k = 360$$

360 trout would need to be added each year to maintain a long-term steady state of 2000.

d 1 A steady state is achieved when the $(n + 1)$th term is the same as the nth term.

$$C_{n+1} = 1.07C_n + u$$

The desired steady state is 800.

$$800 = 1.07 \times 800 + u$$

Substitute the desired steady state into the equation.

2 Use the solve function on CAS to find the value of u.

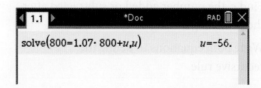

3 State the value of u and interpret the result.

$$u = -56$$

Removing 56 trout at the end of each year.

WACE QUESTION ANALYSIS

© SCSA MA2017 Q10 Calculator-assumed (12 marks)

In a laboratory experiment, the population of a particular bacteria began with 400 present. The population grew at a rate of 35% each week, where P is the number of bacteria and t is the number of weeks from the start of the experiment.

a Four possible equations were produced to model this experiment:

$P = 400(1.35)^t$

$P = 400(0.35)^t$

$P = 540(1.35)^{t-1}$

$P = 540(1.35)^{t+1}$

Identify the correct equation(s). (2 marks)

b Calculate the population of bacteria after three weeks. (1 mark)

c During which week did the population of bacteria first reach 1800? (2 marks)

d After eight weeks the growth rate slowed to 20% each week. How many weeks in total did it take for the population of bacteria to reach 15 812? (3 marks)

e What constant weekly growth rate would produce the same change in population from 400 to 15 812 in the same time as found in part **d**? (2 marks)

f Once the bacteria population reached 15 812 it began to die out at a rate of 250 each day. Approximately how many weeks did it take for the bacteria to die out completely? (2 marks)

Video
WACE question analysis: Growth and decay in sequences

Reading the question

- t is 'the number of weeks from the start of the experiment'.
- 'Identify the correct equation(s)' indicates more than one answer is possible.
- The bacteria are counted in whole numbers. Rounding will be required.
- 'as found in part **d**' indicates that the answer from part **d** will be needed to answer part **e**.

Thinking about the question

- You will need to calculate values using the different models. The value from one model might be needed to set up another model.
- Part **e** gives a rate per day but requires an answer in weeks. You will need to convert between days and weeks.
- The question relates to geometric sequences, except part **f**, which relates to an arithmetic sequence.

Worked solution ($\checkmark$ = 1 mark)

a   An increase of 35% corresponds to a ratio of 1.35.

$$r = 1 + \frac{35}{100} = 1.35$$

For a geometric sequence, the rule for the nth term is usually written as $T_n = ar^{n-1}$, where $a = T_1$. However, when T_0 is available, the rule for the nth term can also be written as $T_n = T_0 r^n$.

In this case, use 'P' instead of 'T' and use 't' instead of 'n'.

$P_0 = 400$

$P_1 = 1.35 \times P_0 = 1.35 \times 400 = 540$

Using the most common form:

$P_t = ar^{t-1}$

$P_t = 540(1.35)^{t-1}$

Using the less common form:

$P_t = P_0 r^t$

$P_t = 400(1.35)^t$

Therefore, there are two correct equations:

$\boldsymbol{P = 400(1.35)^t}$

$\boldsymbol{P = 540(1.35)^{t-1}}$

states first correct equation $\checkmark$

states second correct equation $\checkmark$

b   Either rule for the nth term can be used.

$P_t = 400(1.35)^t$

$P_3 = 400(1.35)^3$

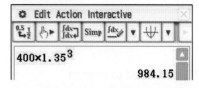

The bacteria are counted in whole numbers.

The population of bacteria after three weeks is **984**. $\checkmark$

9780170476959

c Substitute the desired value of P into either rule for the nth term.

$$P_t = 400(1.35)^t$$

$$1800 = 400(1.35)^t$$

ClassPad

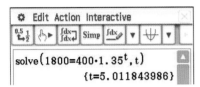

TI-Nspire

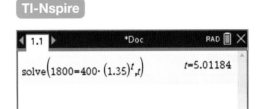

$t = 5.0118$ ✓

As the population is growing, this indicates that when $t = 5$, the number of bacteria was not yet 1800. '$t = 5$' is equivalent to 'after 5 weeks'. Therefore, a value bigger than 5 means that we are in the 6th week.

During the 6th week. ✓

d This part involves changing conditions. First, find the number of bacteria after eight weeks.

$$P_8 = 400(1.35)^8 = 4412.9615\ldots$$

4413 bacteria after 8 weeks.

Write a rule for the nth term under the new growth rate. Use B_t to represent the number of bacteria after t weeks, from the beginning of the new growth rate.

$$r = 1 + \frac{20}{100} = 1.2$$

4413 is the value of B_0, so we use the $T_n = T_0 r^n$ form of the rule.

$$B_t = 4413(1.2)^t$$

Substitute the value of 15 812 and solve for t.

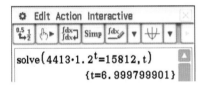

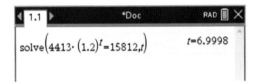

$t = 7$

In total, it took 8 weeks to reach 4413 and then another 7 weeks to reach 15 812 under the new model.

$$8 + 7 = 15$$

It took **15 weeks** for the population of bacteria to reach 15 812.

correctly calculates population after 8 weeks ✓

correctly solves equation using new growth rate ✓

correctly states total number of weeks ✓

3.3

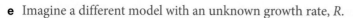

e Imagine a different model with an unknown growth rate, R.

$$T_0 = 400$$

$$T_n = 400(R)^n$$

Using the answer to part **d**, $T_{15} = 15\,812$, substitute the desired value into the equation and solve for R.

$$15\,812 = 400(R)^{15}$$

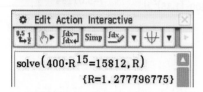

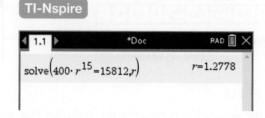

$R \approx 1.28$. Therefore, the new constant growth rate is **28%**.

correctly solves for R ✓

correctly states the new growth rate ✓

f A reduction by 250 each day refers to an arithmetic sequence. The question asks for a number of weeks, so set up a rule for the nth term, in weeks.

$$T_n = a + (n - 1)d$$

Just like in the geometric sequence, a refers to T_1. Here, however, $15\,812$ is T_0. Using T_0, the rule becomes:

$$T_n = T_0 + nd$$

d is the amount lost per week. There are 7 days in a week.

$$250 \times 7 = \textbf{1750}$$

$$d = -1750$$

$$T_n = 15\,812 - 1750n$$

When the bacteria die out completely, $T_n = 0$. Solve $0 = 15\,812 - 1750n$ for n.

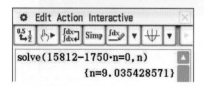

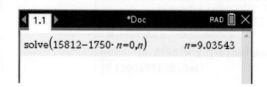

The question says 'approximately', which allows us to round to the nearest whole number.

Time taken is approximately **9 weeks**.

correctly multiplies 250 by 7 ✓

correctly solves equation and states approximate time ✓

Recap

1 Consider the sequence generated by the recursive rule

$$T_{n+1} = \frac{1}{3} T_n, \; T_1 = 3.$$

If n is plotted on the horizontal axis and T_n on the vertical axis, which is the correct graph of the terms of the sequence?

A T_n

B T_n

C T_n

D T_n

E T_n

2 Consider the arithmetic sequence 30, 38, 46, 54 … and the geometric sequence 2, 3, 4.5, 6.75 … Which sequence has the larger value at position 13? What is the difference between the larger and smaller value, correct to two decimal places?

Mastery

3 [Using CAS 4] Write the first four terms generated by each of these first-order linear recurrence relations.

a $T_{n+1} = 3T_n + 5$, $T_1 = 1$

b $T_{n+1} = \frac{1}{2} T_n - 7$, $T_1 = 26$

c $T_{n+1} = 1.5T_n + 10$, $T_1 = 20$

d $T_{n+1} = 0.9T_n + 7$, $T_1 = 70$

4 [WORKED EXAMPLE 12] Find the value of k in each of these first-order linear recurrence relations, given the first three terms of the sequence.

a 350, 360, 369 …

$T_{n+1} = 0.9T_n + k$, $T_1 = 350$

b 100, 80, 58.4 …

$T_{n+1} = 1.08T_n + k$, $T_1 = 100$

c 1050, 2150, 5450 …

$T_{n+1} = 3T_n + k$, $T_1 = 1050$

5 [Using CAS 5] [Using CAS 6] For each of the following first-order linear recurrence relations

   i graph the values of M_{10}, M_{20}, M_{30}, M_{40} and M_{50}. Label the vertical axis M_n and the horizontal axis n. Label the coordinates of each point correct to one decimal place.

   ii identify whether the long-term trend is increasing, is decreasing or levels out to a steady state.

a $M_{n+1} = 0.875M_n + 4$, $M_1 = 1$

b $M_{n+1} = 1.05M_n - 2$, $M_1 = 51$

6 ⊘ WORKED EXAMPLE 13 What is the long-term steady state of the sequence defined by the following recurrence relations?

a $T_{n+1} = \dfrac{1}{4} T_n + 24, T_1 = 10$

b $D_{n+1} = \dfrac{3}{5} D_n + 14, D_1 = 250$

c $J_{n+1} = \dfrac{4}{5} J_n + 3, J_1 = 1000$

7 ⊘ WORKED EXAMPLE 14 Consider the sequence generated by the first-order linear recurrence relation $F_{n+1} = \dfrac{2}{3} F_n + k, F_1 = 25$. What value of k is needed to produce a long-term steady state of

a 15   **b** 33?

8 ⊘ WORKED EXAMPLE 15 Consider the sequence generated by the first-order linear recurrence relation $T_{n+1} = 0.63T_n + k, T_1 = 68$.

a If k is equal to 7, what is the long-term steady state of the sequence, correct to two decimal places?

b What value of k would produce a long-term steady state of

  i 30   **ii** 75   **iii** −8?

9 ⊘ WORKED EXAMPLE 16 In a region of the Kimberley, a colony of red-collared lorikeets experiences population decline due to gradual loss of habitat. During each year, 4% of the population is lost. At the end of each year, 30 red-collared lorikeets from more adversely affected areas migrate to the region. There are 980 birds initially. Let L_n represent the number of birds in the region at the beginning of the nth year.

a Determine a recursive rule for the red-collared lorikeet population.

b How many birds will there be in the long run, to the nearest whole number?

c How many birds would need to migrate to this region each year to maintain a steady state of 1100?

Some years after achieving the steady state, the loss of habitat is reversed and the population exhibits an annual natural growth rate of 8%. The new population model becomes $C_{n+1} = 1.08C_n + u, C_1 = v$.

d What is the value of v?

e What value of u would be required to maintain the constant red-collared lorikeet population found in part **b**? What does this value represent in the context of the question?

Calculator-free

10 © SCSA MA2019 Q6 (6 marks) The population of turtles in an artificial lake at a wildlife sanctuary is initially 32 and research has shown a natural decrease in population of 50% each year. Twenty extra turtles are introduced to the lake at the end of each year.

a Determine a recursive rule for the turtle population. (2 marks)

b Determine the long-term steady state of the turtle population. (2 marks)

c If the wildlife sanctuary preferred a long-term steady state of 80 turtles, what yearly addition of turtles would be required to produce this steady state? Assume all other conditions remain the same. (2 marks)

11 © SCSA | MA2017 Q16 | (8 marks) In a Northern Territory river, the crocodile population is dropping by 7.5% each year. The current population is 200. A scheme is being trialled under which 20 crocodiles are introduced to the river each year.

The population of crocodiles in the river can be modelled by the first-order linear recurrence relation $T_{n+1} = 0.925T_n + b$, $T_1 = 200$, where T_n is the number of crocodiles in the river at the beginning of the nth year.

a **i** Interpret the coefficient 0.925 in the context of the question. (1 mark)

ii State the value of b. (1 mark)

b Copy the axes and graph the number of crocodiles in the river for every five-year period (commencing at $n = 5$), up to the 30th year. (2 marks)

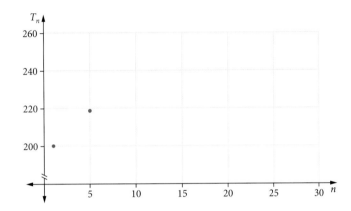

c Using your graph, comment on how the population of crocodiles is changing over time. (2 marks)

d To the nearest whole number, what is the long-term effect on the crocodile population? (2 marks)

Arithmetic sequences

An **arithmetic sequence** is defined by a starting term, $a = T_1$, and a common difference, d, between consecutive terms.

- The **recurrence relation** for an arithmetic sequence is given by two statements: $T_{n+1} = T_n + d$ and $T_1 = a$.
- The rule for the nth term of an arithmetic sequence is $T_n = a + (n - 1)d$.
- Growth in an arithmetic sequence occurs when $d > 0$ and is represented by a positive slope on the graph.

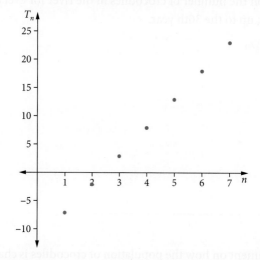

- Decay in an arithmetic sequence occurs when $d < 0$ and is represented by a negative slope on the graph.

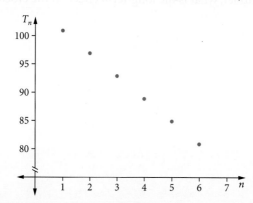

- A shortcut for summing the first n terms of an arithmetic sequence is $\dfrac{T_1 + T_n}{2} \times n$.

Geometric sequences

A **geometric sequence** is defined by a **common ratio**, r, between consecutive terms. Each new term is equal to its preceding term multiplied by this constant ratio. a represents the first term, T_1.

- The common ratio, r, can be found by dividing any term by its previous term: $r = \dfrac{\text{any term}}{\text{previous term}}$.

- The recurrence relation for a geometric sequence is given by two statements: $T_{n+1} = rT_n$ and $T_1 = a$.

- The rule for the nth term of a geometric sequence is $T_n = ar^{n-1}$, where $a = T_1$.

- An alternative form for the rule for the nth term of a geometric sequence is $T_n = T_0r^n$.

- Growth in a geometric sequence occurs when $r > 1$ and is represented by a positive slope on the graph.

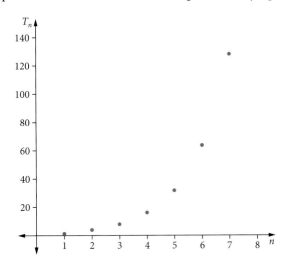

- Decay in a geometric sequence occurs when $0 < r < 1$ and is represented by a negative slope on the graph

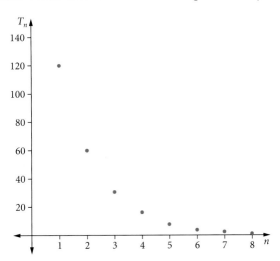

First-order linear recurrence relations

- A **first-order linear recurrence relation** is given by two statements: $T_{n+1} = bT_n + c$ and $T_1 = a$.

- If a **steady state** exists, the steps to find it are:

    1 take the recursive part of the recurrence relation and set both T_{n+1} and T_n to x

    2 solve for x, by hand or by CAS.

- The steps to find the value of c that produces a desired steady state are:

    1 take the recursive part of the recurrence relation and set both T_{n+1} and T_n to the desired steady state

    2 solve for c, by hand or by CAS.

Cumulative examination: Calculator-free

Total number of marks: 19   Reading time: 2 minutes   Working time: 19 minutes

1 © SCSA MA2016 Q4 (12 marks)

   a Given the sequence 256, 128, 64, 32 …

      i Write a recursive rule for the sequence. (2 marks)

      ii Deduce a rule for the nth term of this sequence. Hence, calculate the 15th term, leaving your answer as a fraction. (3 marks)

   b Use the recursive definitions given to state the first three terms of each of the following sequences.

      i $T_{n+1} = T_n + 7$, $T_1 = 11$ (2 marks)

      ii $T_{n+1} = 1.5T_n$, $T_2 = 7.5$ (2 marks)

   c Consider the sequence 12, 7, 2, –3 …

      By deducing a rule for the nth term, or otherwise, determine which term of the sequence is –168. (3 marks)

2 © SCSA MA2021 Q6 (7 marks) A television network programmer was analysing the number of viewers for two children's programs over a period of several weeks, to decide which program should be given the better time slot. The viewing numbers, displayed on the graph below, formed an arithmetic sequence and a geometric sequence.

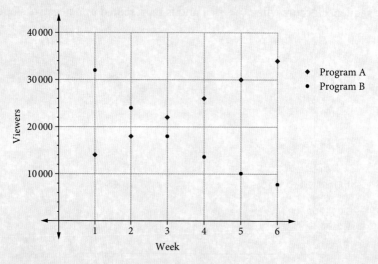

   a Write a recursive rule for the arithmetic sequence. (2 marks)

   b Using the first two data points, deduce a rule for the nth term of the geometric sequence. (2 marks)

   c Explain which program should be given the better time slot. (2 marks)

   d Determine the number of viewers for the more successful program in Week 8. (1 mark)

Cumulative examination: Calculator-assumed

Total number of marks: 26   Reading time: 3 minutes   Working time: 26 minutes

1 (4 marks) A sample of 96 birds are grouped according to their beak size (small, medium, large). The percentage of birds in each group is calculated. The results are displayed in Table 1.

Table 1

Beak size	Percentage (%)
small	25
medium	44
large	31
Total	100

a How many of the 96 birds have small beaks?   (1 mark)

b Use the percentages in Table 1 to construct a percentage segmented bar chart. Copy the template below to assist you in completing this task. Use the key to indicate the segment of your bar chart that corresponds to each beak size.   (1 mark)

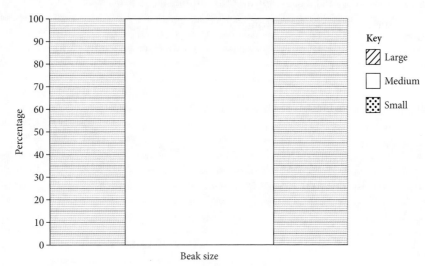

c In order to investigate a possible association between beak size and sex, the same birds are grouped by both their beak size (small, medium, large) and their sex (male, female). The results of this grouping are shown in Table 2.

Table 2

Beak size	Sex	
	Male	Female
small	1	23
medium	26	16
large	27	3
Total	54	42

Does the information provided above support the contention that *beak size* is associated with *sex*? Justify your answer by quoting appropriate percentages. It is sufficient to consider one *beak size* only when justifying your answer.   (2 marks)

2 (5 marks) The congestion level in a city can be recorded as the percentage increase in travel time due to traffic congestion in peak periods (compared to non-peak periods). This is called the percentage congestion level. The percentage congestion levels for the morning and evening peak periods for 19 large cities are plotted on the scatterplot.

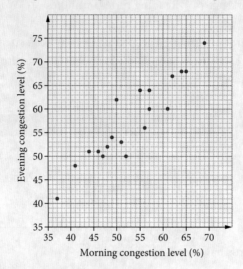

A least-squares line is to be fitted to the data with the aim of predicting *evening congestion level* from *morning congestion level*. The equation of this line is

 evening congestion level = 0.922 × *morning congestion level* + 8.48

a Name the response variable in this equation. (1 mark)

b Use the equation of the least-squares line to predict the evening congestion level when the morning congestion level is 60%. (1 mark)

c Determine the residual value when the equation of the least-squares line is used to predict the evening congestion level when the morning congestion level is 47%. Round your answer to one decimal place. (2 marks)

d The value of the correlation coefficient *r* is 0.92. What percentage of the variation in the evening congestion level can be explained by the variation in the morning congestion level? Round your answer to the nearest whole number. (1 mark)

3 © SCSA MA2020 Q8 (9 marks) A farmer has a large lake on his farm and has started stocking it with fish of a variety that will flourish in the conditions in this lake. Monitoring has shown that the number of adult fish is increasing at a consistent rate of 9% per month and at the beginning of 2020 the lake holds 660 of the adult fish.

a Write a recursive rule to give the number of adult fish in the lake at the end of each month from the beginning of 2020. (2 marks)

b Deduce a rule for the *n*th term of this sequence. (2 marks)

The farmer plans to allow the general public to pay to fish in the lake. This will commence at the beginning of the next month after the adult fish population first reaches 4000.

c Determine how many months after the beginning of 2020 fishing will commence. (2 marks)

d The farmer wishes to maintain a steady state in the adult fish population once fishing commences. Calculate how many adult fish can be taken from the lake each month. (3 marks)

4 © SCSA MA2018 Q9 (8 marks) Deborah is purchasing mealworms for her pet lizard, Lizzy, to eat. Deborah starts by buying 50 mealworms. She then buys an additional 15 at the start of each subsequent week. She feeds 12 mealworms to Lizzy each week, and each week a certain percentage of the mealworms dies.

Deborah has found that the approximate number of mealworms at the start of the nth week can be modelled by M_n where $M_{n+1} = 0.9(M_n - 12) + 15$, $M_1 = 50$.

a What percentage of the mealworms dies each week? (1 mark)

b Determine the approximate number of mealworms Deborah has at the start of the fifth week. (1 mark)

c Deborah claims that she will never run out of mealworms using this model. Justify her claim. (2 marks)

After 10 weeks, hot weather results in a larger percentage of the mealworms dying, so Deborah alters the model to

$$N_{n+1} = 0.8(N_n - 12) + 15, N_1 = c$$

d **i** Determine the value of c. (1 mark)

  ii Determine the approximate number of mealworms Deborah has at the start of the thirtieth week. (1 mark)

Deborah's vet recommends feeding Lizzy 10 mealworms a week. She would also like to maintain a constant number of 30 mealworms at the start of each week, so she changes the above model to:

$$P_{n+1} = 0.8(P_n - 10) + k$$

e Determine the value of k, the number of mealworms she must buy each week, to ensure this occurs. (2 marks)

CHAPTER

4 GRAPHS AND NETWORKS

TOPIC 3.3: GRAPHS AND NETWORKS

The definition of a graph and associated terminology

3.3.1   demonstrate the meanings of, and use, the terms: graph, edge, vertex, loop, degree of a vertex, subgraph, simple graph, complete graph, bipartite graph, directed graph (digraph), arc, weighted graph, and network

3.3.2   identify practical situations that can be represented by a network, and construct such networks

3.3.3   construct an adjacency matrix from a given graph or digraph and use the matrix to form multi-stage matrices to solve associated problems

Planar graphs

3.3.4   demonstrate the meanings of, and use, the terms: planar graph and face

3.3.5   apply Euler's formula, $v + f - e = 2$ to solve problems relating to planar graphs

Paths and cycles

3.3.7   investigate and solve practical problems to determine the shortest path between two vertices in a weighted graph (by trial-and-error methods only)

Mathematics Applications ATAR Course Year 12 syllabus pp. 10–11 © SCSA

4.1 Graphs, networks and their features

Graphs, vertices and edges

A **graph** is a diagram consisting of a set of points, called **vertices**, that are connected by a set of lines called **edges**. A graph is often named using a capital letter; for example, G. The vertices of a graph are typically labelled using a single letter, either capital or lower case, but they can also appear without labels.

An edge typically joins two vertices and is named by the two vertices it connects. For example, in graph G shown, the edge connecting vertices A and B is called edge AB. Any two vertices connected by an edge are called **adjacent vertices**. When an edge connects a vertex of a graph to itself, it is called a **loop**, as seen at vertex C. When two adjacent vertices are connected by two or more edges, the edges are called **multiple edges**. For example, there are multiple edges AE in the graph shown.

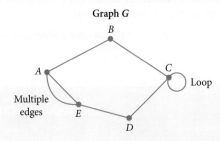

The **degree** of a vertex V is the number of edges connected to that vertex and can be denoted as $\deg(V)$. For graph G:

- B and D have degree 2, i.e. $\deg(B) = \deg(D) = 2$
- A and E have degree 3, i.e. $\deg(A) = \deg(E) = 3$
- C has degree 4, i.e. $\deg(C) = 4$, as a loop counts as two because it is connected to the vertex twice.

We can see that for graph G above containing 7 vertices, the sum of the degrees of all the vertices is $3 + 2 + 4 + 2 + 3 = 14$. This value is called the **degree sum**, S, and is twice the number of edges in a graph, meaning that the degree sum is always even.

$$S = 2 \times \text{number of edges}$$

This is because for any given edge, there will always be two vertices for which it connects to.

Suppose a graph contained a vertex that did not connect to any other vertices. This vertex is called an **isolated vertex** and has degree 0. For example, vertex D in the graph on the right is an isolated vertex.

It is also useful to note that the way in which a graph can be drawn is not necessarily unique. For example, the graph shown with five vertices and three edges below is equivalent to the graph drawn so that BE does not cross edges AD and CD.

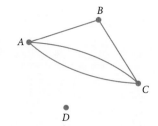

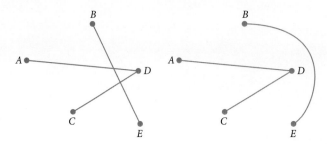

Sometimes drawing a graph so that the edges are not crossing can be helpful in identifying the number of vertices in the graph.

Features of a graph

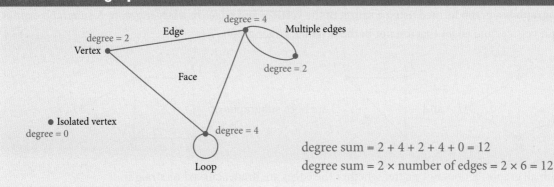

degree sum = 2 + 4 + 2 + 4 + 0 = 12
degree sum = 2 × number of edges = 2 × 6 = 12

WORKED EXAMPLE 1 | **Identifying the features of graphs**

For each of the following graphs

  i count and list the vertices and edges

  ii show that the degree sum is twice the number of edges.

a

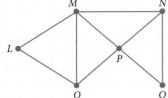

b

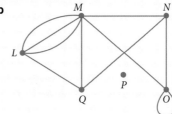

Steps	**Working**
a **i** Count and list the number of vertices and edges.	6 vertices: *L, M, N, O, P, Q* 9 edges: *LM, LQ, MQ, MP, MN, NP, NO, PO, PQ*

ii Find the degree of each vertex and then add them.
 	
Show that multiplying the number of edges by 2 gives the same result.
	<table><tr><td>**Vertex**</td><td>*L*</td><td>*M*</td><td>*N*</td><td>*O*</td><td>*P*</td><td>*Q*</td><td>**Sum**</td></tr><tr><td>**Degree**</td><td>2</td><td>4</td><td>3</td><td>2</td><td>4</td><td>3</td><td>**18**</td></tr></table>
	degree sum = 2 × number of edges
	  = 2 × 9
	  = 18

b **i** **1** Redraw the graph to uncross the intersecting edges to show that there is no vertex at the point of intersection.	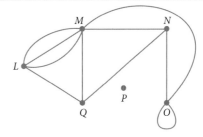
2 Count and list the number of vertices and edges.	6 vertices: *L, M, N, O, P, Q* 10 edges: *LM* × 3, *LQ, MQ, MN, MO, NO, NQ, OO*

ii Find the degree of each vertex and then add them.
 	
Show that multiplying the number of edges by 2 gives the same result.
	<table><tr><td>**Vertex**</td><td>*L*</td><td>*M*</td><td>*N*</td><td>*O*</td><td>*P*</td><td>*Q*</td><td>**Sum**</td></tr><tr><td>**Degree**</td><td>4</td><td>6</td><td>3</td><td>4</td><td>0</td><td>3</td><td>**20**</td></tr></table>
	degree sum = 2 × number of edges
	  = 2 × 10
	  = 20

Subgraphs

A **subgraph** is a graph formed using a subset of the vertices and edges in a larger graph. A subgraph cannot contain vertices and edges that are not in the original, larger graph.

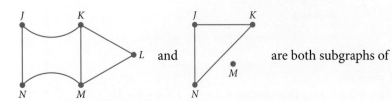

As with all graphs, it does not matter whether the edges are drawn curved or straight.

> ### Subgraphs
> Any selection of vertices and edges chosen from this graph makes up a **subgraph**.

WORKED EXAMPLE 2 | **Identifying and drawing subgraphs**

Consider the graph X.

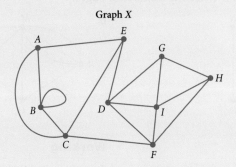

a State, with reasons, whether the following graph is a subgraph of graph X.

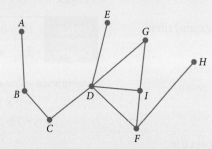

b Draw a subgraph of X containing one vertex of degree 2 and one edge.

Steps	Working
a Does the graph contain *only* vertices and edges from the original graph?	It is not a subgraph because the larger graph does not have the edge CD.
b Identify that for a single vertex to have a degree 2, it must be a loop.	B ⬭

9780170476959

Networks: applying graphs and subgraphs

A **network** is a group of interconnected elements such as people, places or things. A graph can be used to show these connections in real-life situations such as road systems, maps, railway networks, friendships and social connections, food webs and round-robin sporting competitions. When applying graphs to practical situations in the form of a network, it is important to identify what the set of vertices and set of edges represent. For example, in a road system, the labelled vertices may represent towns and the edges may represent the routes between each of the connected towns.

WORKED EXAMPLE 3	Representing road systems as a network

The road system shows how roads connect the three towns of Springfield (*S*), Hawkins (*H*) and Rosewood (*R*).

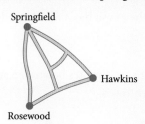

a Determine whether the network representing this road system contains:

  i loops.

  ii multiple edges.

Justify your answer.

b Hence, construct the network representing the road system between *S*, *H* and *R*.

c Is it possible to construct a subgraph containing *S*, *H* and *R*, with three edges and one isolated vertex? If so, draw this subgraph. If not, justify why.

Steps	Working
a **i** Find the loops by identifying a route from a town back to the same town that does not go through another town.	There is one route from *S* to *S*. 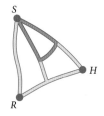 So, the graph has a loop at *S*. There is one route from *H* to *H* without going through another town. 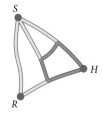 So, the graph has a loop at *H*.

ii Find the multiple edges by identifying routes between two towns that do not go through another town.

There are 3 routes between S and R that do not go through one of the other towns.

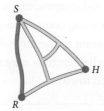 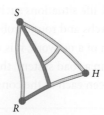 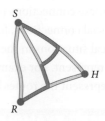

So, the graph has 3 multiple edges joining S and R.

There are 4 routes between S and H that do not go through one of the other towns.

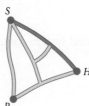 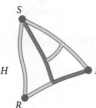 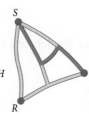 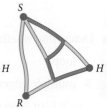

So, the graph has 4 multiple edges joining S and H.

There are 2 routes between R and H that do not go through one of the other towns.

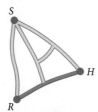 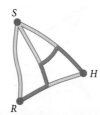

So, the graph has 2 multiple edges joining R and H.

b Draw the network.

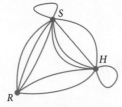

c 1 In order for one of the three vertices to be isolated, the subgraph must contain either a loop or multiple edges.

2 Isolate one vertex and connect the other two, using three edges.

Note: Other answers are possible.

WORKED EXAMPLE 4   Representing a region as a network

A small suburban park space has five areas, labelled as A to E on the map. Area D is surrounded by a small pond and only accessible by water. Draw a network showing the land connections between the five areas.

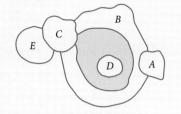

Steps	**Working**
1 List all the land connections.	E and C have 1 land connection: 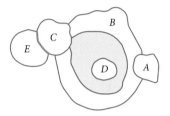
	C and B have 2 land connections: 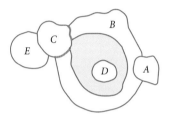
	A and B have 1 land connection: 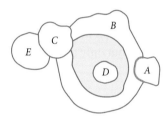
	D has no land connections.
2 Let A, B, C, D and E be the vertices of the graph. The land connections are the edges. Draw the vertices and connect them with the number of edges.	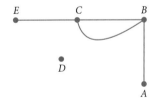

Mastery

1 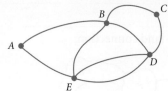 For each of the following graphs

   i count and list the vertices and edges

   ii show that the degree sum is twice the number of edges.

a

b

2 WORKED EXAMPLE 2 Consider graph H.

Graph H

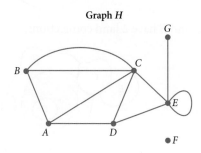

  a State, with reasons, whether the graph on the right is a subgraph of H.

  b Draw a subgraph of H containing five vertices and four edges, but only one isolated vertex.

3 WORKED EXAMPLE 3 The road system shows how roads connect the four towns Karton (K), Lucton (L), Michelton (M) and Nellton (N).

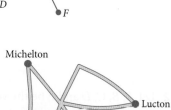

  a Determine whether the network representing this road system contains:

    i loops.   **ii** multiple edges.

    Justify your answer.

  b Hence, construct the network representing the road system between K, L, M and N.

  c Construct a possible subgraph containing three edges and all vertices, but where L is the only isolated vertex.

4 WORKED EXAMPLE 4 The city of Freshwater is divided into six suburbs, labelled as A to F on the map. A lake in the middle of the city is shown on the map in blue. Draw a network diagram showing the land connections between the six suburbs.

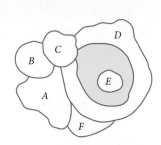

5 (2 marks) Consider the five graphs below.

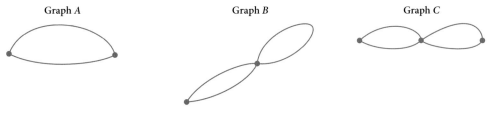

Graph A   Graph B   Graph C

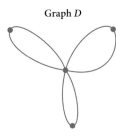

Graph D   Graph E

    a Identify which of the graphs contains a loop.   (1 mark)

    b Identify which of the graphs contains multiple edges.   (1 mark)

6 (5 marks)

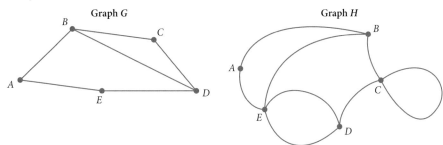

Graph G   Graph H

    a State the degree sum of graph G shown.   (2 marks)

    b Graph H has five vertices and eight edges.

        i Identify the pair of adjacent vertices that are connected by multiple edges.   (1 mark)

        ii List the vertices that have an even degree.   (1 mark)

        iii State the degree sum of graph H.   (1 mark)

7 (4 marks) Consider the graph.

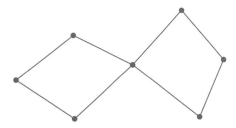

    a Label the vertices of the graph using the letters A to G, so that vertex A has
the greatest degree.   (2 marks)

    b Is it possible to draw a subgraph of the above graph containing all seven vertices,
so that the degree of each vertex is 2? Justify your answer.   (2 marks)

▶ **8** (2 marks)

a Consider the five graphs shown.

Graph G

Graph H

Graph I

Graph J

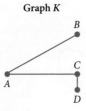

Graph K

Identify which of the graphs contains the following subgraph. (1 mark)

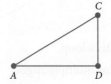

b Consider the five graphs shown.

Graph L

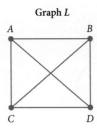

Graph M

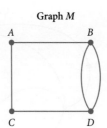

Graph N

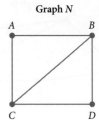

Graph O

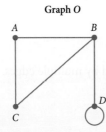

Graph P

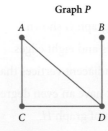

Identify which of the graphs contains the following subgraph. (1 mark)

9780170476959

9 (3 marks) The West Coast Cricket Club has five new players join its team: Alex, Bo, Cameron, Dale and Emerson. The graph shows the players who have played cricket together before joining the team. For example, the edge between Alex and Bo shows that they have previously played cricket together.

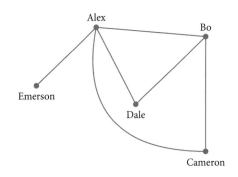

a List the players that have previously played cricket with

   i Emerson. (1 mark)

   ii both Alex and Bo. (1 mark)

b During the season, another new player, Finn, joined the team.

  Finn had not played cricket with any of these players before. State the term used to describe Finn in the context of graph theory. (1 mark)

10 (3 marks) A map of the roads connecting five suburbs of a city, Appletown (A), Belfort (B), Canningside (C), Dalton (D) and East Swan (E), is shown.

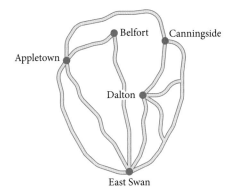

a Starting at Belfort, which **two** suburbs can be driven to using only one road? (1 mark)

b A graph that represents the map of the roads is shown. One of the edges that connects to vertex E is missing from the graph.

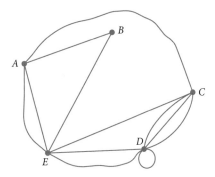

   i Copy the graph and add the missing edge. (1 mark)

   ii Explain what the loop at D represents in terms of a driver who is departing from Dalton. (1 mark)

11 (8 marks) The map of Australia shows the six states, the Northern Territory and the Australian Capital Territory (ACT). In the network diagram, each of the vertices *A* to *H* represents one of the states or territories shown on the map of Australia. The edges represent a border shared by two states or by a state and territory.

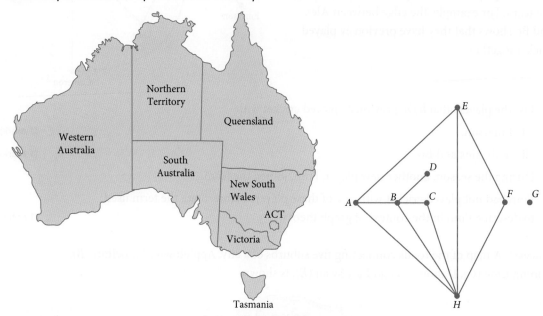

a Identify the vertex that represents Tasmania. Justify your answer. (2 marks)

b Identify the vertex that represents Western Australia. Justify your answer. (2 marks)

c Identify the vertex that represents the Australian Capital Territory (ACT) and state its degree. (2 marks)

d State the degree sum of the above network and explain its significance in the context of the map of Australia. (2 marks)

Video playlist
Types of undirected graphs and their applications

Worksheet
Planar graphs

(4.2) Types of undirected graphs and their applications

Graphs can be directed or undirected. In this section, we will explore different types of **undirected graphs,** which are graphs containing edges that do not indicate a direction between two adjacent vertices. In section 4.3, we will introduce the concept of a directed graph.

Simple and complete graphs

A **simple graph** is a graph without loops or multiple edges between adjacent vertices.

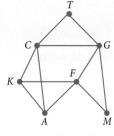

simple graph

not a simple graph
because it has two edges
between vertices *T* and *G*

A **complete graph** is a simple graph where every vertex is connected to every other vertex. A complete graph with n vertices can be called K_n. The complete graph, graph G, is an example of K_4. Suppose graph H contained the edge BE, then it would be an example of K_5.

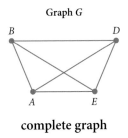

Graph G

complete graph

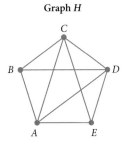

Graph H

not a complete graph because
B and E are not connected

Complete graphs

Given that each vertex in a complete graph, K_n, with n vertices is connected to every other vertex, then the degree of each vertex is $n - 1$.

WORKED EXAMPLE 5 | Simple and complete graphs

Justify whether the graph on the right is a

a simple graph.

b complete graph.

Steps	Working
a Identify the existence of any loops or multiple edges.	The graph contains loops at A and G, and multiple edges between vertices F and H. Therefore, it is not a simple graph.
b Recall that the first condition of a complete graph is that it must be simple.	Given that the graph is not simple, it therefore cannot be complete.

Connected graphs and bridges

A graph is said to be a **connected graph** if any vertex can be reached using a sequence of edges, starting from any other vertex. Otherwise, the graph is called disconnected. An edge that keeps the graph connected is called a **bridge**.

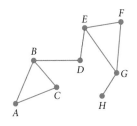

The graph shown on the right is a connected graph that has three bridges BD, DE and GH, shown in red.

Removing edge:	*BD*	*DE*	*GH*
Results in a disconnected graph			

9780170476959 **Chapter 4** | Graphs and networks **155**

Determine the number of bridges in each of these connected graphs. Copy the graphs and indicate the bridges by drawing them in red.

a

b

c

d

Steps	Working
Decide which edge(s) will make the graph disconnected if deleted.	**a** one bridge 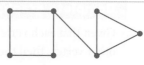
	b two bridges 
	c no bridges
	d four bridges 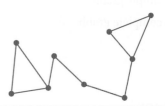

Planar graphs and Euler's formula

Planar graphs are connected graphs that can be drawn so that they don't have any edges crossing. It doesn't matter whether the graph is actually drawn with crossed edges. What's important is how it *can* be drawn.

K_4 is an example of a planar graph. Although two edges are crossing, it can be redrawn so that no edges are crossing:

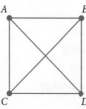 Redraw

However, K_5 is *not* a planar graph. It's possible to redraw it so that *some* of the edges aren't crossing, but it's impossible to redraw it with *no* edges crossing:

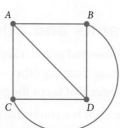 Redraw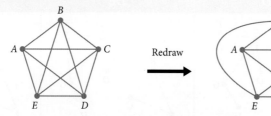

Trying to uncross the last two edges always results in a different cross.

9780170476959

Faces are the regions of a planar graph that are enclosed or bounded by edges. Before we count faces, we need to check that no edges are crossing. If they are, redraw the graph with no crossings where possible. This graph has six faces.

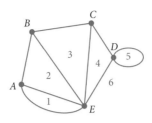

Note that:

- the loop at D, where an edge starts and ends at the same vertex, creates a face
- two multiple edges between A and E create a face
- the infinitely large region outside the graph counts as a face.

Euler's (pronounced 'oiler's') **formula** applies to graphs that are planar and, hence by definition, must be connected. The formula states

    number of vertices + number of faces − number of edges = 2

That is, $v + f - e = 2$.

Graph	Connected/Planar	Euler's formula
	Connected ✓ Planar ✓	$v = 4, f = 4, e = 6$ $v + f - e$ $= 4 + 4 - 6$ $= 2$ Euler's formula works for this graph.
	Connected ✓ Planar ✗ It's not possible to redraw the graph with all edges uncrossed so it's not planar.	If we can't uncross all the edges, we can't identify all the faces, so Euler's formula doesn't work for this graph.
	Connected ✗ Planar ✗	Euler's formula doesn't work for this graph as it is not planar.

Euler's formula

For planar graphs

    $v + f - e = 2$

where v = the number of vertices

    f = the number of faces

    e = the number of edges.

WORKED EXAMPLE 7	Planar graphs, connected graphs and Euler's formula

For the graph shown

a redraw it to show it is a planar graph

b show that Euler's formula holds true.

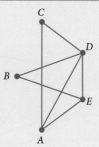

Steps	Working
a To uncross the edges, move edges AC and AD to the outside of the graph.	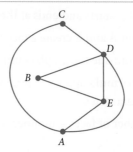 The graph can be redrawn without any edges crossing, so it is a planar graph.
b Count the number of vertices, faces and edges, and substitute into Euler's formula to see if the result is 2.	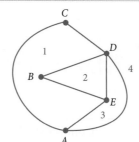 $v = 5, f = 4, e = 7$ $v + f - e = 5 + 4 - 7$ $\qquad\qquad = 2$ Euler's formula works for this graph.

WORKED EXAMPLE 8	Applying Euler's formula

a A planar graph has 14 edges and 5 faces. Determine the number of vertices.

b A planar graph has 10 vertices and 4 faces. Determine the number of edges.

Steps	Working
a Substitute the known values into Euler's formula and solve to find the number of vertices, v.	Substitute $e = 14$ and $f = 5$ into $v + f - e = 2$: $v + 5 - 14 = 2$ $\qquad v = 2 - 5 + 14$ $\qquad v = 11$ The number of vertices is 11.
b Substitute the known values into Euler's formula and solve to find the number of edges, e.	Substitute $v = 10$ and $f = 4$ into $v + f - e = 2$: $10 + 4 - e = 2$ $\qquad e = 10 + 4 - 2$ $\qquad e = 12$ The number of edges is 12.

9780170476959

Bipartite graphs

A **bipartite graph** is a graph with vertices that can be separated into two distinct sets, for example, set X and set Y, so that each edge of the graph only connects a vertex from set X to set Y.

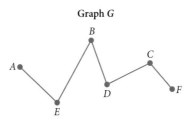

Graph G

For example, graph G is an example of a bipartite graph because vertices A, B and C could form one set of vertices and vertices D, E and F could form another, so that no vertex is connected to any other vertex from the same set.

A very simple example of a bipartite graph is the connected graph with two vertices and one edge.

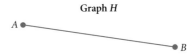

Graph H

A useful way of checking whether a graph is bipartite is to use a vertex two-colouring technique. That is, if the vertices of a graph can be coloured in two distinct colours so that no two adjacent vertices are of the same colour, then the graph is bipartite. For example, graphs G and H can be shown to be bipartite using a blue–red colouring technique.

Graph G Graph H

Graph H is also an example of a **complete bipartite graph**, as every vertex in the first distinct set is connected to every vertex in the second distinct set.

Suppose a complete bipartite graph has two sets of vertices with m and n vertices respectively. Then this graph can be denoted by $K_{m,n}$. Graph H above is the example $K_{1,1}$.

The graphs of $K_{2,3}$ and $K_{3,1}$ are shown below.

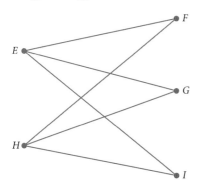

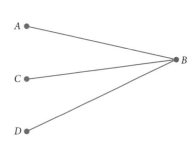

Consider the graph.

Determine, using an appropriate technique, whether the graph is bipartite. If so, identify the edges that need to be added in order for the graph to be complete bipartite. If not, identify the feature of the graph that makes it not bipartite.

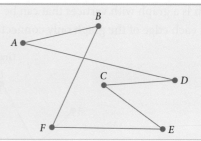

Steps	Working
1 Starting at any vertex, use a two-colour technique to colour the adjacent vertices alternating colours. **2** If no two adjacent vertices share the same colour, it is bipartite.	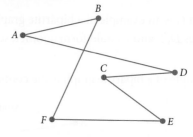 No two adjacent vertices share the same colour. Therefore, the graph is bipartite.
3 Redraw the graph so that adjacent vertices are in different 'columns' or sets of vertices. **4** To make the graph complete bipartite, add edges to ensure that every red vertex is connected to every other blue vertex, and that every blue vertex is connected to every other red vertex.	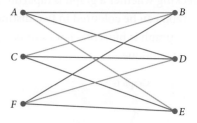 Using the redrawn graph, the required edges for a complete bipartite graph are AE, CB and FD.

EXERCISE 4.2 Types of undirected graphs and their applications

ANSWERS p. 435

Recap

1 The degree sum in the network diagram is

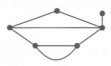

A 6   **B** 7   **C** 8   **D** 15   **E** 16

2 The map on the right shows the road connections between three towns, P, Q and R. The network that could be used to model these road connections is

9780170476959

3 🔗 WORKED EXAMPLE 5 Justify whether each of the following graphs is a

   i simple graph

   ii complete graph.

a
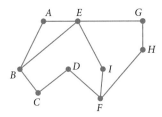

b

c
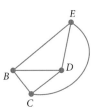

d
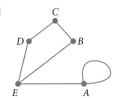

4 🔗 WORKED EXAMPLE 6 Determine the number of bridges in each of these connected graphs. Copy the graphs and indicate the bridges by highlighting them or drawing them in red.

a

b

c

d
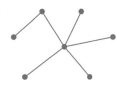

5 🔗 WORKED EXAMPLE 7 For each of the following graphs

   i redraw it to show that it is a planar graph

   ii verify if Euler's formula holds true.

a

b
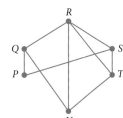

c
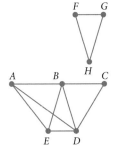

d
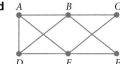

6 🔗 WORKED EXAMPLE 8

   a A planar graph has 9 edges and 6 faces. Determine the number of vertices.

   b A planar graph has 9 vertices and 7 faces. Determine the number of edges.

   c A planar graph has 5 vertices and 7 edges. Determine the number of faces.

7 🔗 WORKED EXAMPLE 9   Consider the graph below.

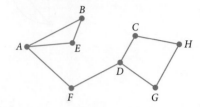

Determine, using an appropriate technique, whether or not the graph is bipartite. If so, identify the edges that need to be added in order for the graph to be complete bipartite. If not, identify the feature of the graph that makes it not bipartite.

Calculator-free

8 (5 marks)

   a Determine the number of edges needed to make

      i K_4 (1 mark)

      ii K_5 (1 mark)

      iii K_n (2 marks)

   b Determine the smallest number of edges that need to be added to the following graph
   to make it connected.   (1 mark)

9 (4 marks)

   a A planar graph has 7 vertices and 9 edges. Determine the number of faces.   (2 marks)

   b Verify that Euler's formula holds true for the following graph.   (2 marks)

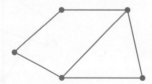

10 (4 marks)

a Identify which of the four graphs below are planar. (2 marks)

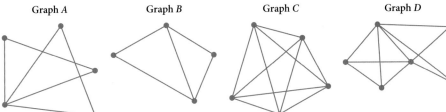

Graph A   Graph B   Graph C   Graph D

b Identify which of the five graphs below are **not** planar. (2 marks)

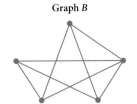

Graph A   Graph B   Graph C

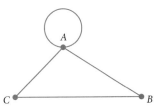

Graph D   Graph E

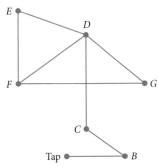

11 © SCSA MA2019 Q1ab (3 marks) The graph represents three buildings *A*, *B* and *C*, with connecting walkways, at a local school.

a Why is the graph planar? (1 mark)

b Show that the graph satisfies Euler's formula. (2 marks)

12 © SCSA MA2021 Q3ad (4 marks) The graph below shows the current network of reticulation pipes in Tarik's garden.

a Using Euler's formula, stating the number of vertices, edges and faces, show that the graph is planar. (2 marks)

b Tarik would like to increase the water pressure by removing one edge (pipe).

  i Identify any edge that cannot be removed. (1 mark)

  ii What is the name given to the type of edge identified in part **b i**? (1 mark)

▶ **13** © SCSA MA2017 Q3a(i-ii)b (8 marks)

  a A planar graph has five faces and five vertices, *A*, *B*, *C*, *D* and *E*.

   i Determine the number of edges for this graph. (2 marks)

   ii Draw the planar graph. (2 marks)

  b **i** A simple connected graph contains five vertices. Determine the minimum and the maximum number of edges it contains. (2 marks)

   ii A simple connected graph contains *n* vertices. Determine the minimum number of edges it contains. (1 mark)

   iii What name is given to the simple connected graph with the maximum number of edges possible? (1 mark)

14 © SCSA MA2020 Q2ab(i) (5 marks) A small research facility consists of five buildings with walkways represented by the edges in this network:

  a Determine the smallest number of edges (walkways) to be added to ensure that the network is

   i connected (1 mark)

   ii complete (1 mark)

   iii planar with 4 regions. (2 marks)

Three different plans for completing the network with the addition of four walkways are shown:

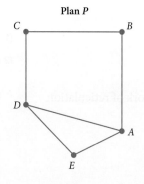

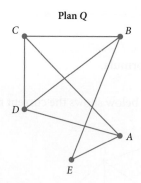

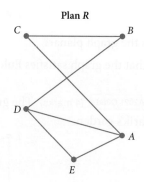

  b State which, if any, of these plans has a graph that is bipartite. (1 mark)

Weighted digraphs and shortest path problems

Directed graphs

The graphs and networks we have looked at so far have been examples of undirected graphs. A **directed graph**, also known as a **digraph**, is a graph containing vertices and edges, where each edge has an indicated direction between the two vertices it connects. These directed edges are commonly known as **arcs** and are shown using an arrowhead. Directed graphs are more commonly used in practical contexts involving graph theory, which will be explored in greater detail in Chapters 8 and 9.

Given that arcs of a directed graph typically indicate a one-way direction, it slightly changes the understanding of the degree of a vertex. In directed graphs, a vertex has an **in-degree** and an **out-degree**. The in-degree of a vertex V is the number of arcs ending at V, whereas the out-degree of a vertex V is the number of arcs starting at V.

Video playlist
Weighted digraphs and shortest path problems

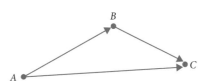

For example, vertex A in the graph above has an out-degree of 2 and an in-degree of 0, whereas vertex C has an in-degree of 2 and an out-degree of 0. Vertex B has both an in-degree and out-degree of 1.

WORKED EXAMPLE 10	Directed graphs

Consider the directed graph.

a Identify the vertex that cannot be reached from vertex A.

b Identify the vertex that cannot be used as a starting point to reach any other vertex. Justify your answer.

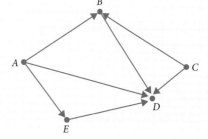

Steps	Working
a Start at A and test to see whether each vertex can be reached.	B can be reached.
	D can be reached.
	E can be reached.
	C cannot be reached.
b Look for the vertex that has an out-degree of 0.	D. As the out-degree of the vertex is 0, i.e. no arcs have D as the starting vertex.

Shortest path problems

A **weighted graph**, sometimes just called a network, is a graph whereby each edge is labelled with a quantity, called the **weight**, to represent extra information between adjacent vertices, such as distances, times or costs.

For example, the weighted graph on the right shows the distances in kilometres by road between several towns, where each vertex represents a town. Note that the lengths of the edges do not need to be drawn to scale to match the weight.

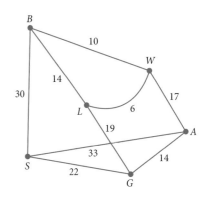

Weights can also be assigned to arcs in a directed graph, as seen in the graph shown.

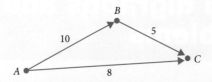

> ## Weighted graphs
>
> A weighted graph is a graph where extra information, such as distances, times or costs, is labelled on the edges. These quantities are the weights and are often used to solve shortest path problems by inspection.

Although we will define a path more formally in Chapter 5, shortest path problems generally involve finding the shortest distance, shortest time or least cost from a starting vertex to an end vertex by following a sequence of edges and vertices. For example, in the weighted digraph above, we could consider vertices *A*, *B* and *C* as destinations within a city and the weights on the arcs as times taken (in minutes) along each route between destination. We may then want to know the shortest time possible to get from *A* to *C*. In this case, the direct route along *AC* of 8 minutes would be the shortest path.

There are formal algorithms that can be used to calculate shortest paths in more complex networks such as Dijkstra's algorithm; however, in many cases an inspection or trial-and-error method will suffice.

WORKED EXAMPLE 11	**Finding the shortest path by inspection**

The network shows the travel times, in minutes, along a series of roads. Find the shortest time, in minutes, that it takes Justin to travel from his house to Ayesha's house by listing all the options.

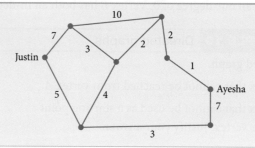

Steps	**Working**
1 Add labels to the vertices.	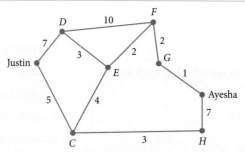
2 List the path options and calculate the total time of each option.	Justin–*D*–*F*–*G*–Ayesha takes 7 + 10 + 2 + 1 = 20 min Justin–*D*–*E*–*F*–*G*–Ayesha takes 7 + 3 + 2 + 2 + 1 = 15 min Justin–*D*–*E*–*C*–*H*–Ayesha takes 7 + 3 + 4 + 3 + 7 = 24 min Justin–*C*–*E*–*D*–*F*–*G*–Ayesha takes 5 + 4 + 3 + 10 + 2 + 1 = 25 min Justin–*C*–*E*–*F*–*G*–Ayesha takes 5 + 4 + 2 + 2 + 1 = 14 min Justin–*C*–*H*–Ayesha takes 5 + 3 + 7 = 15 min
3 Conclude your answer.	The shortest time needed for Justin to travel to Ayesha's house is 14 minutes using the path Justin–*C*–*E*–*F*–*G*–Ayesha.

Recap

4.3

1 Consider the planar graph shown.

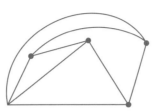

Which one of the following graphs can be redrawn as the planar graph above?

Graph *A*	Graph *B*	Graph *C*	Graph *D*
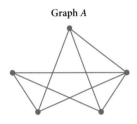	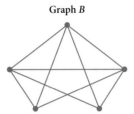	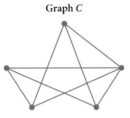	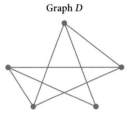

2 True or false? The following graph is bipartite.

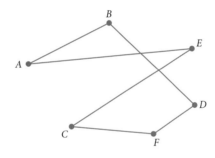

Mastery

3 🔗 WORKED EXAMPLE 10   Consider the directed graph on the right.

  a Identify the vertex that cannot be reached from vertex *A*.

  b Identify the vertices that cannot be used as a starting points to reach any other vertex. Justify your answer.

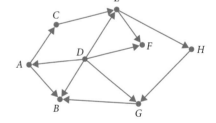

4 🔗 WORKED EXAMPLE 11   The network shows the travel times, in minutes, along a series of roads. By first listing all possible options, find the shortest time, in minutes, that it takes to travel from Riverville (*R*) to Midvale (*M*). State the corresponding shortest path.

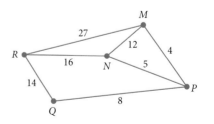

5 Hunter rides his bike to school each day. The edges of the network shown represent the roads that Hunter can use to ride to school. The numbers on the edges give the distance, in kilometres, along each road.

Determine the shortest distance that Hunter can ride between home and school, and state the path that Hunter must take to achieve this shortest distance.

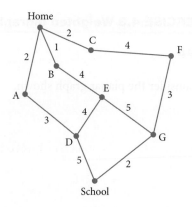

6 The network shows the distances, in kilometres, along a series of roads that connect Town A to Town B.

Determine the shortest distance, in kilometres, from Town A to Town B.

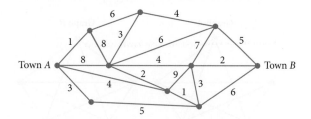

Calculator-free

7 (5 marks) The vertices in the network diagram show the entrance to a wildlife park and six picnic areas in the park: $P1$, $P2$, $P3$, $P4$, $P5$ and $P6$. The numbers on the edges represent the lengths, in metres, of the roads joining these locations.

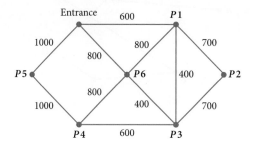

a State the degree of the vertex at the entrance to the wildlife park. (1 mark)

b Determine the shortest distance, in metres, from the entrance to picnic area $P3$, stating the corresponding path. (2 marks)

c Suppose the wildlife park manager decided to make the walkways one-way only, and the entrance is also used as the exit. The above network can now be represented by the following weighted digraph.

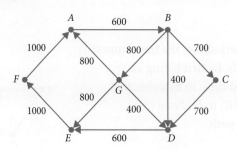

Justify whether the shortest distance in part **b** still holds true. (2 marks)

8 © SCSA MA2018 Q2c (2 marks) A bus company conducts 'jump on, jump off' sightseeing tours, during which tourists can get on and off buses at any of the designated attractions as many times as they like during the same day. The weighted digraph shows Attractions *A* to *F*, along with the time (in minutes) that a bus takes to travel between the attractions. The bus company operates two different circuits around the city, each shown differently on the right.

Vinh has just visited Attraction *C* and wishes to visit Attraction *B* next. Determine the route he should take to arrive at *B* in the shortest travelling time. State the time taken.

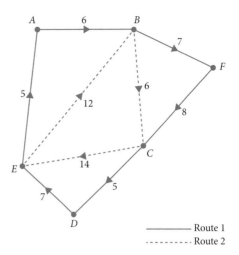

——— Route 1
- - - - - Route 2

9 © SCSA MA2017 Q6a MODIFIED (3 marks) In the network below, the vertices represent towns and the weights on the edges represent the time taken (in minutes) to travel between them.

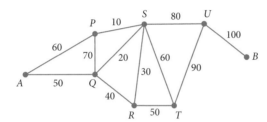

A driver leaves Town *A* and must deliver goods to all the other towns in the shortest time possible, finishing at Town *B*. Determine this shortest time. (Assume that a town may be visited more than once.)

10 © SCSA MA2019 Q4a (3 marks) A marine park has attractions with paths connecting them. The vertices on the graph represent the attractions and the numbers on the edges represent the path distances (km) between the attractions. Visitors can either walk around the park or take one of the many shuttle buses that run between attractions.

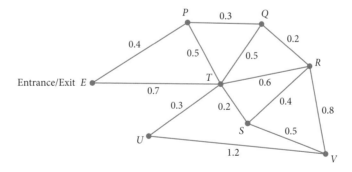

The manager of the marine park leaves his office, which is located at the entrance/exit (*E*) and walks to attraction *V*.

a Determine the shortest distance from *E* to *V*. (1 mark)

b If the manager needs to pick up some tools left at *U* on the way, determine the route he should take and the corresponding distance, given he wants to take the shortest route from *E* to *V*. (2 marks)

▶ **11** ©SCSA MA2021 Q5a (3 marks) The network below shows the relative distances, in hundreds of metres, between Wi-Fi hotspots around a university campus.

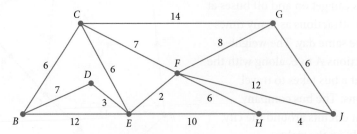

The control room for this system of hotspots is located at *B*. A problem has been reported with the hotspot at *J*.

A technician is sent from the control room to solve the problem at *J*. To get there as quickly as possible, she wants to use the shortest path, travelling from hotspot to hotspot. Determine the required path and its length.

Calculator-assumed

12 ©SCSA MA2020 Q13bc (6 marks) The graph below represents a road transport network from a warehouse at *A* to seven retail outlets *B*, *C*, *D*, *E*, *F*, *G* and *H*. The number on each edge represents the distance, in kilometres, along each road.

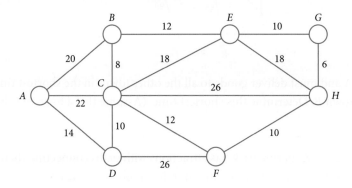

A special delivery must be made from the warehouse to retail outlet *H*.

a Copy the network to determine the shortest path and the distance travelled for this delivery. Working **must** appear on the network to show an appropriate method has been used. (3 marks)

Road *CH* presently goes around what is now a dry salt lake. It is proposed that a direct road be constructed that will reduce the distance between retail outlets *C* and *H*.

b By how much can the direct road between *C* and *H* be reduced, so that the shortest path from the warehouse to *H* includes the direct road *CH*? (3 marks)

Graphs and matrices

Adjacency matrices

An **adjacency matrix** for a non-directed graph with n vertices is an $n \times n$ matrix, where the entry in the ith row and jth column is the number of edges joining the vertices i and j.

Recall from Unit 1 of Mathematics Applications, the **leading diagonal** of a square matrix is the diagonal line of entries that runs from the top left corner to the bottom right corner of the matrix. It is important to note that in an adjacency matrix, a loop is counted as one edge and is shown in the leading diagonal. For example:

Graph	Adjacency matrix	Description
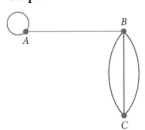	$\begin{array}{c} \\ A \\ B \\ C \end{array} \begin{array}{ccc} A & B & C \\ \left[\begin{array}{ccc} 1 & 1 & 0 \\ 1 & 0 & 3 \\ 0 & 3 & 0 \end{array}\right] \end{array}$	• one loop at A to A • no loops at B to B and C to C • one edge between A and B • three edges between B and C • no edges between A and C

Number of vertices = 3   Number of rows = 3

Number of edges = 5

Video playlist
Graphs and matrices

Worksheets
Adjacency matrices 1

Adjacency matrices 2

Comparing the graph and the adjacency matrix:

Number of vertices in the graph = number of rows in the matrix
$$= 3$$

Number of edges in the graph = sum of elements on and below the **leading diagonal** in the matrix
$$= 1 + 1 + 3$$
$$= 5$$

Degree of a vertex = sum of the elements of the row (+1 for each loop)

$$\begin{array}{c} \\ A \\ B \\ C \end{array} \begin{array}{ccc} A & B & C \\ \left[\begin{array}{ccc} 1 & 1 & 0 \\ 1 & 0 & 3 \\ 0 & 3 & 0 \end{array}\right] \end{array}$$

	A	B	C	Degree
A	1	1	0	$1 + 1 + 1 = 3$
B	1	0	3	$1 + 3 = 4$
C	0	3	0	3

The degree of vertex A is 3.

The degree of vertex B is 4.

The degree of vertex C is 3.

Adjacency matrices of undirected graphs

An adjacency matrix is a square matrix representing a graph where:

- each row and column is labelled as a vertex
- the elements show the number of edges between vertices
- the elements are symmetric about its leading diagonal
- the leading diagonal shows the number of loops.

Number of vertices = number of rows

Number of edges = sum of the elements on and below the leading diagonal

Degree of a vertex = sum of the elements of the row (+1 for each loop)

WORKED EXAMPLE 12 | Constructing an adjacency matrix from an undirected graph

Represent the graph using an adjacency matrix.

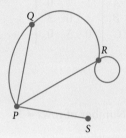

Steps	Working
1 Label the rows and columns of the matrix to match the graph.	$\begin{array}{c} \\ P \\ Q \\ R \\ S \end{array}\begin{array}{cccc} P & Q & R & S \\ \left[\begin{array}{cccc} & & & \\ & & & \\ & & & \\ & & & \end{array} \right] \end{array}$
2 List the connections in terms of the number of edges between vertices.	Two edges between P and Q One edge between P and R One edge between P and S One edge between Q and R One loop from R to R
3 Fill in the matrix based on the number of edges. An edge from A to B is also an edge from B to A, so the matrix is symmetrical around the leading diagonal.	$\begin{array}{c} \\ P \\ Q \\ R \\ S \end{array}\begin{array}{cccc} P & Q & R & S \\ \left[\begin{array}{cccc} & 2 & 1 & 1 \\ 2 & & 1 & \\ 1 & 1 & 1 & \\ 1 & & & \end{array} \right] \end{array}$
4 Complete the matrix by writing 0 for all the remaining elements.	$\begin{array}{c} \\ P \\ Q \\ R \\ S \end{array}\begin{array}{cccc} P & Q & R & S \\ \left[\begin{array}{cccc} 0 & 2 & 1 & 1 \\ 2 & 0 & 1 & 0 \\ 1 & 1 & 1 & 0 \\ 1 & 0 & 0 & 0 \end{array} \right] \end{array}$

9780170476959

WORKED EXAMPLE 13 Constructing a graph from an adjacency matrix

$$
\begin{array}{c}
\begin{array}{ccc} A & B & C \end{array} \\
\begin{array}{c} A \\ B \\ C \end{array}
\left[
\begin{array}{ccc}
1 & 2 & 0 \\
2 & 0 & 1 \\
0 & 1 & 1
\end{array}
\right]
\end{array}
$$

a Justify whether the above adjacency matrix represents a simple graph.

b Use the adjacency matrix provided to construct a possible graph.

Steps	Working
a Observe any entries in the leading diagonal, which indicate loops, or any entries larger than 1, indicating multiple edges.	Given that the entries from A to A and C to C both have 1s, the graph has loops and, hence, is not a simple graph.
b 1 Construct and label the three vertices of the graph.	
2 Identify and draw the loops, as shown in the leading diagonal.	
3 Use the symmetry of the matrix to complete the remaining edges.	Multiple edges (two edges) between A and B One edge between B and C

For a directed graph, the entry in the ith row and jth column of the adjacency matrix is the number of arcs (directed edges) joining the vertices i and j, in the specific direction from i to j. That is, the vertices labelled along the rows are the 'from' vertices and the vertices labelled in the columns are the 'to' vertices. As a result, the adjacency matrix of a directed graph may not necessarily be symmetrical about the leading diagonal. For example:

Graph	Adjacency matrix
Number of vertices = 4 Number of arcs = 5	$\begin{array}{c}\begin{array}{cccc} A & B & C & D \end{array} \\ \begin{array}{c} A \\ B \\ C \\ D \end{array} \left[\begin{array}{cccc} 0 & 1 & 0 & 0 \\ 0 & 0 & 0 & 1 \\ 0 & 0 & 1 & 1 \\ 1 & 0 & 0 & 0 \end{array}\right]\end{array}$ Number of rows = 4

Adjacency matrices of directed graphs

An adjacency matrix is a square matrix representing a graph where:

- each row and column is labelled as a vertex, whereby the rows are the 'from' vertices and the columns are the 'to' vertices
- the elements show the number of arcs between vertices
- the elements are not necessarily symmetrical about its leading diagonal
- the leading diagonal shows the number of loops.

In-degree of a vertex = sum of the elements of the respective column

Out-degree of a vertex = sum of the elements of the respective row

---

WORKED EXAMPLE 14 | Constructing an adjacency matrix from a digraph

Consider the digraph.

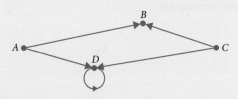

a Represent the digraph using an adjacency matrix.

b Describe how the adjacency matrix would change if an arc from *B* to *C* was added.

Steps	Working
a 1 Count the number of vertices, *n*, and construct and label an *n* × *n* matrix.	$\begin{array}{c} \\ A \\ B \\ C \\ D \end{array} \begin{array}{cccc} A & B & C & D \\ \left[\begin{array}{cccc} & & & \\ & & & \\ & & & \\ & & & \end{array}\right] \end{array}$
2 Start from vertex *A* and systematically complete each row of the adjacency matrix.	$\begin{array}{c} \\ A \\ B \\ C \\ D \end{array} \begin{array}{cccc} A & B & C & D \\ \left[\begin{array}{cccc} 0 & 1 & 0 & 1 \\ 0 & 0 & 0 & 0 \\ 0 & 1 & 0 & 1 \\ 0 & 0 & 0 & 1 \end{array}\right] \end{array}$
b 1 Draw the additional arc onto the digraph.	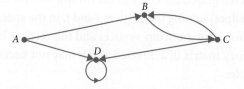
2 Identify the corresponding cell that would change in the matrix and describe the change.	The cell in the 2nd row, 3rd column representing 'from *B* to *C*' would now become 1 instead of 0.

Applying adjacency matrices

Recall from Unit 1 of Mathematics Applications that square matrices can be raised to a power. Squaring a square matrix is relatively simple as a non-calculator process using standard matrix multiplication; however, powers greater than 2 are generally computed by using technology.

9780170476959

USING CAS 1 | Raising a matrix to a power and evaluating

Find $\begin{bmatrix} 1 & 0 & 2 \\ 0 & 1 & 1 \\ 2 & 1 & 0 \end{bmatrix}$ raised to the power of 2 and 3.

ClassPad

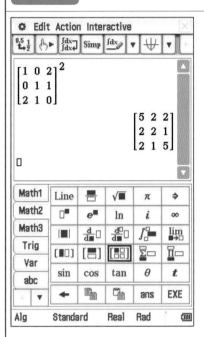

1 Open the **Main** application.

2 Open **Keyboard > Math2**.

3 Tap the **2×2** matrix template twice to create a 3×3 matrix.

4 Enter the values into the template as shown above, then raise the matrix to the power of **2**.

5 Press **EXE**.

6 The square of the matrix will be displayed.

7 Change the power from **2** to **3**.

8 Press **EXE**.

9 The cube of the matrix will be displayed.

TI-Nspire

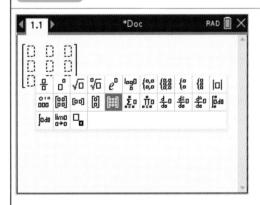

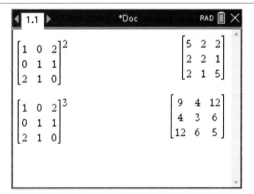

1 Open a **Calculator** application.

2 Press the **template** key.

3 Select the **3×3** template.

4 In the dialogue box, keep the default settings of 3 rows and 3 columns.

5 Press **enter** to insert a 3×3 matrix.

6 Enter the values into the template as shown above, then raise the matrix to the power of **2**.

7 The square of the matrix will be displayed.

8 Copy the matrix and change the power from **2** to **3**.

9 The cube of the matrix will be displayed.

In the context of graphs and networks, raising adjacency matrices to an integer power k will produce a **multi-stage matrix**. Let M be an adjacency matrix. Then M^k is a k-stage matrix, whereby the entry in the ith row and jth column of M^k represents the numbers of ways vertex j can be reached from vertex i using k edges.

For example, consider the undirected graph represented by the adjacency matrix:

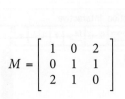

$$M = \begin{bmatrix} 1 & 0 & 2 \\ 0 & 1 & 1 \\ 2 & 1 & 0 \end{bmatrix}$$

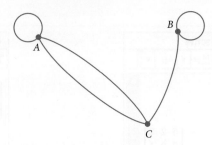

Raising M to the power of 2 produces the matrix $M^2 = \begin{bmatrix} 1 & 0 & 2 \\ 0 & 1 & 1 \\ 2 & 1 & 0 \end{bmatrix}\begin{bmatrix} 1 & 0 & 2 \\ 0 & 1 & 1 \\ 2 & 1 & 0 \end{bmatrix} = \begin{bmatrix} 5 & 2 & 2 \\ 2 & 2 & 1 \\ 2 & 1 & 5 \end{bmatrix}$.

Observing the entry $m^2_{11} = 5$, means that there are 5 two-stage journeys starting from A and finishing at A. Inspecting the graph, we can see that there are the following five possibilities:

- A–A–A (direction along the loop is not important)
- A–C (bottom edge)–A (bottom edge)
- A–C (bottom edge)–A (top edge)
- A–C (top edge)–A (top edge)
- A–C (top edge)–A (bottom edge).

Instead, if the graph was directed as shown below, then the multi-stage matrix would change, given that the adjacency matrix would no longer be symmetrical about the leading diagonal. That is:

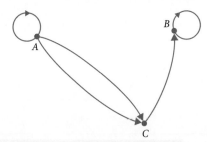

$$M = \begin{bmatrix} 1 & 0 & 2 \\ 0 & 1 & 0 \\ 0 & 1 & 0 \end{bmatrix}$$

Then $M^2 = \begin{bmatrix} 1 & 2 & 2 \\ 0 & 1 & 0 \\ 0 & 1 & 0 \end{bmatrix}$. Now observing the entry $m^2_{11} = 1$, means that there is only 1 two-stage journey

starting from A and finishing at A. Inspecting the graph, we can see that the only possibility is to use the loop twice: A–A–A. Interpreting multi-stage matrices such as these can be used to solve practical problems with both undirected and directed graphs.

The graph shows the possible roads a bus can take connecting four towns: Kelly, Lindon, Milton and Nate.

Show the use of an adjacency matrix and any relevant multi-stage matrices to determine the number of routes the bus could take to travel from Kelly to Milton, using at most three roads. Hence, state the possible routes.

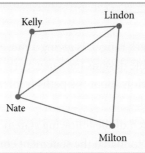

Steps	Working
1 Label the rows and columns of the adjacency matrix to match the graph and call the matrix A.	$A = \begin{array}{c} \\ K \\ L \\ M \\ N \end{array} \begin{array}{cccc} K & L & M & N \\ \left[\begin{array}{cccc} & & & \\ & & & \\ & & & \\ & & & \end{array} \right] \end{array}$
2 Complete the adjacency matrix.	$A = \begin{array}{c} \\ K \\ L \\ M \\ N \end{array} \begin{array}{cccc} K & L & M & N \\ \left[\begin{array}{cccc} 0 & 1 & 0 & 1 \\ 1 & 0 & 1 & 1 \\ 0 & 1 & 0 & 1 \\ 1 & 1 & 1 & 0 \end{array} \right] \end{array}$
3 Interpret 'at most three roads' as requiring the sum of the one-stage matrix, the two-stage matrix and the three-stage matrix. Compute A^2 and A^3 and, hence, $A + A^2 + A^3$.	$A^2 = \begin{array}{c} \\ K \\ L \\ M \\ N \end{array} \begin{array}{cccc} K & L & M & N \\ \left[\begin{array}{cccc} 2 & 1 & 2 & 1 \\ 1 & 3 & 1 & 2 \\ 2 & 1 & 2 & 1 \\ 1 & 2 & 1 & 3 \end{array} \right] \end{array}$ $A^3 = \begin{array}{c} \\ K \\ L \\ M \\ N \end{array} \begin{array}{cccc} K & L & M & N \\ \left[\begin{array}{cccc} 2 & 5 & 2 & 5 \\ 5 & 4 & 5 & 5 \\ 2 & 5 & 2 & 5 \\ 5 & 5 & 5 & 4 \end{array} \right] \end{array}$ $A + A^2 + A^3 = \begin{array}{c} \\ K \\ L \\ M \\ N \end{array} \begin{array}{cccc} K & L & M & N \\ \left[\begin{array}{cccc} 4 & 7 & 4 & 7 \\ 7 & 7 & 7 & 8 \\ 4 & 7 & 4 & 7 \\ 7 & 8 & 7 & 7 \end{array} \right] \end{array}$
4 Identify the entry corresponding to 'Kelly to Milton' and state the number of possible routes.	Kelly to Milton is the 1st row, 3rd column. Therefore, there are 4 possible routes.
5 List the possible routes.	No one-stage routes Two two-stage routes: • $K\text{–}N\text{–}M$ • $K\text{–}L\text{–}M$ Two three-stage routes: • $K\text{–}N\text{–}L\text{–}M$ • $K\text{–}L\text{–}N\text{–}M$

Video
WACE
question
analysis:
Graphs and
networks

WACE QUESTION ANALYSIS

© SCSA MA2019 Q12 Calculator-assumed (6 marks)

Jake, a park ranger, is giving a presentation at a National Park and Wildlife Conference on possible designs for a new park. Unfortunately, Jake made mathematical errors in his presentation about the paths (represented by edges) and shelter huts (represented by vertices) in the park.

a For each of the following statements, the graph drawn by Jake was incorrect. Redraw the graph to match the statement correctly.

    i This park plan has been drawn as a connected planar graph containing six vertices.   (2 marks)

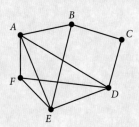

    ii This park plan has been drawn as a bipartite graph.   (3 marks)

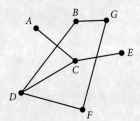

Jake also makes the following incorrect statement in his presentation: 'A park plan can be a complete graph with 21 paths and six shelter huts'.

b If the plan must be a complete graph with 21 paths, how many shelter huts should Jake have quoted?   (1 mark)

Reading the question

- Highlight what the vertices and edges represent in the context of the question.
- Highlight the command words, e.g. redraw, meaning that you cannot just modify the given graph.
- Highlight the key graph theory terminology.

Thinking about the question

- Which of the conditions in part **a i** is not visible? Is the graph connected? Is it planar? Does it already have six vertices?
- What is an appropriate technique that can be used to verify whether the graph in part **a ii** is bipartite?
- What is the relationship between the number of edges and vertices in a complete graph?

Worked solution ($\checkmark$ = 1 mark)

a **i** The graph is currently not visibly planar because of the crossed edges. Redraw the graph so that the edges do not cross to demonstrate that it is planar.

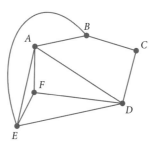

the graph is drawn with no two edges crossing $\checkmark$

the graph is drawn with all connections $\checkmark$

ii Given that the graph can be verified to be bipartite using a two-colouring technique, it is best to demonstrate that the vertices can be separated into two disjoint sets, with no vertex from either set connected to a vertex within the same set.

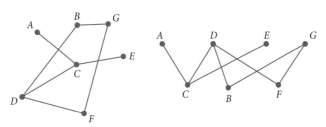

correctly arranges the groups $\checkmark$

correctly joins at least 4 vertices $\checkmark$

correctly joins all vertices $\checkmark$

b The relationship between the number of vertices (n) and edges in the complete graph K_n can be shown in the table of values below.

n	1	2	3	4	5	6	7
Number of edges	0	1	3	6	10	15	21

Therefore, **7 shelter huts** $\checkmark$ should have been quoted.

EXERCISE 4.4 Graphs and matrices

ANSWERS p. 436

Recap

1 The network shows the times taken, in minutes, to walk along a series of tracks in a botanical garden from the Entrance to the Exit.

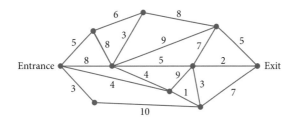

The shortest distance, in kilometres, to travel from the Entrance to the Exit is

A 9   **B** 10   **C** 11   **D** 12   **E** 13

2 The network shows the travel times, in minutes, along a series of roads that connect a student's home to school.

The shortest time, in minutes, for this student to travel from home to school is

A 22

B 23

C 24

D 25

E 26

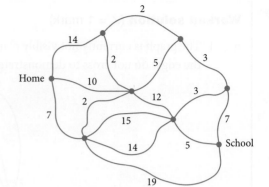

Mastery

3 🔗 WORKED EXAMPLE 12 | Represent each graph using an adjacency matrix.

a

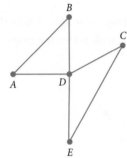

b

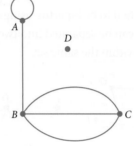

c

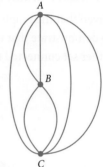

4 🔗 WORKED EXAMPLE 13 | For each of the following adjacency matrices

   i justify whether the matrix represents a simple graph

   ii construct a possible graph.

a

$$
\begin{array}{c c c c c}
 & A & B & C & D \\
A & 0 & 0 & 1 & 1 \\
B & 0 & 0 & 2 & 1 \\
C & 1 & 2 & 0 & 1 \\
D & 1 & 1 & 1 & 1
\end{array}
$$

b

$$
\begin{array}{c c c c c c}
 & A & B & C & D & E \\
A & 0 & 0 & 1 & 0 & 3 \\
B & 0 & 1 & 1 & 0 & 0 \\
C & 1 & 1 & 0 & 1 & 0 \\
D & 0 & 0 & 1 & 0 & 1 \\
E & 3 & 0 & 0 & 1 & 0
\end{array}
$$

5 🔗 WORKED EXAMPLE 14 | Consider the digraph.

  a Represent the digraph using an adjacency matrix.

  b Describe how the adjacency matrix would change if a loop at E was added.

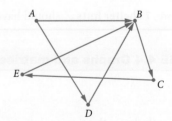

6 🔗 Using CAS 1 | Find $\begin{bmatrix} 1 & 0 & 3 \\ 0 & 0 & 2 \\ 1 & 1 & 0 \end{bmatrix}$ raised to the power of 3 and 4.

7 🔗 WORKED EXAMPLE 15 | The graph shows the possible walking tracks between four landmarks: J, K, L and M.

Show the use of an adjacency matrix and any relevant multi-stage matrices to determine the number of routes from Landmark M to Landmark K, using exactly two tracks. Hence, state the possible routes.

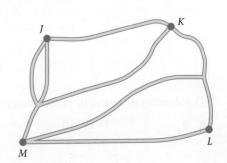

9780170476959

8 © SCSA MA2019 Q1c (3 marks) The graph shown represents three buildings A, B and C, with connecting walkways, at a local school. Construct the adjacency matrix for the graph.

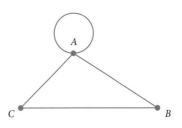

9 © SCSA MA2018 Q2a (2 marks) A bus company conducts 'jump on, jump off' sightseeing tours, during which tourists can get on and off buses at any of the designated attractions as many times as they like during the same day. The weighted digraph below shows Attractions A to F, along with the time (in minutes) that a bus takes to travel between the attractions. The bus company operates two different circuits around the city, each shown differently below.

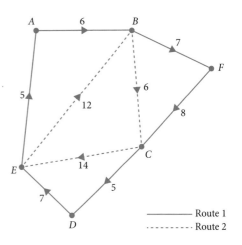

———— Route 1
- - - - - Route 2

Copy and complete the adjacency matrix for the digraph.

$$
\begin{array}{c c}
 & \begin{array}{c c c c c c} A & B & C & D & E & F \end{array} \\
\begin{array}{c} A \\ B \\ C \\ D \\ E \\ F \end{array} &
\left[\begin{array}{c c c c c c}
0 & 1 & 0 & 0 & 0 & 0 \\
0 & 0 & 1 & 0 & 0 & 1 \\
0 & 0 & 0 & 1 & 1 & 0 \\
0 & 0 & 0 & 0 & 1 & 0 \\
1 & 1 & 0 & 0 & 0 & 0 \\
 & & & & &
\end{array} \right]
\end{array}
$$

10 (5 marks) The adjacency matrix shows the number of road connections between towns F, G, H and I.

$$
\begin{array}{c c}
 & \begin{array}{c c c c} F & G & H & I \end{array} \\
\begin{array}{c} F \\ G \\ H \\ I \end{array} &
\left[\begin{array}{c c c c}
0 & 0 & 1 & 1 \\
0 & 1 & 1 & 2 \\
1 & 1 & 0 & 0 \\
1 & 2 & 0 & 0
\end{array} \right]
\end{array}
$$

a Justify whether or not any of the road connections between towns are one-way. (2 marks)

b Construct a graph showing these road connections. (3 marks)

11 (3 marks) The friendships between five children are summarised in the graph.

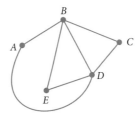

The vertices A, B, C, D and E in the graph represent these children. Each edge between two vertices indicates that the two children are friends. For example, the edge between vertex B and vertex C shows that child B and child C are friends.

a Construct an adjacency matrix that summarises the friendships whereby a '0' indicates no friendship and a '1' indicates a friendship. (2 marks)

b Explain, with reference to the context, why there are no entries in the leading diagonal of this adjacency matrix. (1 mark)

12 (2 marks) In a competition, members of a team work together to complete a series of challenges. The members of one team are Kristy (*K*), Lyn (*L*), Mike (*M*) and Neil (*N*).

In one of the challenges, these four team members are only allowed to communicate directly with each other, as indicated by the edges of the following network diagram.

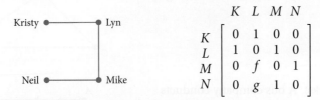

$$\begin{array}{c} \\ K \\ L \\ M \\ N \end{array} \begin{array}{cccc} K & L & M & N \\ \begin{bmatrix} 0 & 1 & 0 & 0 \\ 1 & 0 & 1 & 0 \\ 0 & f & 0 & 1 \\ 0 & g & 1 & 0 \end{bmatrix} \end{array}$$

The adjacency matrix also shows the allowed lines of communication.

a Explain the meaning of a zero in the adjacency matrix. (1 mark)

b State the values of *f* and *g* in the adjacency matrix. (1 mark)

13 © SCSA MA2017 Q7b (3 marks) The adjacency matrix *Q* represents the raised paths connecting the observation platforms in the safari section at the zoo. Copy the vertices and draw a planar graph for the adjacency matrix.

$$Q = \begin{array}{c} \\ A \\ B \\ C \\ D \end{array} \begin{array}{cccc} A & B & C & D \\ \begin{bmatrix} 0 & 2 & 1 & 1 \\ 2 & 0 & 1 & 1 \\ 1 & 1 & 0 & 0 \\ 1 & 1 & 0 & 1 \end{bmatrix} \end{array}$$

A •   *B* •

C •   • *D*

14 © SCSA MA2020 Q6 (8 marks) A small business office has five separate areas connected by doorways shown as gaps in this diagram.

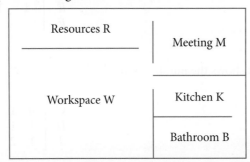

This adjacency matrix below represents the number of doorways directly between each area:

$$\begin{array}{c} \\ B \\ K \\ M \\ R \\ W \end{array} \begin{array}{ccccc} B & K & M & R & W \\ \begin{bmatrix} 0 & 1 & 0 & 0 & 0 \\ 1 & 0 & 0 & 0 & 1 \\ 0 & 0 & 0 & 1 & 1 \\ 0 & 0 & 1 & 0 & Y \\ 0 & X & 1 & 2 & 0 \end{bmatrix} \end{array}$$

a State the meaning of the zero entries in the matrix. (1 mark)

b Determine the value of *X* and *Y*. (2 marks)

c Describe how the total number of doorways for each area can be found from the adjacency matrix. (1 mark)

d Complete this network with vertices corresponding to the office areas and the edges representing the doorways.

K •   • *W*

M •——————• *R* (3 marks)

e Determine how many different routes there are between the meeting room and the workspace that pass through exactly two doorways. (1 mark)

15 (3 marks) Consider the graph.

The adjacency matrix, A, for this graph, with some elements missing, is shown.

$$A = \begin{array}{c} \\ W \\ X \\ Y \\ Z \end{array} \overset{\displaystyle W \ \ X \ \ Y \ \ Z}{\begin{bmatrix} 1 & - & - & - \\ - & 0 & - & - \\ - & - & 0 & - \\ - & - & - & 1 \end{bmatrix}}$$

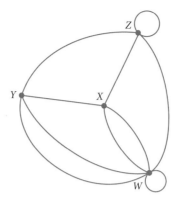

a Copy and complete the adjacency matrix. (1 mark)

b Evaluate A^3 and interpret the entry a_{33}^3. (2 marks)

16 © SCSA | MA2016 Q15 (8 marks) An express bus service runs between seven adjacent shopping centres in the city. Below is an adjacency matrix of the seven shopping centres, A to G.

$$\begin{array}{c} \\ A \\ B \\ C \\ D \\ E \\ F \\ G \end{array} \overset{\displaystyle A \ \ B \ \ C \ \ D \ \ E \ \ F \ \ G}{\begin{bmatrix} 0 & 1 & 0 & 0 & 1 & 0 & 1 \\ 1 & 0 & 1 & 1 & 1 & 0 & 0 \\ 0 & 1 & 0 & 1 & 0 & 1 & 0 \\ 0 & 1 & 1 & 0 & 0 & 0 & 0 \\ 1 & 1 & 0 & 0 & 0 & 0 & 0 \\ 0 & 0 & 1 & 0 & 0 & 0 & 1 \\ 1 & 0 & 0 & 0 & 0 & 1 & 0 \end{bmatrix}}$$

a Copy the vertices below and draw the network diagram associated with the adjacency matrix, assuming the arcs are undirected. (3 marks)

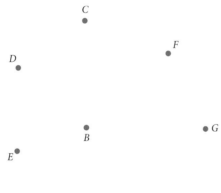

b The buses only run between adjacent shopping centres. However, a passenger can buy a multi-stage ticket at any shopping centre. A one-stage ticket means a passenger can travel from one shopping centre to an adjacent shopping centre, such as:

- $A \to B$ or $A \to E$ etc.

Similarly for a two-stage ticket:

- $A \to B \to A$ which is a return journey

- $A \to B \to C$ which is a one-way journey.

    i What feature on the adjacency matrix tells us that the buses run in both directions between adjacent shopping centres? (2 marks)

    ii How many different one-stage journeys are available from shopping centre B? (1 mark)

    iii List **all** the different two-stage, one-way journeys available from shopping centre B. (2 marks)

④ Chapter summary

Features of graphs

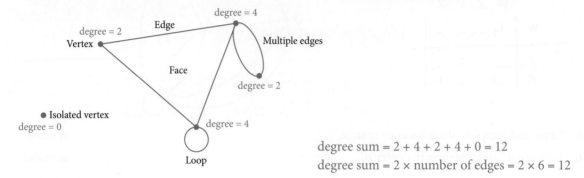

degree sum = 2 + 4 + 2 + 4 + 0 = 12
degree sum = 2 × number of edges = 2 × 6 = 12

- Any selection of vertices and edges chosen from this graph makes up a **subgraph**.

Types of graphs

- Graphs can be **undirected** (no direction indicated on each edge) or **directed** (directed edges are called **arcs**).
- Vertices of directed graphs have **in-degrees** and **out-degrees** due to the direction.
- Graphs when used to represent a context are often called **networks**.
- A **simple graph** is a graph without any loops or multiple edges.
- A **complete graph** is a simple graph where every vertex is connected to every other vertex. A complete graph with n vertices is denoted as K_n.
- Every vertex in a complete graph has degree = number of vertices − 1.
- A **connected graph** is a graph where any vertex is reachable from any other vertex.
- A **bridge** is any edge that keeps a graph connected; otherwise, it becomes disconnected.
- A **planar graph** is a connected graph that can be drawn so that it does not have any edges crossing. **Euler's formula** holds true for planar graphs such that $v + f - e = 2$, where

  v = the number of vertices

  f = the number of faces

  e = the number of edges.

- A **bipartite graph** can be drawn so that the set of vertices can be separated into two distinct sets, e.g. set X and set Y, so that each edge of the graph only connects a vertex from set X to set Y. Bipartite graphs can also be **complete**, denoted as $K_{m,n}$.
- A **weighted graph** is a graph where extra information, such as distances, times or costs, is labelled on the edges. These quantities are the **weights** and are often used to solve **shortest path problems by inspection**.

Graphs and matrices

- An **adjacency matrix** is a square matrix representing a graph where:
  - each row and column is labelled as a vertex
  - the elements show the number of edges between vertices
  - for undirected graphs, the elements are symmetrical about its leading diagonal
  - the leading diagonal shows the number of loops.
- Adjacency matrices can be raised to the power of k to create a **multi-stage matrix** to solve problems involving k stages.

Total number of marks: 20   Reading time: 2 minutes   Working time: 20 minutes

1 (7 marks)

   a An arithmetic sequence is defined by the recursive rule

$$A_{n+1} = A_n + 15, A_1 = 4$$

     i Determine the first five terms of this sequence. (2 marks)

     ii Deduce a rule for the nth term in the sequence and, hence, state A_{50}. (2 marks)

   b A geometric sequence is defined by the general rule

$$B_n = 100(0.75)^{n-1}$$

     i Describe the long-term behaviour of this sequence. (1 mark)

     ii Express the sequence in the form of a recursive rule. (2 marks)

2 © SCSA MA2016 Q5ab (7 marks)

   a Redraw the following graphs as planar graphs.

     i (1 mark)

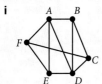

     ii (2 marks)

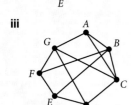

     iii (2 marks)

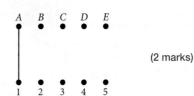

   b Verify Euler's formula for the planar graph obtained in part **a i**. (2 marks)

3 © SCSA MA2021 Q2abc (6 marks) A construction company uses five different machines and has five employees who operate those machines. The adjacency matrix on the right shows each of the five employees (A, B, C, D, E) and the five machines they are trained to operate. These are the only machines they may use.

Employee

$$\text{Machine} \begin{array}{c} 1 \\ 2 \\ 3 \\ 4 \\ 5 \end{array} \begin{array}{ccccc} A & B & C & D & E \\ \left[\begin{array}{ccccc} 1 & 0 & 0 & 1 & 0 \\ 0 & 0 & 0 & 0 & 1 \\ 1 & 1 & 0 & 0 & 0 \\ 0 & 0 & 1 & 0 & 1 \\ 0 & 0 & 1 & 1 & 0 \end{array} \right] \end{array}$$

   a Draw the adjacency matrix as a bipartite graph by copying the start of the graph on the right. The $A1$ connection has already been drawn on the graph. (2 marks)

   b Does the bipartite graph in part **a** represent a

     i planar graph? (1 mark)

     ii connected graph? (1 mark)

   c Explain why the bipartite graph in part **a** is a simple graph. (2 marks)

Cumulative examination: Calculator-assumed

Total number of marks: 22   Reading time: 2 minutes   Working time: 22 minutes

1 (4 marks) A large population of birds lives on a remote island. As part of the winter migration process, the birds gradually leave the island. Let P_n be the population of birds on the island n days since the start of the migration process. The population of birds on the island can be modelled by the first order linear recurrence relation

$$P_{n+1} = 0.8P_n + 10, P_1 = 250$$

a Determine the population of birds, rounded to the nearest whole, left on the island after

   i five days   (1 mark)

   ii two weeks.   (1 mark)

b Describe the long-term behaviour of the population of birds on the island.   (2 marks)

2 (5 marks) The scatterplot provided shows the *rainfall* (in mm) and the *percentage of clear days* for each month of 2022 in Perth.

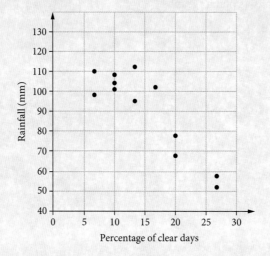

An equation of the least-squares line for this data set is

rainfall = −2.68 × *percentage of clear days* + 131

a Use the equation of the least-squares line to predict the rainfall for a month with 35% of clear days. Write your answer in mm correct to one decimal place.   (1 mark)

b The coefficient of determination for this data set is 0.8081.

   i Interpret the coefficient of determination in terms of the variables *rainfall* and *percentage of clear days*.   (1 mark)

   ii Determine the value of the correlation coefficient, correct to three decimal places and, hence, describe the correlation between the two variables.   (3 marks)

3 (6 marks) A zoo has an entrance, a café and nine animal exhibits: bears (B), elephants (E), giraffes (G), lions (L), monkeys (M), penguins (P), seals (S), tigers (T) and zebras (Z). The edges on the graph below represent the paths between the entrance, the café and the animal exhibits. The numbers on each edge represent the length, in metres, along that path. Visitors to the zoo can use only these paths to travel around the zoo.

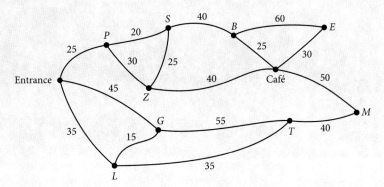

a Fred, a visitor to the zoo, wants to visit the seal exhibit (S) as quickly as possible upon entering the zoo. State the shortest path from the entrance to S and the distance travelled. (2 marks)

b Determine the shortest distance from the entrance of the zoo to the monkey exhibit (M), stating the path taken. (2 marks)

A reptile exhibit (R) will be added to the zoo. A new path of length 20 m will be built between the reptile exhibit (R) and the giraffe exhibit (G). A second new path, of length 35 m, will be built between the reptile exhibit (R) and the café.

c Justify whether this new path will reduce the shortest distance between the entrance and the monkey exhibit (M) found in part **b**. (2 marks)

4 © SCSA | MA2018 Q5ab (7 marks) Consider two country towns in which roads connect the local attractions. The adjacency matrix (Town 1) and graph (Town 2), shown below, represent the road connections between attractions (vertices) within each town.

Town 1

$$\begin{array}{c c} & \begin{matrix} A & B & C & D \end{matrix} \\ \begin{matrix} A \\ B \\ C \\ D \end{matrix} & \begin{bmatrix} 0 & 1 & 2 & 0 \\ 1 & 0 & 2 & 0 \\ 2 & 1 & 1 & 1 \\ 0 & 0 & 2 & 0 \end{bmatrix} \end{array}$$

Town 2

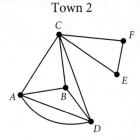

a Consider the adjacency matrix for Town 1.

  i Explain why the network represented by this matrix is a directed graph. (1 mark)

  ii Give **two** reasons why the network represented by this matrix is not a simple graph. (2 marks)

b Consider the graph shown for Town 2. The adjacency matrix for Town 2 has been squared, and is shown below.

$$\begin{array}{c c} & \begin{matrix} A & B & C & D & E & F \end{matrix} \\ \begin{matrix} A \\ B \\ C \\ D \\ E \\ F \end{matrix} & \begin{bmatrix} 6 & 3 & 3 & 2 & 1 & 1 \\ 3 & 3 & 2 & 3 & 1 & 1 \\ 3 & 2 & 5 & 3 & 1 & 1 \\ 2 & 3 & 3 & 6 & 1 & 1 \\ 1 & 1 & 1 & 1 & 2 & 1 \\ 1 & 1 & 1 & 1 & 1 & 2 \end{bmatrix} \end{array}$$

  i Explain the significance of the element in Row 3, Column 4. (2 marks)

  ii Draw a connected subgraph containing only vertices A, C and D. (2 marks)

CHAPTER

5

PATHS AND CYCLES

Syllabus coverage

TOPIC 3.3: GRAPHS AND NETWORKS

Paths and cycles

3.3.6   demonstrate the meanings of, and use, the terms: walk, trail, path, closed walk, closed trail, cycle, connected graph, and bridge

3.3.7   investigate and solve practical problems to determine the shortest path between two vertices in a weighted graph (by trial-and-error methods only)

3.3.8   demonstrate the meanings of, and use, the terms: Eulerian graph, Eulerian trail, semi-Eulerian graph, semi-Eulerian trail and the conditions for their existence, and use these concepts to investigate and solve practical problems

3.3.9   demonstrate the meanings of, and use, the terms: Hamiltonian graph and semi-Hamiltonian graph, and use these concepts to investigate and solve practical problems

Mathematics Applications ATAR Course Year 12 syllabus p. 11 © SCSA

Video playlists (4):

5.1   Exploring and travelling with graphs
5.2   Eulerian and semi-Eulerian graphs
5.3   Hamiltonian and semi-Hamiltonian graphs
WACE question analysis Paths and cycles

Worksheets (4):

5.2   Eulerian graphs • Eulerian circuits • Eulerian trails and circuits
5.3   Hamiltonian paths and cycles

Nelson MindTap

To access resources above, visit
cengage.com.au/nelsonmindtap

5.1 Exploring and travelling with graphs

Types of walks

In the previous chapter, we introduced the idea of shortest path problems. These problems are examples of how graphs can be used to solve exploring and travelling problems, involving moving from one vertex to another via an edge or sequence of edges. However, there are broader definitions that we now need to know when dealing with other types of exploring and travelling problems.

A **walk** in a graph is a sequence of connected vertices such that from each vertex there is an edge to the next vertex in the sequence. Walks can include repeated edges and vertices. An example of a walk in the graph on the right is $A-B-E-C-E-D$. The edge EC is travelled along twice, meaning that the vertex E is visited twice. In undirected graphs, the direction of travel along a repeated edge is not important. That is, EC and CE are considered the same edge.

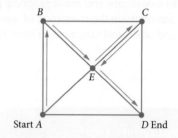

Each walk has a **length**, which is defined as the number of edges it includes. For example, the walk $A-B-E-C-E-D$ has a length of 5.

Given that this walk starts and finishes at different vertices, then it is said to be an **open walk**. However, if the walk became $A-B-E-C-E-D-A$ such that it now finished back at A, it would be considered a **closed walk**.

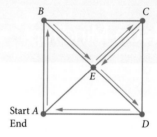

When a walk does not contain repeated edges, it is called a **trail**. In a trail, vertices can be repeated. An example of a trail in this graph is $A-B-E-C-D-E$.

In this example, the vertex E is visited twice.

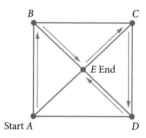

Similar to a walk, if the trail starts and finishes at different vertices, then it is said to be an **open trail**, whereas if it starts and finishes at the same vertex, then it is said to be a **closed trail** (also called a **circuit**). For example, $A-B-E-C-D-E-A$ is a closed trail in this graph, whereby A and E are visited twice.

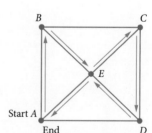

A **path** in a graph is a walk with no repeated edges and no repeated vertices. If the path starts and finished at different vertices, then it is said to be an **open path**. For example, A–B–E–C–D is an open path in this graph.

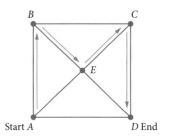

If the path finishes at the same vertex at which it starts, then it is said to be a **closed path**. A closed path can also be referred to as a **cycle**. Hence, a cycle can also be defined as a closed walk in which all of the edges and vertices are different, except for the first and last vertex. For example, A–B–E–C–D–A is a cycle in this graph.

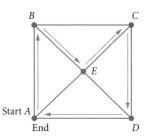

Types of walks

Walks that:	Open trail	Closed trail	Path	Cycle
have no repeated edges	✓	✓	✓	✓
have no repeated vertices			✓	✓
start and finish at the same vertex		✓		✓

Walk classification chart

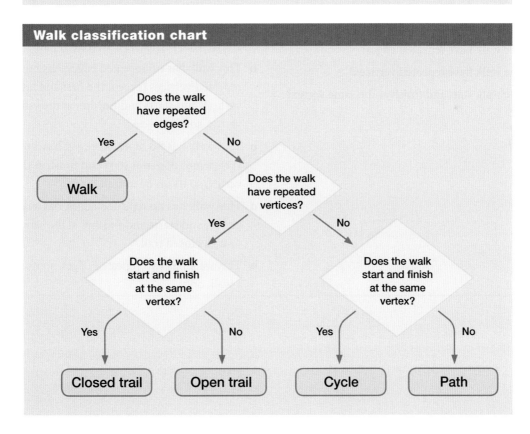

For each of the following walks, state whether it is an open trail, closed trail, path, cycle or walk only. Justify your answer.

a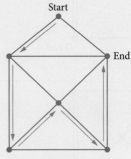

b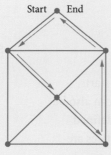

c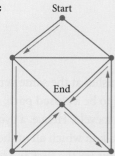

d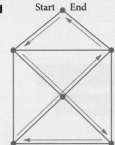

e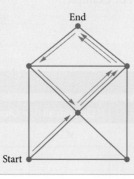

Steps	Working
Use the three classification questions to help classify the walks, in this order: Does the walk have repeated edges? Does the walk have repeated vertices? Does the walk start and finish at the same vertex?	**a** This walk has no repeated edges, has no repeated vertices, and does not start and finish at the same vertex, so it's a (open) path. **b** This walk has no repeated edges, has no repeated vertices (except the first and last vertex), and starts and finishes at the same vertex, so it's a cycle. **c** This walk has no repeated edges, has a repeated vertex, and does not start and finish at the same vertex, so it's an open trail. **d** This walk has no repeated edges, has a repeated vertex, and starts and finishes at the same vertex, so it's a closed trail. **e** This walk has two repeated edges, so it's a walk only.

9780170476959

WORKED EXAMPLE 2 **Classifying walks from a list of vertices**

Rebekah is hiking along tracks in the John Forrest National Park. For each of the following tracks described as lists of vertices, state whether the track is considered an open trail, a closed trail, a path, a cycle or a walk only. Justify your answer.

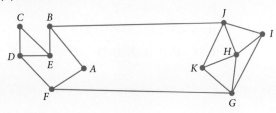

a C–E–B–A–F–D–C

b A–F–G–K–H

c G–K–H–I–J–K–H–G

d D–E–C–D–F–G

e G–K–J–I–H–K–G

f K–H–I–J–H–G–K

Steps	Working
Use the three classification questions to help classify the walks, in this order: Does the walk have repeated edges? Does the walk have repeated vertices? Does the walk start and finish at the same vertex?	**a** This walk has no repeated edges, has no repeated vertices (except the first and last vertex), and starts and finishes at the same vertex, so it's a cycle. **b** This walk has no repeated edges, has no repeated vertices, and does not start and finish at the same vertex, so it's a (open) path. **c** This walk has a repeated edge *KH*, so it's a walk only. **d** This walk has no repeated edges, has a repeated vertex *D*, and does not start and finish at the same vertex, so it's an open trail. **e** This walk has a repeated edge (*GK* is the same edge as *KG*), so it's a walk only. **f** This walk has no repeated edges, has a repeated vertex *H*, and starts and finishes at the same vertex, so it's a closed trail.

EXERCISE 5.1 Exploring and travelling with graphs   ANSWERS p. 438

Mastery

1 [WORKED EXAMPLE 1] For each of the following walks, state whether it's an open trail, a closed trail, a path, a cycle or a walk only. Justify your answer.

a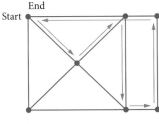

b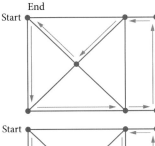

c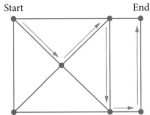

d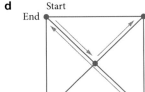

e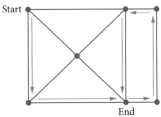

2 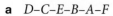 **WORKED EXAMPLE 2** Alex is trekking along tracks on Tropicana Island. For each of the following tracks described as lists of vertices, state whether the track is considered an open trail, a closed trail, a path, a cycle or a walk only. Justify your answer.

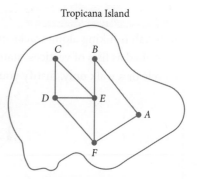

Tropicana Island

a D–C–E–B–A–F

b D–C–E–B–A–F–D

c D–C–E–B–A–F–E–D

d C–D–E–F–A–B–E–D

e F–E–B–A–F–E

f C–E–B–A–F–E–D

Calculator-free

3 © SCSA MA2019 Q1d MODIFIED (3 marks) The graph shown represents three buildings A, B and C, with connecting walkways, at a local school.

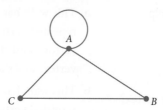

a A student wishes to carry out closed walks of length two from Building A. List all his possible walks. (2 marks)

b Another student starts at Building B, travels to Building C and then back to Building B. Explain why this walk cannot be called a closed trail nor a cycle. (1 mark)

4 © SCSA MA2018 Q2b MODIFIED (6 marks) A bus company conducts 'jump on, jump off' sightseeing tours, during which tourists can get on and off buses at any of the designated attractions as many times as they like during the same day. The weighted digraph below shows Attractions A to F, along with the time (in minutes) that a bus takes to travel between the attractions. The bus company operates two different routes around the city, each shown differently below. Mai gets on a bus at E. She travels directly to B, changes bus, and continues travelling BFCDEA.

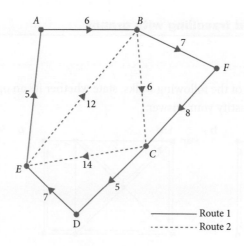

——— Route 1
- - - - - Route 2

a Explain why her route is an open trail. (2 marks)

b State the length of the trail. (1 mark)

c State the amount of time the bus takes to complete this trail. (1 mark)

d Find a closed path of length 4 that commences at A and, hence, state the amount of time the bus takes to complete this closed path. (2 marks)

5 (9 marks) The West Water Aquarium installs new one-way travellators between each of the aquatic exhibitions according to the directed graph shown.

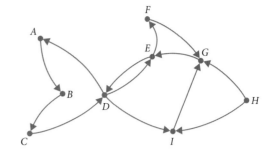

Visitors to West Water Aquarium enter at *H* and can choose their own journey through the aquarium. They can leave the aquarium from any of the other exhibitions.

a A family travels through the aquarium along the walk *H–G–E–D–A–B–C–D–I* and then leaves.

    i State the length of the walk. (1 mark)

    ii State the appropriate mathematical term that can be used to describe their walk. (1 mark)

b A couple travel through the aquarium along the walk *H–I–G–E–D–A–B–C* and then leave.

    i State the length of the walk. (1 mark)

    ii State the appropriate mathematical term that can be used to describe their walk. (1 mark)

c A group of teenagers enter at *H* and wish to complete a cycle through the aquarium. Explain why this is not possible. (1 mark)

d If the entrance was at *I*, a cycle through the aquarium would be possible. State the cycle and its corresponding length. (2 marks)

e If the entrance was at *I*, a closed trail through the aquarium would also be possible. State the closed trail and its corresponding length. (2 marks)

Eulerian and semi-Eulerian graphs

Video playlist
Eulerian and semi-Eulerian graphs

Worksheets
Eulerian graphs

Eulerian circuits

Eulerian trails and circuits

Open and closed Eulerian trails

In 1736, mathematician Leonhard Euler proposed a solution to the well-known real-world problem, the Königsberg bridge problem. The problem asks: can the seven bridges of the city of Königsberg (a city in Prussia, now Kaliningrad, Russia) all be crossed in a single trip that starts and finishes at the same place?

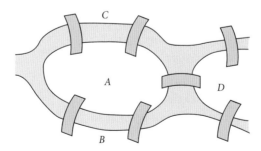

This problem laid the foundations for a type of graph theory problem, named after Euler himself.

An **Eulerian trail** is a trail that includes *every* edge in a graph, but only once. An Eulerian trail may include repeated vertices. If the Eulerian trail starts and finishes at the same vertex, then it can be called a **closed Eulerian trail**. If a connected graph contains a closed Eulerian trail, then it is called an **Eulerian graph**. If the Eulerian trail starts and finishes at different vertices, then it can be called an **open Eulerian trail**. If a connected graph contains an open Eulerian trail (also called a **semi-Eulerian trail**), then it is called a **semi-Eulerian graph**.

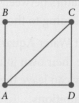

Identify whether the graph is Eulerian, semi-Eulerian or neither. If Eulerian or semi-Eulerian, state the corresponding closed or open trail found.

Steps	Working
Pick a starting vertex and test if it is possible to use every edge without repeating.	Starting at *A*, the following walk is possible: *A–B–C–A–D–C*. Other trails are possible, such as *A–C–D–A–B–C*. Given that this is an open Eulerian trail, the graph is semi-Eulerian.

Existence conditions and traversability of graphs

Consider the following problem.

Can you trace every edge of this graph without lifting your pencil and without repeating any edges? Do you start and finish at the same vertex?

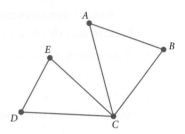

Such a problem is called a **traversability problem**. A graph is said to be **traversable** if there exists an open or closed Eulerian trail; that is, it is a semi-Eulerian or an Eulerian graph. As a result, the graph on the right can be called traversable, because there exists a closed Eulerian trail (e.g. *A–B–C–D–E–C–A*), but what is it about the features of the graph that allows it to be traversed?

Existence condition for an Eulerian graph

A graph can be called Eulerian if every vertex of the graph has an even degree; that is, a closed Eulerian trail exists in a graph if the graph has 0 vertices of odd degree.

What if we were to ask the same question for the slightly more complex graph on the right?

Can you trace every edge of this graph without lifting your pencil, without repeating any edges? Do you start and finish at the same vertex?

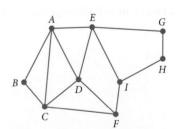

Hopefully after a couple of tries, you can see that it is possible to traverse every edge without repeats, but it is not possible to finish at the same starting vertex without repeating an edge. For example, *F–C–B–A–C–D–A–E–D–F–I–E–G–H–I* is an Eulerian trail and so the graph can be called semi-Eulerian. It cannot be called Eulerian as the edge *IF* has already been traversed. What is it about the features of the graph that allows it to be traversed? Identifying the degree of each of the vertices, we should see the following result.

Existence condition for a semi-Eulerian graph

A graph can be called semi-Eulerian if exactly 2 vertices in the graph have an odd degree; that is, an open Eulerian trail exists in a graph if the graph has exactly 2 vertices of odd degree.

 Exam hack

When a question asks whether a closed or open Eulerian trail exists, always find the degree of every vertex to quickly identify whether the correct existence conditions apply.

 Exam hack

To find an open Eulerian trail, start at one of the two vertices of odd degree and finish at the other vertex of odd degree.

By first considering the degree of each of the vertices in the following graphs, justify whether the graphs are Eulerian, semi-Eulerian or neither. If Eulerian or semi-Eulerian, state the corresponding closed or open Eulerian trail found.

a

b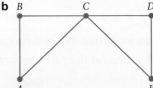

Steps	Working
a 1 Label each of the vertices with its corresponding degree.	
2 Highlight and count the number of odd vertices.	$\deg(A) = 3$ $\deg(B) = 3$ $\deg(C) = 3$ $\deg(D) = 3$ $\deg(E) = 4$
3 Identify which, if any, of the existence conditions are met and conclude.	There are neither 0 nor 2 vertices of odd degree, so the graph is neither Eulerian nor semi-Eulerian.
b 1 Label each of the vertices with its corresponding degree.	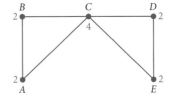
2 Highlight and count the number of odd vertices.	$\deg(A) = 2$ $\deg(B) = 2$ $\deg(C) = 4$ $\deg(D) = 2$ $\deg(E) = 2$
3 Identify which, if any, of the existence conditions are met and conclude.	All vertices have an even degree; that is, there are 0 vertices of odd degree, so the graph is Eulerian. A possible closed Eulerian trail is $A–B–C–D–E–C–A$.

Practical problems

In some problems, we may need to consider a real-life, practical context where *every edge* in a graph is included only once. This contextual problem would suggest we need to look for the existence of open or closed Eulerian trails.

<table>
<tr><td colspan="2">WORKED EXAMPLE 5   Solving a practical problem involving Eulerian graphs</td></tr>
<tr><td colspan="2">

A local high school has six buildings. The network shows these buildings represented by vertices. The edges of the network represent the paths between the buildings.

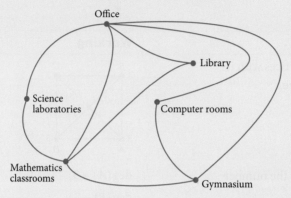

The organiser of a school tour is trying to plan the tour such that it starts from the Office and finishes at the Office. The tour is allowed to visit the same building more than once, but cannot walk along the same path more than once. Justify whether such a school tour is possible. If so, state the order of the tour.

</td></tr>
<tr><td>Steps</td><td>Working</td></tr>
<tr><td>

1   Label each vertex with its corresponding degree.

</td><td>

</td></tr>
<tr><td>

2   Highlight and count the number of odd vertices.

</td><td>

$\deg(\text{Office}) = 5$

$\deg(\text{Science}) = 2$

$\deg(\text{Maths}) = 4$

$\deg(\text{Library}) = 2$

$\deg(\text{Computer}) = 2$

$\deg(\text{Gym}) = 3$

</td></tr>
<tr><td>

3   Identify which, if any, of the existence conditions are met and conclude.

</td><td>

There are exactly 2 vertices of odd degree, and so there exists an Eulerian trail.

A school tour that walks along every path exactly once without repeats is possible, but it will not finish back at the Office.

</td></tr>
</table>

Features of Eulerian and semi-Eulerian graphs

A trail:
- has no repeated edges.

An open Eulerian trail:
- has no repeated edges
- includes every edge exactly once
- includes every vertex.

A closed trail:
- has no repeated edges
- starts and finishes at the same vertex.

A closed Eulerian trail:
- has no repeated edges
- starts and finishes at the same vertex
- includes every edge exactly once
- includes every vertex.

EXERCISE 5.2 Eulerian and semi-Eulerian graphs

ANSWERS p. 438

Recap

1 Consider the graph. Which one of the following is **not** a path for this graph?

A $P–R–Q–T–S$

B $P–Q–R–T–S$

C $P–R–T–S–Q$

D $P–T–Q–S–R$

E $P–T–R–Q–S$

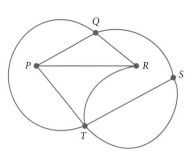

2 Consider the graph.

Which of the following terms best describes the following sequence of vertices: $A–B–C–D–E–F–D–A$?

A walk

B path

C trail

D closed trail

E cycle

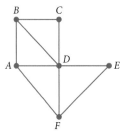

Mastery

3 [WORKED EXAMPLE 3] Identify whether the graph is Eulerian, semi-Eulerian or neither. If Eulerian or semi-Eulerian, state the corresponding closed or open trail found.

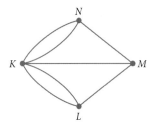

4 🔗 WORKED EXAMPLE 4   By first considering the degree of each of the vertices in the following graphs, justify whether the graphs are Eulerian, semi-Eulerian or neither. If Eulerian or semi-Eulerian, state the corresponding closed or open Eulerian trail found.

a

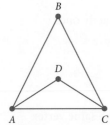

b

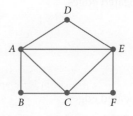

5 🔗 WORKED EXAMPLE 5   The suburb of Rampsville has a skateboard park with seven ramps. The ramps are shown as vertices *T*, *U*, *V*, *W*, *X*, *Y* and *Z* on the graph.

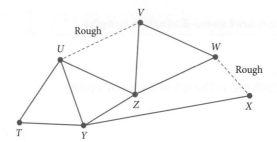

The tracks between ramps *U* and *V* and between ramps *W* and *X* are rough, as shown on the graph, and so not all skaters can travel along them. Nathan is a skater who can travel over all tracks, including the rough ones. He wants to begin skating at ramp *W*, travel along all tracks exactly once and finish back at *W*. Justify whether such a skate is possible. If so, state the order of the skate.

Calculator-free

6 (2 marks)   An open Eulerian trail for the graph shown will be possible if only one edge is removed. Determine the number of ways this can be achieved, stating the edge removed in each case.

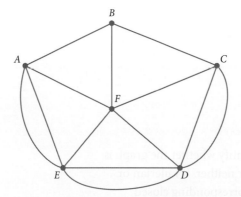

7 © SCSA   MA2016 Q5c   (2 marks)   Consider the three planar graphs below.

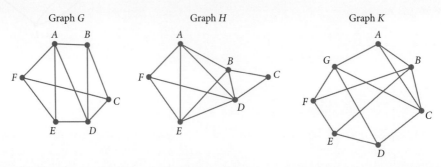

One of the planar graphs is semi-Eulerian. State which graph it is, giving a reason for your choice.

8 © SCSA MA2017 Q7a (3 marks) The graph below shows the paths connecting the exhibits at a zoo.

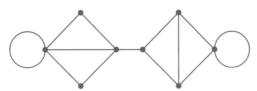

    a Explain why the graph is not semi-Eulerian. (1 mark)

    b Copy the graph and draw one edge on the graph so that it becomes semi-Eulerian and does not contain a bridge. (2 marks)

9 © SCSA MA2018 Q5c (2 marks) Consider a country town, labelled Town 1, in which roads connect the local attractions. The graph of Town 1 represents the road connections between attractions (vertices) within each town.

The local council of Town 1 wants to add one extra road so that an Eulerian trail is possible.

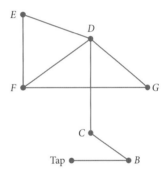

    a Copy the graph of Town 1 and draw an edge on the graph that allows this to occur. (1 mark)

    b Explain why a closed Eulerian trail is now possible. (1 mark)

10 © SCSA MA2021 Q3bc (2 marks) The graph shows the current network of reticulation pipes in Tarik's garden.

The water needs to travel from the tap and through all pipes.

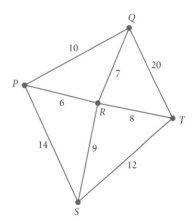

    a List a possible route for the water. (1 mark)

    b What is the mathematical term for the route listed in part **a**? (1 mark)

Calculator-assumed

11 (3 marks) Parcel deliveries are made between five nearby towns, P to T. The roads connecting these five towns are shown on the graph. The distances, in kilometres, are also shown. A road inspector will leave from town P to check all the roads and return to town P when the inspection is complete. He will travel the minimum distance possible.

    a Justify why the inspector's journey cannot be considered a closed Eulerian trail. (1 mark)

    b State the number of roads that the inspector will have to travel more than once. (1 mark)

    c Determine the minimum distance, in kilometres, that the inspector will travel. (1 mark)

5.3 Hamiltonian and semi-Hamiltonian graphs

Hamiltonian paths and cycles

In 1857, a game was created by William Rowan Hamilton that involved finding a path along the edges of a dodecahedron so that every vertex was visited exactly once and the path started and finished at the same vertex. This game, called the icosian game, led to a type of problem in graph theory named after Hamilton.

A **Hamiltonian path** in a graph is a walk that includes every vertex of a graph exactly once; that is, there are no repeat vertices. A Hamiltonian path that starts and finishes at the same vertex can be called a **Hamiltonian cycle** and the connected graph that contains a Hamiltonian cycle is called a **Hamiltonian graph**. If a connected graph contains an open Hamiltonian path, it is called a **semi-Hamiltonian graph**. Unlike with open and closed Eulerian trails, Hamiltonian paths and cycles do not have any existence conditions, and so to find whether a graph is Hamiltonian, semi-Hamiltonian or neither, trial and error or inspection methods need to be used.

 Exam hack

To remember the difference between Eulerian and Hamiltonian, remember open and closed Eulerian trails must travel on every **E**dge, and **H**amiltonian paths and cycles are like visiting every **H**ouse (vertex) on a map.

WORKED EXAMPLE 6 | Identifying Hamiltonian paths and cycles

For each of the graphs provided, where possible, state a

  i Hamiltonian path

  ii Hamiltonian cycle.

a   **b**   **c**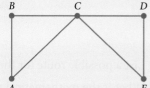

Steps	Working
1 Pick a starting vertex.	**a** **i** *A–B–C–D*
2 Identify an open path visiting every vertex with no repeated edges or vertices, but not finishing at the starting vertex.	Other answers are possible.
	ii *A–B–C–D–A*
	Other answers are possible.
3 Identify a closed path visiting every vertex with no repeated edges or vertices, but finishing at the starting vertex.	**b** **i** *A–B–E–C–D*
	Other answers are possible.
Trial and error/inspection is the only method that can be used.	**ii** *A–B–E–C–D–A*
	Other answers are possible.
	c **i** *A–B–C–D–E*
	Other answers are possible.
	ii No Hamiltonian cycle exists.

Features of Hamiltonian and semi-Hamiltonian graphs

A path:

- has no repeated edges
- has no repeated vertices.

A Hamiltonian path:

- has no repeated edges
- has no repeated vertices
- includes every vertex exactly once.

A cycle:

- has no repeated edges
- has no repeated vertices
- starts and finishes at the same vertex.

A Hamiltonian cycle:

- has no repeated edges
- has no repeated vertices
- starts and finishes at the same vertex
- includes every vertex exactly once.

Practical problems

In some problems, we may need to consider a real-life, practical context where *every vertex* in a graph is included only once. This contextual problem would suggest we need to look for the existence of Hamiltonian paths and cycles.

WORKED EXAMPLE 7 Solving a practical problem involving Hamiltonian graphs

Simon's holiday home has four rooms, A, B, C and D. The floor plan shows these rooms and the outside area. There are 12 doors, as shown on the floor plan. Only room C and the outside are labelled. A graph of the floor plan is also shown. On this graph, vertices represent the rooms and the outside area, and edges represent the doors.

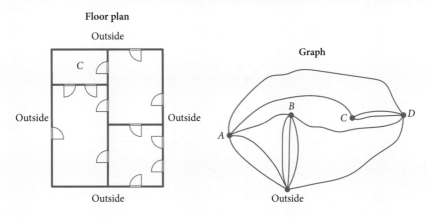

a Room C has already been labelled on the floor plan. Copy the floor plan and label the other three rooms A, B and D.

b Simon is in room C and his daughter Zofia is outside. Simon calls Zofia to see him in room C. Zofia visits every other room once on her way to room C. Give the mathematical term that describes Zofia's journey.

Steps	Working
a **1** Count the degree of each vertex. **2** Match the degrees to the number of doors for each room.	$\deg(A) = 5$ $\deg(B) = 5$ $\deg(C) = 3$ $\deg(D) = 5$ $\deg(\text{Outside}) = 6$ **Floor plan** Outside *C* *A* Outside Outside *D* *B* Outside
b Identify the critical features of Zofia's journey.	Not all edges are traversed, but every vertex is visited once. Given that Zofia starts Outside and finishes at *C*, her journey is a Hamiltonian path.

Video
WACE question analysis: Paths and cycles

WACE QUESTION ANALYSIS

© SCSA   MA2016 Q14bcd   Calculator-assumed   **(5 marks)**

Therese, a mathematics student at Trinity College, Dublin was employed as a guide for a cultural tour of Dublin. She decided to use graph theory to plan the walking tour.

Below is a network she constructed in which the:

- vertices represent the points of interest to be visited, and
- edges represent the most direct route between adjacent vertices.

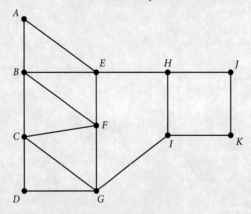

a Therese planned to take the group on a closed walk. Explain the meaning of a closed walk.   **(1 mark)**

b She also stated that the walk would qualify as a Hamiltonian cycle. State the **two** properties that makes the walk a Hamiltonian cycle.   **(2 marks)**

c Given that the walk started at *G* (Trinity College, Dublin), copy the network above and then mark the Hamiltonian cycle.   **(2 marks)**

Reading the question

- Highlight what the vertices and edges represent in the context of the question.
- Highlight the command words, e.g. explain and mark.
- Highlight the key graph theory terminology, for example, closed walk and Hamiltonian cycle.

Thinking about the question

- 'Explain' requires a written response that contains the key mathematical language.
- What are the conditions for a Hamiltonian cycle?

Worked solution (✓ = 1 mark)

a Define what it means for a walk to be 'closed' in terms of the vertices.

A closed walk is a walk which starts and ends at the same vertex. ✓

b Define what it means to be a 'cycle', i.e. a closed path, and define what it means to be 'Hamiltonian'.

A Hamiltonian cycle is a walk that

passes through all vertices only once (except for the start and finish vertex) ✓

starts and finishes the walk at the same vertex (a closed path) ✓

c Use the fact that the start and end vertex must be G to construct the Hamiltonian cycle.

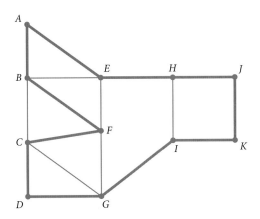

correctly draws a closed walk through G ✓

correctly visits all vertices exactly once with no repeated edges ✓

 Exam hack

When marking and annotating graphs, working out needs to be clearly indicated or highlighted on the graph, or the graph should be redrawn if unclear.

Recap

1 For the network diagram, an Eulerian trail can be found

  A without altering the network diagram.

  B by adding an edge that joins A to H.

  C by adding an edge that joins C to F.

  D by removing the edge that joins B to C.

  E by removing the edge that joins D to E.

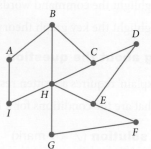

2 Consider the graph. The minimum number of extra edges that are required so that a closed Eulerian trail is possible in this graph is

  A 0   **B** 1   **C** 2

  D 3   **E** 4

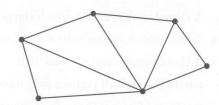

Mastery

3 [WORKED EXAMPLE 6] For each of the graphs provided, where possible, state a

  i Hamiltonian path

  ii Hamiltonian cycle.

  a 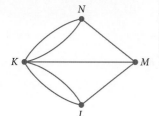   **b** 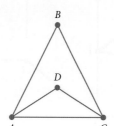   **c**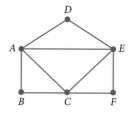

4 [WORKED EXAMPLE 7] Maggie's house has five rooms, A, B, C, D and E, and eight doors. The floor plan of these rooms and doors is shown below. The outside area, F, is shown shaded on the floor plan. The floor plan is represented by the graph to the right of the floor plan. On this graph, vertices represent the rooms and the outside area. Edges represent direct access to the rooms through the doors.

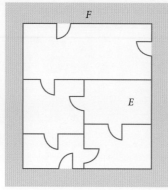

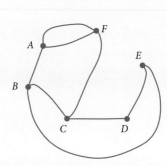

  a Room E and the outside region F have already been labelled on the floor plan. Copy the floor plan and label the other four rooms A, B, C and D.

  b Maggie hires a cleaner to clean the house. It is possible for the cleaner to enter the house from the outside area, F, and walk through each room only once, cleaning each room as they go and finishing in the outside area, F. Give the mathematical term that describes the cleaner's route.

5 (2 marks) For each of the graphs shown, state a Hamiltonian cycle.

5.3

a

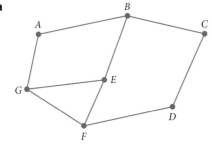

b

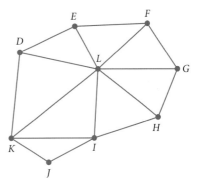

6 © SCSA MA2018 Q2d (2 marks) A bus company conducts 'jump on, jump off' sightseeing tours, during which tourists can get on and off buses at any of the designated attractions as many times as they like during the same day. The weighted digraph below shows Attractions *A* to *F*, along with the time (in minutes) that a bus takes to travel between the attractions. The bus company operates two routes around the city, each shown differently on the right.

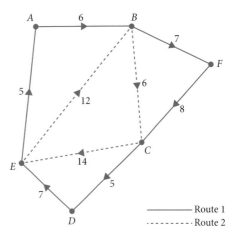

The bus company plans to include another attraction, *G*. Instead of adding another route, they will provide shuttle buses between *F* and *G* (four minutes in each direction) and *D* and *G* (five minutes in each direction), as shown on the right.

Toshi is at Attraction *D*. He wants to complete a Hamiltonian path. State the route he should take.

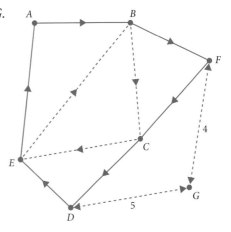

7 © SCSA MA2019 Q4b (2 marks) A marine park has attractions with paths connecting them. The vertices on the graph represent the attractions and the numbers on the edges represent the path distances (km) between the attractions. Visitors can either walk around the park or take one of the many shuttle buses that run between attractions.

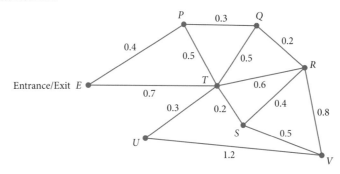

Rachel arrives at the entrance. She wants to complete a Hamiltonian cycle. State the route she should take.

8 © SCSA | MA2021 Q5c | (4 marks) The network below shows the relative distances, in hundreds of metres, between Wi-Fi hotspots around a university campus.

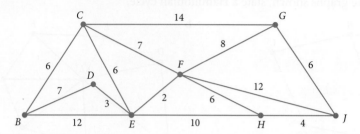

The control room for this system of hotspots is located at *B*. A problem has been reported with the hotspot at *J*.

After repairs have been made, all hotspots need to be checked. A technician is sent from the control room, travelling to all hotspots once only, and finishing back at the control room.

a State the name given to this type of path. (2 marks)

b Determine the length of the shortest path possible and state the path used. (2 marks)

Calculator-assumed

9 © SCSA | MA2020 Q13a | (3 marks) The graph below represents a road transport network from a warehouse at *A* to seven retail outlets *B*, *C*, *D*, *E*, *F*, *G* and *H*. The number on each edge represents the distance, in kilometres, along each road.

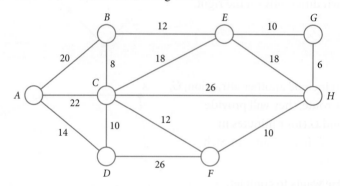

Identify the shortest Hamiltonian path from the warehouse and state its length.

10 (2 marks) A factory supplies groceries to stores in five towns, *Q*, *R*, *S*, *T* and *U*, represented by vertices on the graph.

The edges of the graph represent roads that connect the towns and the factory. The numbers on the edges indicate the distance, in kilometres, along the roads. Vehicles may only travel along the road between towns *S* and *Q* in the direction of the arrow due to temporary roadworks. Each day, a van must deliver groceries from the factory to the five towns. The first delivery must be to town *T*, after which the van will continue on to the other four towns before returning to the factory.

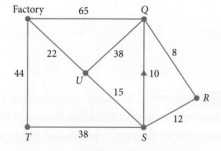

a The shortest possible route from the factory for this delivery run, starting with town *T*, is not Hamiltonian. Copy and complete the order in which these deliveries would follow this shortest possible route:

factory – *T* – _____ – factory (1 mark)

b With reference to the town names in your answer to part **a**, explain why this shortest route is not a Hamiltonian cycle. (1 mark)

Types of walks

- A **walk** is a sequence of connected vertices.
- A **trail** is a walk with no repeated edges.
- A **closed trail** is a walk with no repeated edges that starts and finishes at the same vertex.
- A **path** is a walk with no repeated edges and no repeated vertices.
- A **cycle** is a walk with no repeated edges and no repeated vertices that starts and finishes at the same vertex.

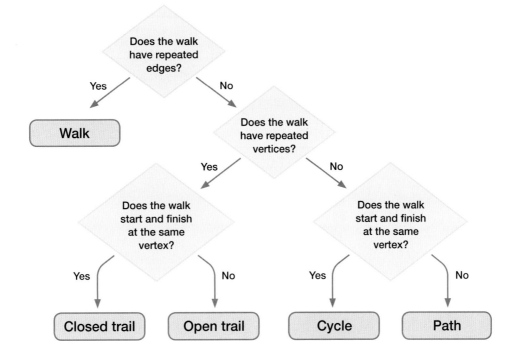

Open and closed Eulerian trails

Eulerian trail

- An **Eulerian trail** is a walk with no repeated edges that includes every *edge* in a graph.
- An **open Eulerian trail** (**semi-Eulerian trail**) will only exist if the graph has *exactly two* vertices of odd degree.
- To find the Eulerian trail, you must start at one of the two vertices of odd degree and finish at the other vertex of odd degree.

Closed Eulerian trail

- A **closed Eulerian trail** is a walk with no repeated edges that includes every *edge* in a graph and starts and finishes at the same vertex.
- A closed Eulerian trail will only exist if the graph has all vertices of even degree.

Hamiltonian paths and cycles

- A **Hamiltonian path** is a walk with no repeated vertices that includes every vertex in a graph.
- A **Hamiltonian cycle** is a walk with no repeated vertices that includes every vertex in a graph and starts and finishes at the same vertex.
- We need to use trial and error to find whether a Hamiltonian path or cycle exists.

Walk summary

Walks that:	Open trail	Open Eulerian trail	Closed trail	Closed Eulerian trail	Path	Hamiltonian path	Cycle	Hamiltonian cycle
have no repeated edges	✓	✓	✓	✓	✓	✓	✓	✓
have no repeated vertices					✓	✓	✓	✓
start and finish at the same vertex			✓	✓			✓	✓
include every edge		✓		✓				
include every vertex		✓		✓		✓		✓

9780170476959

Cumulative examination: Calculator-free

Total number of marks: 20   Reading time: 2 minutes   Working time: 20 minutes

1 (6 marks) Consider the scatterplot below showing a sample of 10 students taken from a Year 12 Mathematics Applications class. The scatterplot shows the students' performances (as percentages) in the first two tests of the year: Bivariate data and Graph theory.

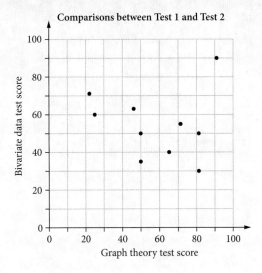

Comparisons between Test 1 and Test 2

A linear regression model is applied to the data set and is found to be $y = -0.05x + 57.6$, with a correlation coefficient of $r = -0.07$.

a Give a reason as to why the correlation coefficient suggests there is no correlation between these two data sets.   (1 mark)

b Predict a student's score in the Bivariate data test, if they scored 20% in the Graph theory test. Comment on the reliability of this prediction.   (2 marks)

c If the student who scored 91% in the Graph theory test and 90% in the Bivariate data test was removed from the data set, explain the effect on the

   i correlation coefficient   (1 mark)

   ii gradient of the least-squares line   (1 mark)

   iii y-intercept of the least-squares line.   (1 mark)

2 (2 marks)

a Consider the following four graphs.

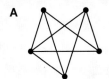 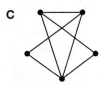

State which of the four graphs have a closed Eulerian trail.   (1 mark)

b Consider Graph E.

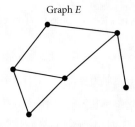

Graph E

Explain whether the graph has a Hamiltonian path.   (1 mark)

3 © SCSA MA2017 Q3a(iii, iv) (3 marks) A planar graph has five faces and five vertices, A, B, C, D and E, as shown in the diagram.

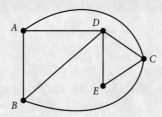

a Determine a Hamiltonian cycle for the graph, giving your answer as a sequence of vertices. (1 mark)

b Is the graph Eulerian, semi-Eulerian or neither? Justify your answer. (2 marks)

4 © SCSA MA2020 Q2b(ii, iii) (3 marks) A small research facility consists of five buildings with walkways represented by the edges in this network:

Three different plans for completing the network with the addition of four walkways are shown.

Plan P Plan Q Plan R

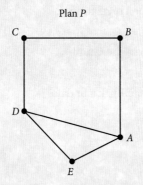 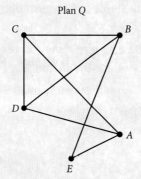 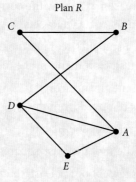

State which, if any, of these plans has a graph that

a is Eulerian (1 mark)

b contains a Hamiltonian cycle. (2 marks)

5 (6 marks) A graph has five vertices, A, B, C, D and E.
The adjacency matrix for this graph is shown.

$$\begin{array}{c} \\ A \\ B \\ C \\ D \\ E \end{array} \begin{array}{ccccc} A & B & C & D & E \\ \left[\begin{array}{ccccc} 0 & 1 & 0 & 1 & 2 \\ 1 & 0 & 1 & 0 & 1 \\ 0 & 1 & 1 & 0 & 1 \\ 1 & 0 & 0 & 0 & 1 \\ 2 & 1 & 1 & 1 & 0 \end{array}\right] \end{array}$$

a Construct a possible graph that represents this adjacency matrix. (2 marks)

b Justify whether the graph is

  i Eulerian, semi-Eulerian or neither (2 marks)

  ii Hamiltonian, semi-Hamiltonian or neither. (2 marks)

Total number of marks: 17   Reading time: 2 minutes   Working time: 17 minutes

1 (7 marks) The surface temperature of a hot boiled egg is measured at regular one-minute intervals after being removed from the boiling water. It is found that its surface temperature (T_n) after n minutes can be modelled by the first order linear recurrence relation:

$$T_{n+1} = 0.65T_n + 10, T_1 = 90$$

a Calculate the surface temperature of the egg five minutes after being removed from the boiling water, correct to two decimal places. (2 marks)

b Suppose room temperature is 30°C. After how many minutes will the surface temperature of the egg reach room temperature? (2 marks)

c Describe the long-term behaviour of the surface temperature of the egg. Support your answer with evidence of a calculation. (3 marks)

2 (3 marks) Bus routes connect six suburbs. The towns are North City (N), Osbornevale (O), Palmira (P), Queens Garden (Q), Riverville (R) and Spearton (S). The graph gives the cost, in dollars, of bus travel along these routes. Bai lives in North City (N) and he is planning on travelling by bus to holiday in Spearton (S). The bus routes and the corresponding distances are shown in the following network.

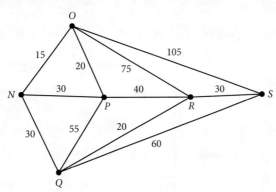

a Bai considers travelling by bus along the route $N–O–S$. State the cost of this journey. (1 mark)

b If Bai takes the cheapest route from N to S, which other town(s) will he pass through? (1 mark)

c Show that Euler's formula, $v + f = e + 2$, holds for this graph. (1 mark)

3 (4 marks) The diagram shows the food web for insects (*I*), birds (*B*) and lizards (*L*). The matrix *M* has been constructed to represent the information in this diagram. In matrix *M*, a '1' is read as 'eat' and a '0' is read as 'do not eat'.

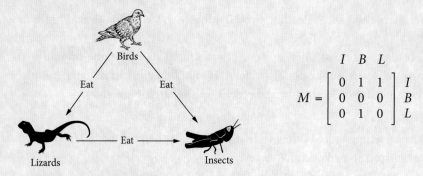

$$M = \begin{bmatrix} 0 & 1 & 1 \\ 0 & 0 & 0 \\ 0 & 1 & 0 \end{bmatrix} \begin{matrix} I \\ B \\ L \end{matrix}$$

with column headers $I \ B \ L$

a Interpret the following features of the matrix M in context of the question.

 i The '1' in M_{13}. (1 mark)

 ii The row of zeros in matrix *M*. (1 mark)

The diagram below shows the food web for insects (*I*), birds (*B*), lizards (*L*) and frogs (*F*). A second matrix *N* has been set up to represent the information in this diagram. This matrix *N* has not been completed.

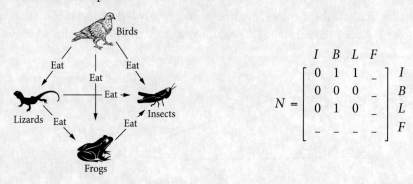

$$N = \begin{bmatrix} 0 & 1 & 1 & - \\ 0 & 0 & 0 & - \\ 0 & 1 & 0 & - \\ - & - & - & - \end{bmatrix} \begin{matrix} I \\ B \\ L \\ F \end{matrix}$$

with column headers $I \ B \ L \ F$

b Copy and complete the matrix *N*. (1 mark)

c Hence, explain the significance of the values in the leading diagonal of *N* in context of the question. (1 mark)

4 (3 marks) George lives in town *G* and Macey lives in town *M*. The diagram shows the network of main roads between town *G* and town *M*. The vertices *G*, *H*, *I*, *J*, *K*, *L*, *M*, *N* and *O* represent towns. The edges represent the main roads. The numbers on the edges indicate the distances, in kilometres, between adjacent towns.

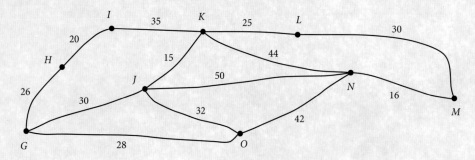

a Determine the shortest distance, in kilometres, between town *G* and town *M*, stating the corresponding path. (2 marks)

b George plans to travel to Macey's house. He will pass through all the towns shown above. George plans to take the shortest route possible. Identify the town that George will pass through twice. (1 mark)

9780170476959

CHAPTER 6

TIME SERIES ANALYSIS

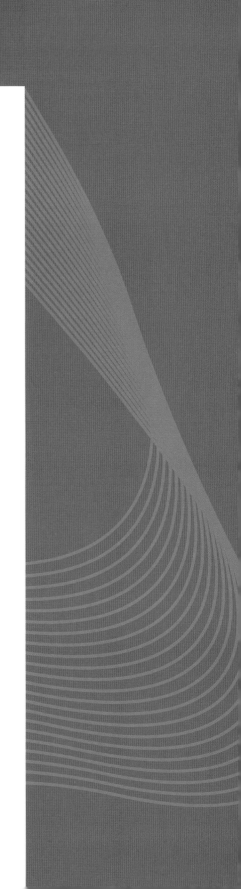

TOPIC 4.1: TIME SERIES ANALYSIS

Describing and interpreting patterns in time series data

4.1.1   construct time series plots

4.1.2   describe time series plots by identifying features such as trend (long term direction), seasonality (systematic, calendar-related movements), and irregular fluctuations (unsystematic, short term fluctuations), and recognise when there are outliers

Analysing time series data

4.1.3   smooth time series data by using a simple moving average, including the use of spreadsheets to implement this process

4.1.4   calculate seasonal indices by using the average percentage method

4.1.5   deseasonalise a time series by using a seasonal index, including the use of spreadsheets to implement this process

4.1.6   fit a least-squares line to model long-term trends in time series data

4.1.7   predict from regression lines, making seasonal adjustments for periodic data

The data investigation process

4.1.8   implement the statistical investigation process to answer questions that involve the analysis of time series data

Mathematics Applications ATAR Course Year 12 syllabus p. 13 © SCSA

Video playlists (5):

6.1   Time series data and time series plots

6.2   Moving averages

6.3   Seasonal adjustment

6.4   Least-squares trend lines

WACE question analysis Time series analysis

Worksheets (5):

6.1   Time series plots

6.2   Smoothing time series data • Moving means

6.3   Seasonal adjustment • Deseasonalisation

Nelson MindTap

To access resources above, visit
cengage.com.au/nelsonmindtap

6.1 Time series data and time series plots

A **time series** involves data recorded over a period of time. The data is usually recorded at regular intervals, such as hours, days, weeks, months, seasons or years. Examples of time series data are daily rainfall recorded by the Bureau of Meteorology, hourly temperature or blood pressure of a patient at a hospital recorded by a nurse, daily share prices, and quarterly sales figures for a company.

A **time series plot** is a scatterplot of a time series where time is plotted on the horizontal axis and the data points are joined in order by straight lines. The time divisions on the horizontal axis are often simplified into numbers 1, 2, 3 … to make them easier to work with.

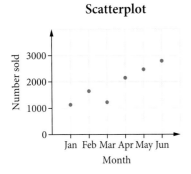

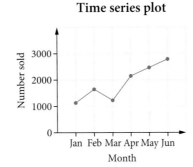

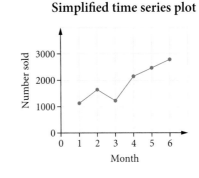

Video playlist
Time series data and time series plots

Worksheet
Time series plots

🔓 Exam hack

In a scatterplot with *time* as the explanatory variable, you can have more than one data point for a specific time value. In a time series plot you can't. For example, you can't draw a time series plot for the data shown on this scatterplot.

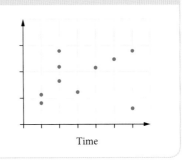

USING CAS 1 | Constructing time series plots

Create a time series plot using this table showing births in a state from 1910 to 1930.

Year	Total live births		Year	Total live births
1910	41104		1921	39576
1911	40623		1922	45623
1912	39200		1923	48789
1913	29239		1924	49792
1914	41300		1925	50124
1915	45601		1926	50678
1916	45122		1927	51237
1917	40123		1928	53456
1918	47563		1929	54789
1919	48323		1930	53627
1920	48124			

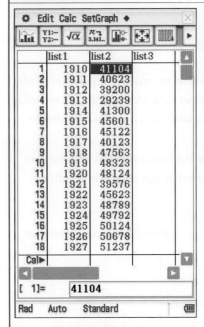

ClassPad

1. Open the **Statistics** application.
2. Clear all lists and enter the data from the table as shown.

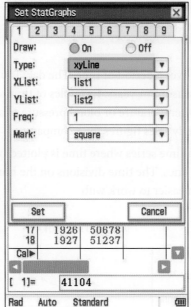

3. Tap **SetGraph > Setting**.
4. Change the **Type:** field to **xyLine**.
5. Tap **Set**.

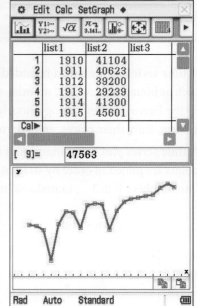

6. Tap **Graph**.
7. The time series plot will be displayed in the lower window.

TI-Nspire

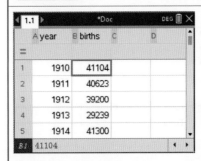

1. Start a new document and add a **Lists & Spreadsheet** page.
2. Label the columns and enter the data from the table, as shown above.

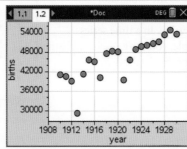

3. Insert a **Data & Statistics** page.
4. For the horizontal axis, select **year**.
5. For the vertical axis, select **births**.

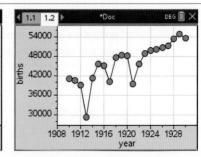

6. Press **menu > Plot Type > XY Line Plot**.
7. The time series plot will be displayed.

Features of time series plots

Time series plots are used to observe the movement and fluctuations of the time series data. A time series plot comprises the following features.

Trend

The **trend** is the general direction of the time series (increasing or decreasing) over a long period of time.

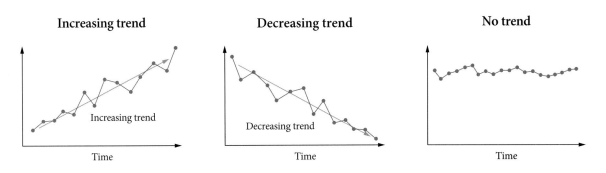

Seasonal variation

Seasonal variation is the regular rise and fall in the time series that recurs in time periods of a year or less, such as days, weeks, months, quarters, or actual seasons. Retail stores, for example, regularly and predictably sell more on Saturday and Sunday than on the other days of the week. Ice-cream sales each year regularly and predictably peak during summer and dip during winter.

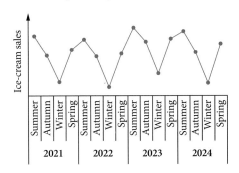

> ## 🔒 Exam hack
>
> To decide if a graph is showing seasonality, check to see if the peaks always occur at the same time; for example, always on a Sunday or always in summer. If they don't, then the graph is not showing seasonality. To determine the number of seasons in a period, locate the peaks (or troughs) of the time series and find the number of intervals between two peaks (or troughs).

It's possible for data to show both seasonality and a trend at the same time. The time series below has regular peaks every four months, as well as having an increasing trend.

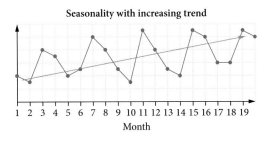

Seasonality with increasing trend

Irregular fluctuations

An **irregular variation** (also called **fluctuation** or **noise**) is an erratic and short-term variation in a time series that is a product of chance occurrences and can't be explained by trends and seasonality. All real-world time series plots will have some irregular fluctuations.

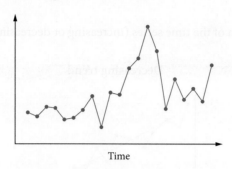

Time

WORKED EXAMPLE 1	Identifying seasonal variations in time series

For each of the following, state whether a time series of the data is likely to show seasonality. Explain your reasoning.

a sales of air conditioners   **b** sales of bread

c sales of dishwashers   **d** number of interstate visitors with school-age children in Perth

Steps	Working
Is there likely to be a pattern throughout the year? If so, what time period would the pattern be based on?	**a** Sales of air conditioners are likely to show seasonality because more would be bought in summer than in winter. **b** Sales of bread are not likely to show seasonality because bread is eaten regularly each day. **c** Sales of dishwashers are likely to show seasonality because more people would have the time to shop for them on weekends than on weekdays. **d** The number of interstate visitors with school-age children to Perth is likely to show seasonality because there would be a larger number during school holidays and fewer during the school term.

Outliers in time series

Outliers in a time series can be easy to identify, as in the example shown below. The difficulty is deciding whether to keep the outliers in the data analysis or to take them out. For example, if the outlier was the result of a one-off freak storm over a weather station, we would remove it, but if it was a measurement on a heart monitor, we wouldn't.

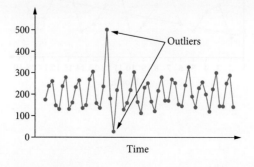

Time

Mastery

1 [Using CAS 1] Create a time series plot using this table showing births by year in a state from 1910 to 1920.

Births by year in a state from 1910 to 1920

Year	Total live births	Year	Total live births
1910	11 124	1916	14 100
1911	12 000	1917	14 127
1912	12 200	1918	15 163
1913	13 329	1919	15 323
1914	13 500	1920	16 224
1915	14 005		

2 [WORKED EXAMPLE 1] For each of the following, state whether a time series of the data is likely to show seasonality. Explain your reasoning.

a sales of umbrellas

b sales of shorts

c sales of television sets

d consumption of milk

e occupancy rates at a holiday resort

f money earned by teachers

g money earned by fruit pickers

h number of people attending AFL matches.

Calculator-free

3 (3 marks) Describe the trend of each of these time series plots.

a

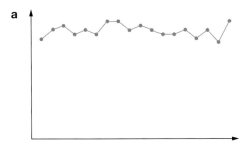

(1 mark)

b

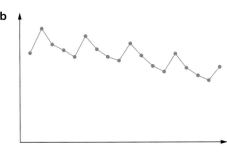

(1 mark)

c

(1 mark)

4 (12 marks) For each of the following sets of time series data

    i construct the time series plot using CAS   (1 mark)

    ii describe the trend of the time series   (1 mark)

    iii identify any possible outliers or irregular variations   (1 mark)

    iv determine the number of seasons for each period.   (1 mark)

a The following table shows the quarterly sales data of swimming pool covers.   (4 marks)

Time	Year	Month	Sales of swimming pool covers ('00)
1	2020	February	3.2
2		May	2.4
3		August	1.4
4		November	2.7
5	2021	February	4.4
6		May	3.3
7		August	2.6
8		November	3.7
9	2022	February	5.5
10		May	4.3
11		August	3.3
12		November	4.6

b The following data shows the quarterly water usage of a typical 4-person household in Melville.   (4 marks)

Time	Year	Month	Water usage (kL)
1	2018	March	168
2		June	85
3		September	76
4		December	136
5	2019	March	173
6		June	88
7		September	78
8		December	140
9	2020	March	176
10		June	89
11		September	80
12		December	143
13	2021	March	183
14		June	93
15		September	83
16		December	241

c The following data shows the prices of unleaded petrol recorded over a 3-week period in Perth.

(4 marks)

Week	Day	Price (cents/litre)
1	Monday	165.0
	Tuesday	163.8
	Wednesday	163.5
	Thursday	163.6
	Friday	165.9
	Saturday	169.1
	Sunday	173.2
2	Monday	163.1
	Tuesday	161.5
	Wednesday	163.7
	Thursday	163.5
	Friday	165.1
	Saturday	168.2
	Sunday	170.9
3	Monday	163.0
	Tuesday	159.0
	Wednesday	160.2
	Thursday	161.1
	Friday	162.3
	Saturday	164.1
	Sunday	169.0

 **Moving averages**

Video playlist
Moving averages

Smoothing time series data is a technique for levelling out fluctuations to produce a smoother graph that lets us see the underlying trend more clearly. One way of achieving this is through **moving averages**.

Worksheets
Smoothing time series data

Moving means

Smoothing using moving averages

Moving averages is the most common numerical smoothing technique. It involves finding a series of averages of a fixed number of data points. The simplest method is to smooth using an odd number of data points, such as 3, 5 or 7.

The following table shows the number of students in a Year 12 Mathematics Applications class over the last 10 years.

Year	1	2	3	4	5	6	7	8	9	10
Number of students	25	18	23	21	19	20	18	16	17	15

a Use a table with three columns to calculate the smoothed data using the method of 3-point moving averages.

b What are the smoothed number of students for the fifth and tenth years?

c Graph the original data and the smoothed data on the same set of axes.

d What does the graph of the smoothed data indicate about the trend in the original data?

Steps	**Working**

a 1 Set up a table with three columns with Year in column 1, Number of students in column 2, and the calculations for 3-point moving averages in column 3.

2 Find the mean for each group of 3 consecutive values for the number of students.

3 Write the mean in the row of the middle number of the 3 consecutive values.

Note: You cannot find a moving average for the first and last value for the number of students.

Year	Number of students	3-point moving averages
1	25	
2	18	$\dfrac{25 + 18 + 23}{3} = 22$
3	23	$\dfrac{18 + 23 + 21}{3} = 20.67$
4	21	$\dfrac{23 + 21 + 19}{3} = 21$
5	19	$\dfrac{21 + 19 + 20}{3} = 20$
6	20	$\dfrac{19 + 20 + 18}{3} = 19$
7	18	$\dfrac{20 + 18 + 16}{3} = 18$
8	16	$\dfrac{18 + 16 + 17}{3} = 17$
9	17	$\dfrac{16 + 17 + 15}{3} = 16$
10	15	

b Read from the table.

The smoothed number of students for the fifth year is 20. There was not enough data to calculate the smoothed number of students for the tenth year.

c Graph the original data and the smoothed data on the same set of axes.

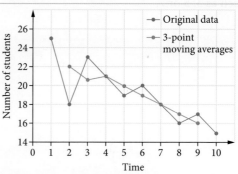

d Identify whether the smoothed graph shows an increasing or a decreasing trend.

The graph of the smoothed data indicates a decreasing trend.

In **3-point moving average smoothing**, the first and last points are lost because there isn't enough data to calculate them. In **5-point moving average smoothing**, the first two and the last two points are lost. This pattern continues as the number of odd points increases.

Using the data in the previous example, we can show a 5-point moving average smoothing in the table below.

Year	Number of students	5-point moving averages
1	25	
2	18	
3	23	$\dfrac{25 + 18 + 23 + 21 + 19}{5} = 21.2$
4	21	$\dfrac{18 + 23 + 21 + 19 + 20}{5} = 20.2$
5	19	$\dfrac{23 + 21 + 19 + 20 + 18}{5} = 20.2$
6	20	$\dfrac{21 + 19 + 20 + 18 + 16}{5} = 18.8$
7	18	$\dfrac{19 + 20 + 18 + 16 + 17}{5} = 18$
8	16	$\dfrac{20 + 18 + 16 + 17 + 15}{5} = 17.2$
9	17	
10	15	

The result of the smoothing using 5-point moving averages can be seen in the below graph. The larger the number of points used for the moving average, the greater the smoothing effect.

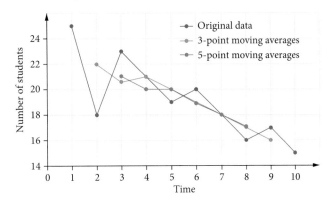

Smoothing with an even number of points

Smoothing with an even number of points is more complicated than smoothing with an odd number of points because the centre of an even number of points isn't at one of the original time values. We deal with this using a process called **centring**, which involves finding the mean of each of the two consecutive smoothed values.

WORKED EXAMPLE 3 | Moving average smoothing with an even number of points

Determine the 4-point centred moving averages for the sales of a company.

Year	Quarter	Period	Sales
2019	January	1	8.95
	April	2	10
	July	3	2.88
	October	4	5.6
2020	January	5	12.5
	April	6	13.38
	July	7	4.8
	October	8	8.98
2021	January	9	17.52
	April	10	18.89
	July	11	5.78
	October	12	10.05

Note:

- If the sales data were to be smoothed using the 4-point moving averages, the moving averages will not coincide with the time period.

- There are two ways of calculating the 4-point centred moving averages. The second method is faster.

▶

Steps

Method 1

1 Set up a table with four columns and extra rows between time intervals.

2 Determine the 4-point moving averages.

3 Determine the 4-point centred moving averages by calculating the means of the 4-point moving averages.

Working

Period	Sales ($'000)	4-point moving averages	4-point centred moving averages
1	8.95		
2	10		
		$\frac{8.95+10+2.88+5.6}{4} = 6.8575$	
3	2.88		$\frac{6.8575+7.745}{2} = 7.301$
		$\frac{10+2.88+5.6+12.5}{4} = 7.745$	
4	5.6		$\frac{7.745+8.59}{2} = 8.168$
		$\frac{2.88+5.6+12.5+13.38}{4} = 8.59$	
5	12.5		$\frac{8.59+9.07}{2} = 8.830$
		$\frac{5.6+12.5+13.38+4.8}{4} = 9.07$	
6	13.38		$\frac{9.07+9.915}{2} = 9.493$
		$\frac{12.5+13.38+4.8+8.98}{4} = 9.915$	
7	4.8		$\frac{9.915+11.17}{2} = 10.543$
		$\frac{13.38+4.8+8.98+17.52}{4} = 11.17$	
8	8.98		$\frac{11.17+12.5475}{2} = 11.859$
		$\frac{4.8+8.98+17.52+18.89}{4} = 12.5475$	
9	17.52		$\frac{12.5475+12.7925}{2} = 12.670$
		$\frac{8.98+17.52+18.89+5.78}{4} = 12.7925$	
10	18.89		$\frac{12.7925+13.06}{2} = 12.926$
		$\frac{17.52+18.89+5.78+10.05}{4} = 13.06$	
11	5.78		
12	10.05		

Method 2

1 Set up a table with three columns.

2 Determine the first 4-point centred moving average by adding half of the first and fifth data points and the sum of the second, third and fourth data points, and then divide by 4.

3 Continue the process until the data points are exhausted.

Period	Sales ($'000)	4-point centred moving averages
1	8.95	
2	10	
3	2.88	$\dfrac{0.5\times8.95+10+2.88+5.6+0.5\times12.5}{4} = 7.301$
4	5.6	$\dfrac{0.5\times10+2.88+5.6+12.5+0.5\times13.38}{4} = 8.168$
5	12.5	$\dfrac{0.5\times2.88+5.6+12.5+13.38+0.5\times4.8}{4} = 8.830$
6	13.38	$\dfrac{0.5\times5.6+12.5+13.38+4.8+0.5\times8.98}{4} = 9.493$
7	4.8	$\dfrac{0.5\times12.5+13.38+4.8+8.98+0.5\times17.52}{4} = 10.543$
8	8.98	$\dfrac{0.5\times13.38+4.8+8.98+17.52+0.5\times18.89}{4} = 11.859$
9	17.52	$\dfrac{0.5\times4.8+8.98+17.52+18.89+0.5\times5.78}{4} = 12.670$
10	18.89	$\dfrac{0.5\times8.98+17.52+18.89+5.78+0.5\times10.05}{4} = 12.926$
11	5.78	
12	10.05	

We also lose data points when smoothing with an even number of points. In **5-point centred moving average smoothing**, the first 2 and the last 2 points are lost and so on. Following the same logic, for a **6-point centred moving average smoothing**, the first 3 points and the last 3 points are lost. This pattern continues as the number of even points increases.

Moving average smoothing with an odd number of points

3-point moving averages

Time	Original data	Smoothed data
1	*	
2	*	average → *
3	*	average → *
4	*	

5-point moving averages

Time	Original data	Smoothed data
1	*	
2	*	
3	*	average → *
4	*	average → *
5	*	
6	*	

9780170476959

4-point moving averages with centring

Time	Original data	Centring	Smoothed data
1	*		
2	*		
		average ►*	
3	*		average ► *
		average ►*	
4	*		
5	*		

Faster method for 4-point moving averages with centring

Time	Original data	Smoothed data
1	*	
2	*	
3	*	$\dfrac{0.5*+*+*+*+0.5*}{4}$
4	*	
5	*	

Choosing the number of points for moving averages

Matching the moving averages to natural cycles

The number of moving average points is usually chosen by looking at the natural cycle of the data being considered.

	Number of points for moving means
Daily sales figures for stores open Monday to Sunday	7
Daily sales figures for stores open only Monday to Friday	5
Monthly figures	12
Quarterly accounts	4

How high can you go?

There are a number of factors to take into account when choosing how many points to use for moving average smoothing. The larger the number of points, the more data points that are lost.

- With 3-point moving averages, we lose 2 data points (one at the start and one at the end).
- With 4-point centred and 5-point moving averages, we lose 4 data points (2 at the start and 2 at the end).
- With 6-point centred and 7-point moving averages, we lose 6 data points and so on.

If we go too high, we could end up losing nearly all of the data points.

Using spreadsheets for smoothing

Spreadsheets can be used to generate the moving averages.

WORKED EXAMPLE 4	Moving averages smoothing with an odd number of points using spreadsheets

The following table shows the total sales for a company for the last 12 periods.

Time period	1	2	3	4	5	6	7	8	9	10	11	12
Total sales ($'000)	20	27	25	22	30	25	24	33	30	25	36	30

Use a spreadsheet to determine the 3-point moving averages.

Steps

Working

1 Type your data into two columns in any spreadsheet (Microsoft Excel is used in this example). The first column should have the time period and the second column the time series data (the total sales figures in this case). Make sure there are no blank rows in your cell data.

2 Calculate the first 3-point moving average for the time series data. For this example, type '=(B2+B3+B4)/3' into cell C3.

	A	B	C
1	Time period	Total sales ($'000)	3-point moving averages
2	1	20	
3	2	27	24.000
4	3	25	
5	4	22	
6	5	30	
7	6	25	
8	7	24	
9	8	33	
10	9	30	
11	10	25	
12	11	36	
13	12	30	

3 Drag the square in the bottom right corner down to move the formula to all cells in the column to calculate the moving averages for the successive time period.

	A	B	C
1	Time period	Total sales ($'000)	3-point moving averages
2	1	20	
3	2	27	24.000
4	3	25	24.667
5	4	22	25.667
6	5	30	25.667
7	6	25	26.333
8	7	24	27.333
9	8	33	29.000
10	9	30	29.333
11	10	25	30.333
12	11	36	30.333
13	12	30	

9780170476959

Recap

1 Describe the following time series plots.

a

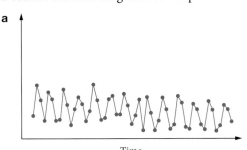

Time

b

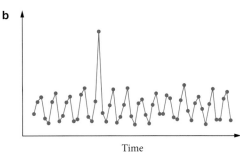

Time

c

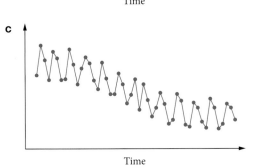

Time

d

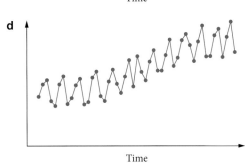

Time

2 The table shows the monthly profit, in dollars, of a new coffee shop for the first nine months of 2018.

Month	Jan	Feb	Mar	Apr	May	June	July	Aug	Sept
Profit ($)	2890	1978	2402	2456	4651	3456		2678	2345

a Determine the smoothed profit for April using the 4-point centred average method.

b The 3-point moving average for July is $2986. Determine the actual profit for July to the nearest dollar.

Mastery

3 WORKED EXAMPLE 2 The introduction of speed cameras in Victoria helped to reduce the number of deaths in road accidents over the period from 1990 to 2000. This data is shown below.

Year	1990	1991	1992	1993	1994	1995	1996	1997	1998	1999	2000
Number of road deaths	225	240	201	192	185	160	172	127	132	101	100

a Use a table with three columns to calculate the smoothed data using the method of 3-point moving averages.

b What are the smoothed number of road deaths in 1994 and 2000?

c Copy the graph on the right and add the time series plot for the smoothed data.

d What does the graph of the smoothed data indicate about the trend in the original data?

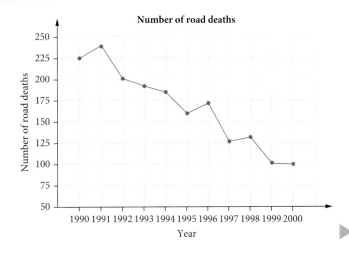

▶ **4** 🔗 WORKED EXAMPLE 3　The table shows the yearly sales of a mathematics textbook from 2010 to 2022.

Year	2010	2011	2012	2013	2014	2015	2016	2017	2018	2019	2020	2021	2022
Number of textbook sales	2250	2230	2000	2010	1990	3000	2045	2989	3000	1950	2120	2255	2297

a Use a table with four columns to calculate the smoothed data using the method of 4-point moving averages with centring.

b What is the smoothed number of sales for 2011 and 2019?

c Graph the smoothed data by hand.

d What does the graph of the smoothed data indicate about the trend in the original data?

5 🔗 WORKED EXAMPLE 4　The table shows the total number of users per week for a local gym for 12 weeks.

Week	1	2	3	4	5	6	7	8	9	10	11	12
Number of users	1788	1420	1100	1690	1398	1010	1619	1305	998	1586	1220	869

Use a spreadsheet to calculate the 3-point moving averages for the number of users.

Calculator-assumed

6 (6 marks)　Copy and complete the following tables by calculating the appropriate moving averages.

a

(3 marks)

Time period	4-monthly sales ($'000)	3-point moving average
1	20	
2	27	24
3	25	
4	22	
5	30	
6	25	
7	24	
8	33	
9	30	
10	25	
11	36	
12	30	
13	28	
14	38	
15	33	
16	32	
17	41	
18	35	

9780170476959

b

Time period	Number of patrons each quarter	4-point centred moving average
1	94	
2	120	
3	103	103
4	97	
5	90	
6	112	
7	98	
8	90	
9	86	
10	104	
11	91	
12	80	
13	74	
14	93	
15	78	
16	70	
17	65	
18	79	

7 (5 marks) The following table shows the quarterly expenditure ($'000) of a small mining company in the south-west of Western Australia.

Year	Quarter	Time period	Expenditure ($'000s)	4-point centred moving average	5-point moving average
2019	March	1	33		
	June	2	40		
	September	3	15	37.625	36.6
	December	4	A	37.25	37.6
2020	March	5	32	37.125	32.8
	June	6	38	37	42
	September	7	16	B	35.4
	December	8	61	36.375	36.8
2021	March	9	30	36.375	32.2
	June	10	39	36	40.8
	September	11	15	35.5	34.2
	December	12	59	34.875	35.4
2022	March	13	28	34.25	C
	June	14	36		
	September	15	13		

a Determine the missing entries A, B and C. (3 marks)

b From the two sets of moving averages, which one is more appropriate for the company's analyst to determine the trend of the expenditure of the company? Justify your answers. (2 marks)

8 © SCSA | MA2017 Q15a | (3 marks) The table below shows some time series data where t represents time.

t	1	2	3	4	5	6	7	8
x	14	17	18	24	21	19	16	13

Calculate at $t = 4$

a the 3-point moving average. (1 mark)

b the 6-point centred moving average. (2 marks)

9 © SCSA | MA2018 Q12 | (6 marks) A service centre manager recorded the number of customers over time periods, t, and produced the following spreadsheet to compare different moving averages.

	A	B	C	D	E	F
1	t	Number of customers	3-point moving average	4-point centred moving average	5-point moving average	6-point centred moving average
2	1	840				
3	2	927	901			
4	3	936	919	902.625	892.8	
5	4	894	899	893.625	890.4	C
6	5	867	863	879	888.6	895
7	6	828	871	880.875	886.8	890.25
8	7	918	891	886.5	883.8	882
9	8	927	908	891	B	874.5
10	9	879	886	881.625	879	877
11	10	852	850	866.625	876.6	883.75
12	11	819	859	869.25	875.4	878.75
13	12	906	882	876	872.4	869.75
14	13	921	897	879.375	868.8	863
15	14	864	873	870	868.2	
16	15	834	838			
17	16	A				

a What is the purpose of calculating moving averages for time series data? (1 mark)

b Determine the values of A, B and C in the above table. (3 marks)

c From those in the table above, which is the most appropriate moving average for the manager of the service centre to consider? Justify your choice. (2 marks)

▷ **10** © SCSA | MA2018 Q13abc | (8 marks) The graph below shows the quarterly retail turnover per capita ($) in Australia, i.e. the average amount spent per person at retail outlets during each quarter.

Quarterly retail turnover per capita, Australia, 2013–2017

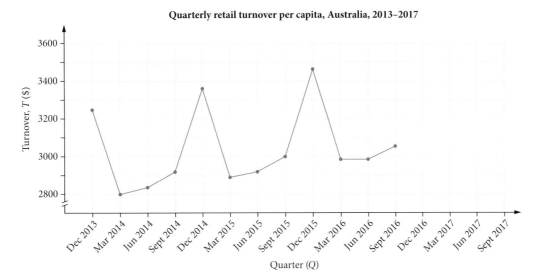

The data for the next four quarters is shown in the following table.

Quarter	December 2016	March 2017	June 2017	September 2017
Quarterly retail turnover per capita ($)	3521.40	2980.10	3045.00	3075.30

a Copy and complete the time series plot by including this additional information. (2 marks)

b The equation of the least-squares line for the above data is $T = 9.6143Q + 2986.50$, where $Q = 1$ for December 2013, $Q = 2$ for March 2014, etc.

  i Fit this line to the graph. (2 marks)

  ii Describe the trend and seasonality of these data. (2 marks)

c The 4-point centred moving average for March 2017 is $3152.78 (correct to two decimal places). Determine the **actual** retail turnover per capita for September 2016. (2 marks)

 6.3 **Seasonal adjustment**

Interpreting seasonal indices

As we have already seen, some data has regular and predictable changes that repeat during a year or less, which we describe as seasonality. Remember, a season can be a day, a week, a month, a quarter, or an actual season.

Moving averages are good for determining trend. We can fit a least-square regression line using the period (i.e. the time component as the explanatory variable). A linear regression equation can be obtained and the gradient of the line can be used to describe the trend. However, it is not possible to use the least-square regression equation to predict the future value.

Video playlist
Seasonal adjustment

Worksheet
Seasonal adjustment

Seasonal indices

- **Seasonal indices** compare each season to an average season.

- A season that is exactly average has a seasonal index of 1.

- An above average season has a seasonal index greater than 1.

- A below average season has a seasonal index less than 1.

- The mean of seasonal indices is 1.

- To interpret seasonal indices, convert them to percentages.

In addition, it's not effective to smooth seasonal data using the moving averages because smoothing two consecutive points might connect two data points with significant seasonal differences. We could end up smoothing out important information rather than an irregular fluctuation. To deal effectively with this sort of data, we need to make **seasonal adjustments**. **Seasonal indices** are used to make seasonal adjustments.

WORKED EXAMPLE 5	Interpreting seasonal indices

The table shows the seasonal indices for the number of pizzas sold on each day of the week by a pizza chain. Use the table to answer the following questions.

	Mon	Tue	Wed	Thu	Fri	Sat	Sun
	0.3	1.05	0.4	0.85	1.4	1.8	1.2

Steps	Working
a How many days had below average sales?	
How many indices are less than 1?	3 days have below average sales
b Which day had the closest to average sales?	
Which of the indices is closest to 1?	Tuesday
c Which day had sales furthest away from the average?	
Which of the indices is furthest from 1?	Saturday
d What do the seasonal indices add to?	
Add the indices.	$0.3 + 1.05 + 0.4 + 0.85 + 1.4 + 1.8 + 1.2 = 7$
e What is the mean of the seasonal indices?	
Divide the sum of the indices by the number of indices.	$\frac{7}{7} = 1$
f Rewrite the table so that the seasonal indices are converted to percentages.	

Convert each decimal to a percentage.	Mon	Tue	Wed	Thu	Fri	Sat	Sun
	30%	105%	40%	85%	140%	180%	120%

Steps	Working
g What is the mean of the percentaged seasonal indices?	
Add the percentages and divide by 7.	$\frac{30 + 105 + 40 + 85 + 140 + 180 + 120}{7} = \frac{700}{7} = 100\%$
h What percentage below average were Wednesday's sales?	
Compare the percentage to 100%.	Wednesday's pizza sales were 60% *below* average.

The seasonal indices in the table in the example above add up to 7. This is because the season we are looking at is the number of days in a week. If the seasonal indices were for months, they would add up to 12, and if they were for quarters, they would add up to 4.

The sum of seasonal indices

The sum of the seasonal indices = the number of seasons.

Type of data	Number of seasons	Cycle	Sum of seasonal indices
Daily figures for data from Monday to Sunday	7	full week	7
Daily figures for data from Monday to Friday	5	working week	5
Monthly figures	12	year	12
Quarterly accounts	4	year	4

Calculating seasonal indices

The seasonal indices can be calculated from one season's data using the formula

$$\text{seasonal index} = \frac{\text{actual figure}}{\text{average for the season}}$$

6.3

WORKED EXAMPLE 6 Calculating seasonal indices

The quarterly sales figures for the number of cars sold were recorded by a car sales yard for 2023.

Quarter 1	Quarter 2	Quarter 3	Quarter 4
5	7	9	3

a Find the seasonal indices for each of the quarters, correct to two decimal places.

b Rewrite the table so that the seasonal indices are converted to percentages.

c Which of the quarters' sales were 50% below average?

Steps	Working
a 1 Find the average for the season.	$\dfrac{5 + 7 + 9 + 3}{4} = \dfrac{24}{4} = 6$
2 For each of the actual figures, use the formula $\text{seasonal index} = \dfrac{\text{actual figure}}{\text{average for the season}}$ Round your answer to two decimal places.	<table><tr><th>Quarter 1</th><th>Quarter 2</th><th>Quarter 3</th><th>Quarter 4</th></tr><tr><td>$\frac{5}{6} = 0.83$</td><td>$\frac{7}{6} = 1.17$</td><td>$\frac{9}{6} = 1.50$</td><td>$\frac{3}{6} = 0.50$</td></tr></table>
b Convert each decimal to a percentage.	<table><tr><th>Quarter 1</th><th>Quarter 2</th><th>Quarter 3</th><th>Quarter 4</th></tr><tr><td>83%</td><td>117%</td><td>150%</td><td>50%</td></tr></table>
c Read from the percentage table.	Quarter 4

Deseasonalising time series data

The most common seasonal adjustment is to deseasonalise data. To do this we use seasonal indices to remove the seasonal component of the time series. **Deseasonalisation** is a form of smoothing, which takes out the seasonal effects of the data so that a line of best fit can be fitted and long-term trends can be predicted.

Worksheet
Deseasonal-
isation

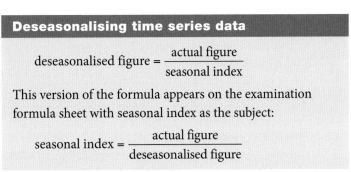

Deseasonalising time series data

$$\text{deseasonalised figure} = \frac{\text{actual figure}}{\text{seasonal index}}$$

This version of the formula appears on the examination formula sheet with seasonal index as the subject:

$$\text{seasonal index} = \frac{\text{actual figure}}{\text{deseasonalised figure}}$$

Realistically, a seasonal index would be calculated from several years' data. Each year, there will be different seasonal indices for the same season due the variability of the data collected. Therefore, the **average percentage method** is used to determine one seasonal index for each season.

The following example shows how to calculate the seasonal index from a table of actual figures across several years using the average percentage method.

The quarterly sales figures for the number of cars sold were recorded by a car sales yard for 2021 to 2023.

Year	Q1	Q2	Q3	Q4
2021	5	7	9	3
2022	4	8	9	4
2023	5	9	10	5

a Calculate the seasonal index for each quarter, correct to four decimal places.

b Use the seasonal indices to deseasonalise the data, correct to two decimal places.

c Plot the original and the deseasonalised data on the same set of axes.

Steps	Working

a 1 Calculate the yearly mean. Calculate the totals for each year, and find the mean for each year by dividing each total by 4.

Year	Q1	Q2	Q3	Q4	Yearly mean
2021	5	7	9	3	$\dfrac{5+7+9+3}{4} = 6$
2022	4	8	9	4	$\dfrac{4+8+9+4}{4} = 6.25$
2023	5	9	10	5	$\dfrac{5+9+10+5}{4} = 7.25$

2 Calculate the quarterly proportions. Divide each quarterly sales figure by the corresponding yearly mean to obtain quarterly proportions. Give answers correct to four decimal places.

Year	Q1	Q2	Q3	Q4
2021	$\dfrac{5}{6}$ = 0.8333	$\dfrac{7}{6}$ = 1.1667	$\dfrac{9}{6}$ = 1.5000	$\dfrac{3}{6}$ = 0.5000
2022	$\dfrac{4}{6.25}$ = 0.6400	$\dfrac{8}{6.25}$ = 1.2800	$\dfrac{9}{6.25}$ = 1.4400	$\dfrac{4}{6.25}$ = 0.6400
2023	$\dfrac{5}{7.25}$ = 0.6897	$\dfrac{9}{7.25}$ = 1.2414	$\dfrac{10}{7.25}$ = 1.3793	$\dfrac{5}{7.25}$ = 0.6897

3 Calculate the seasonal indices by finding the mean of the quarterly proportions. Give answers correct to four decimal places.

Year	Q1	Q2	Q3	Q4
2021	0.8333	1.1667	1.5000	0.5000
2022	0.6400	1.2800	1.4400	0.6400
2023	0.6897	1.2414	1.3793	0.6897
Total	2.1630	3.6881	4.3193	1.8297
Seasonal index	$\dfrac{2.1630}{3}$ = 0.7210	$\dfrac{3.6881}{3}$ = 1.2294	$\dfrac{4.3193}{3}$ = 1.4398	$\dfrac{1.8297}{3}$ = 0.6099

4 Add the four seasonal index values to check the sum is 4. Note that the values may not add exactly to 4 because of rounding errors.

$0.7210 + 1.2294 + 1.4398 + 0.6099 = 4.0001$

5 It may be useful to write a summary table for the seasonal indices.

Year	Q1	Q2	Q3	Q4
Seasonal index	0.7210	1.2294	1.4398	0.6099

b 1 Deseasonalise the original time series data using the formula

deseasonalised figure

$= \dfrac{\text{actual figure}}{\text{seasonal index}}$

Year	Q1	Q2	Q3	Q4
2021	$\dfrac{5}{0.7210}$ $= 6.94$	$\dfrac{7}{1.2294}$ $= 5.69$	$\dfrac{9}{1.4398}$ $= 6.25$	$\dfrac{3}{0.6099}$ $= 4.92$
2022	$\dfrac{4}{0.7210}$ $= 5.55$	$\dfrac{8}{1.2294}$ $= 6.51$	$\dfrac{9}{1.4398}$ $= 6.25$	$\dfrac{4}{0.6099}$ $= 6.56$
2023	$\dfrac{5}{0.7210}$ $= 6.94$	$\dfrac{9}{1.2294}$ $= 7.32$	$\dfrac{10}{1.4398}$ $= 6.95$	$\dfrac{5}{0.6099}$ $= 8.20$

2 Write a summary table for the deseasonalised values, correct to two decimal places.

Year	Q1	Q2	Q3	Q4
2021	6.94	5.69	6.25	4.92
2022	5.55	6.51	6.25	6.56
2023	6.94	7.32	6.95	8.20

c Plot the original and the deseasonalised time series data on the same set of axes.

Use $t = 1$ to represent the first quarter of 2021, $t = 2$ to represent the second quarter of 2021, etc.

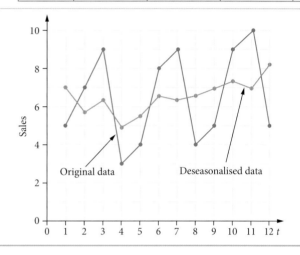

Original data

Deseasonalised data

The analysis in the example above can be summarised in the table below.

Year	Data number (n)	Quarter	No. of cars sold	Quarterly mean	Percentage of quarterly mean	Deseasonalised figure
2021	1	1	5		83.33	6.94
	2	2	7	6	116.67	5.69
	3	3	9		150.00	6.25
	4	4	3		50.00	4.92
2022	5	1	4		64.00	5.55
	6	2	8	6.25	128.00	6.51
	7	3	9		144.00	6.25
	8	4	4		64.00	6.56
2023	9	1	5		68.97	6.94
	10	2	9	7.25	124.14	7.32
	11	3	10		137.93	6.95
	12	4	5		68.97	8.20

Using the averaged percentage method, the seasonal indices for the four quarters are shown in the table below.

Quarter	1	2	3	4
Seasonal index	72.10	122.94	143.98	60.99

The seasonal indices are used to calculate the deseasonalised figures.

Re-seasonalising time series data

Sometimes we are asked to re-seasonalise data. **Re-seasonalisation** involves finding the actual figure given the deseasonalised figure and seasonal index. To do this we rearrange the seasonal index formula so that the actual figure is the subject.

Re-seasonalising time series data

actual figure = deseasonalised figure × seasonal index

WORKED EXAMPLE 8 Deseasonalising and re-seasonalising time series data

The following table shows the quarterly seasonal indices for revenue to a publishing company from the sales of mathematics textbooks.

Quarter	1	2	3	4
Seasonal index	0.7		0.6	1.9

a What is the missing Quarter 2 seasonal index?

b To correct for seasonality, by what percentage should the sales for Quarter 2 be increased?

c The company predicts that its deseasonalised quarterly sales will be $1 000 000 for each quarter. Based on this, what would you predict the actual sales for Quarter 2 to be?

Steps	Working
a Use the fact that the seasonal indices need to add to 4 for quarterly data.	$4 - 0.7 - 0.6 - 1.9 = 0.8$ The Quarter 2 seasonal index is 0.8.
b Use the deseasonalising formula $\text{descasonalised figure} = \dfrac{\text{actual figure}}{\text{seasonal index}}$	$\text{deseasonalised figure} = \dfrac{\text{actual figure}}{0.8}$ $= \dfrac{1}{0.8} \times \text{actual figure}$ $= 1.25 \times \text{actual figure}$ $= 125\% \times \text{actual figure}$ So to correct for seasonality, the sales for Quarter 2 should be increased by 25%.
c Use the re-seasonalising formula actual figure = deseasonalised figure × seasonal index	actual figure = $1 000 000 \times 0.8$ $= \$800\,000$ So, the actual sales for Quarter 2 are $800 000.

Recap

Use the following information to answer Questions 1–3.

The time series plot shows the number of passengers who flew with an airline each month over the period 1955–1960.

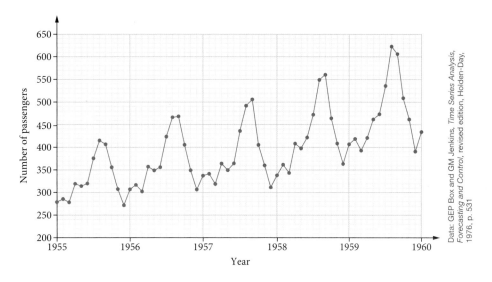

Data: GEP Box and GM Jenkins, *Time Series Analysis, Forecasting and Control*, revised edition, Holden-Day, 1976, p. 531

1 From the graph, identify the period of seasonal variation.

2 Comment on the long-term trend of the data.

3 What moving average would you use to smooth this set of data? Give reason(s) for your choice.

Mastery

4 [WORKED EXAMPLE 5] The table shows the seasonal indices for the number of ice creams sold each month by an ice-cream franchise.

Jan	Feb	Mar	Apr	May	Jun	Jul	Aug	Sep	Oct	Nov	Dec
2.5	2.1	1.6	0.7	0.2	0.2	0.1	0.0	0.6	0.9	1.2	1.9

a How many months had below average sales?

b Which month had the closest to average sales?

c Which month had sales furthest away from the average?

d What do the seasonal indices add to?

e What is the mean of the seasonal indices?

f Rewrite the table so that the seasonal indices are converted to percentages.

g What is the mean of the percentaged seasonal indices?

h What percentage above average were January's sales?

5 [WORKED EXAMPLE 6] The quarterly sales figures for the number of king-sized beds sold were recorded by a furniture shop for 2023.

Q1	Q2	Q3	Q4
11	8	1	3

a Find the seasonal indices for each of the quarters, correct to two decimal places.

b Rewrite the table so that the seasonal indices are converted to percentages.

c Which of the quarters' sales were 39% above average?

6 [WORKED EXAMPLE 7] The quarterly sales figures for the number of king-sized beds sold were recorded by a furniture shop for 2023 to 2025.

Year	Q1	Q2	Q3	Q4
2023	11	8	1	3
2024	9	9	3	1
2025	4	9	1	7

a Calculate the seasonal index for each quarter, correct to three decimal places.

b Use the seasonal indices to deseasonalise the data.

c Plot the original and the deseasonalised data on the same set of axes.

7 [WORKED EXAMPLE 8] The following table shows the quarterly seasonal indices for revenue to a company from the sales of a brand of fruit juice.

Quarter	1	2	3	4
Seasonal index	1.4	0.8		1.1

a What is the missing Quarter 3 seasonal index?

b To correct for seasonality, by what percentage should the sales for Quarter 3 be increased? Round your answer to the nearest percentage.

c The company predicts that its deseasonalised quarterly sales will be $100 000 for each quarter. Based on this, what would you predict the actual sales for Quarter 3 to be?

Calculator-assumed

8 [© SCSA] [MA2018 13d] (5 marks) The seasonal indices (correct to two decimal places) are shown in the table below.

Quarter	Seasonal index
December	110.76%
March	95.00%
June	
September	98.20%

a Copy and complete the table by determining the seasonal index for June. (1 mark)

b Use the seasonal index to determine the deseasonalised retail turnover per capita for December, which is $3521.40. (2 marks)

c The deseasonalised retail turnover per capita for March is $3142.42. Determine the **actual** retail turnover per capita for this quarter. (2 marks)

9780170476959

9 © SCSA MA2016 10ab (7 marks) A school canteen manager recorded the number of ice-creams sold for three weeks. The data are recorded in the table below, together with some calculations.

	Sales day (d)	Ice-cream sales	Weekly mean	Percentage of weekly mean
Monday	1	210		132.9%
Tuesday	2	230		145.6%
Wednesday	3	100	B	63.3%
Thursday	4	90		57.0%
Friday	5	160		101.3%
Monday	6	190		128.4%
Tuesday	7	230		155.4%
Wednesday	8	90	148	60.8%
Thursday	9	80		54.1%
Friday	10	150		101.4%
Monday	11	180		126.8%
Tuesday	12	220		154.9%
Wednesday	13	A	142	C
Thursday	14	70		49.3%
Friday	15	150		105.6%

a Determine the value of A, B and C, giving the value of C correct to one decimal place. (4 marks)

b **i** Use the average percentage method to complete the table below by calculating the seasonal index for Wednesday. (1 mark)

Day	Seasonal index
Monday	129.4% = 1.294
Tuesday	152.0% = 1.520
Wednesday	
Thursday	56.8% = 0.568
Friday	102.8% = 1.028

ii Use the seasonal index to determine the deseasonalised number of ice-cream sales for Tuesday of Week 3, correct to the nearest 10. (2 marks)

A retailer in a shopping centre sells mobile phones. The data of its quarterly sales, together with some calculations, are shown in the table below.

Year	Data number (n)	Quarter	Mobile phone sales	Quarterly mean	Percentage of quarterly mean	Deseasonalised figure (D)
2013	1	March	901		99.56	915
	2	June	802	905	88.62	914
	3	September	A		97.68	900
	4	December	1033		114.14	894
2014	5	March	973		98.83	988
	6	June	863	984.5	C	984
	7	September	964		97.92	981
	8	December	1138		115.59	985
2015	9	March	1049		98.45	1065
	10	June	932	1065.5	87.47	E
	11	September	1049		98.45	1068
	12	December	1232		115.63	1066
2016	13	March	1119		97.01	1136
	14	June	1006	B	87.21	1147
	15	September	1142		99.00	1162
	16	December	1347		116.78	1166

a Determine the value of A, B and C in the table above. (3 marks)

b Copy and complete the seasonal index table below. (1 mark)

Quarter	March	June	September	December
Seasonal index	0.9846	0.8774	0.9826	

c Determine the value of E in the table above. (2 marks)

6.4 Least-squares trend lines

As with other associations between two numerical variables we have looked at, we can use a least-squares line of best fit (often called a **trend line** for time series) to model time series trends, as long as the data appears to be linear.

Seasonality and forecasting

If there is seasonality in the time series, then we usually need to go through the extra step of deseasonalising the data before fitting the least-squares line. The least-squares line based on the deseasonalised data can be used to make predictions; however, the result will give a deseasonalised figure. This value then needs to be re-seasonalised to give the actual figure using:

actual figure = deseasonalised figure × seasonal index

When we use the least-squares line to make predictions outside of the data range, the same issues of extrapolation that we have discussed in Chapter 2 apply. In the case of time series, this involves extending into the future, which is called **trend line forecasting**. We can never be certain that the equation of the line will apply in the future, and the further into the future we are trying to predict, the less reliable the equation of the least-squares line will be.

In forecasting using the least-squares equation based on the deseasonalised data, if the forecasted data is within one period of the whole set of data, we can consider the forecast to be reliable. However, any forecast away from the one period would be considered as unreliable.

Video playlist
Least-squares
trend lines

WORKED EXAMPLE 9 | Working with trend lines for deseasonalised data

The following table lists the deseasonalised number of sales of a particular joke coffee mug in a novelty store for each quarter in 2022–2023 and the seasonal indices.

Quarter	1	2	3	4
Deseasonalised number of sales in 2022	5	10	12	28
Deseasonalised number of sales in 2023	26	25	31	27
Seasonal index	1.8	1	0.5	0.7

a Find the equation of the least-squares trend line for the deseasonalised time series data for 2022–2023. Round the slope and intercept to three decimal places.

b Plot the time series and draw the trend line for the deseasonalised data on the same axes. Comment on the trend by interpreting the slope of the trend line equation.

c Use the trend line equation to forecast the deseasonalised number of sales for Quarter 3 2024.

d Use the trend line equation to forecast the actual number of sales for Quarter 3 2024.

Steps		Working

a 1 Rewrite the deseasonalised number of sales in a table that represents the quarters from 1 to 8.

Quarter number	1	2	3	4	5	6	7	8
Deseasonalised number of sales 2022–2023	5	10	12	28	26	25	31	27

2 Use CAS to find a and b for the least-squares line of best fit equation, rounding to the required decimal places.

$a = 3.523\,81$ $b = 4.642\,86$

deseasonalised number of sales
$= 3.524 \times quarter\ number + 4.643$

ClassPad

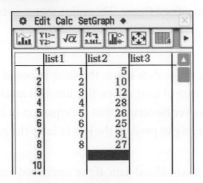

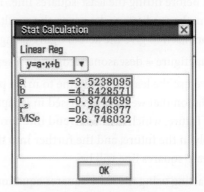

1 Tap **Menu > Statistics**.

2 Enter the values into **list1** and **list2**.

3 Tap **Calc > Regression > Linear Reg.**

4 On the next screen, keep the default settings of **XList: list1** and **YList: list2** and tap **OK**.

5 Select **y=ax+b** from the dropdown menu.

TI-Nspire

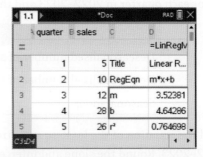

1 Open a **Lists & Spreadsheet** page.

2 Enter the appropriate headings and the values in columns **A** and **B**.

3 Press **menu > Statistics > Stat Calculations > Linear Regression (mx+b)**.

4 Select the headings then select **OK**.

b Use CAS to plot the time series and draw the line of best fit.

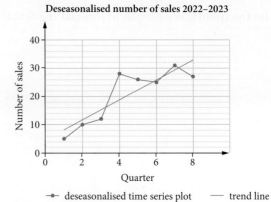

Deseasonalised number of sales 2022–2023

→ deseasonalised time series plot   — trend line

Refer to the slope from the trend line equation in your comment.

During 2022–2023 the sales of the mug increased on average by 3.52 per quarter.

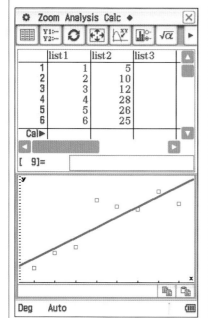
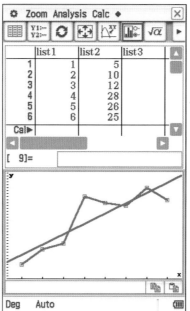
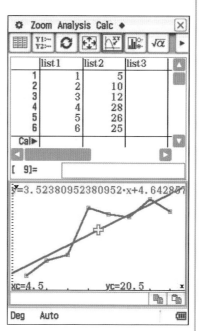

1 Tap **OK** from the previous screen to display the scatterplot and least-squares regression line in the lower window.

2 Tap the **Set StatGraphs** tool.

3 Change the **Type:** from **Scatter** to **xyLine** and tap **Set**.

4 The time series and trend line will both be displayed.

5 Tap the **Equation** tool to display the equation.

TI-Nspire

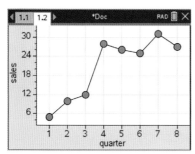
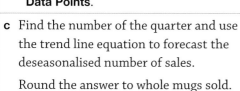
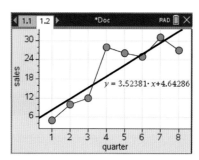

1 Add a **Data & Statistics** page.

2 Select the headings from the **Lists & Spreadsheet** columns.

3 Press **menu > Plot Properties > Connect Data Points**.

4 Press **menu > Analyze > Regression > Show Linear (mx+b)**.

5 The time series and trend line will both be displayed.

c Find the number of the quarter and use the trend line equation to forecast the deseasonalised number of sales.

Round the answer to whole mugs sold.

If Q4 2023 is quarter number 8, then

Q3 2024 is quarter number 8 + 3 = 11.

deseasonalised number of sales = 3.52 × 11 + 4.64

= 43.36

Deseasonalised number forecast to be sold in Q3 2024 is 43 mugs.

d Re-seasonalise by using:

actual figure = deseasonalised figure × seasonal index

Use the **unrounded** deseasonalised figure, then round the answer to whole mugs sold.

actual figure = 43.36 × 0.5

= 21.68

Actual number forecast to be sold in Q3 2024 is 22 mugs.

Video
WACE
question
analysis:
Time series
analysis

WACE QUESTION ANALYSIS

© SCSA MA2019 Q11 Calculator-assumed (13 marks)

Data for the total occupancy of rooms for each season of the year at a Perth hotel is shown below.

n	Year	Season	Total rooms occupied	Seasonal mean	4-point centred moving average	Total rooms occupied as a per cent of seasonal mean
1	2015/16	Spring	1770	1660.5		106.59
2		Summer	1904			B
3		Autumn	1591		1644.375	95.81
4		Winter	1377		1622.5	82.93
5	2016/17	Spring	1641	1610.25	1618	101.91
6		Summer	1858		1614.75	115.39
7		Autumn	1601		1602.25	99.43
8		Winter	1341		1584.75	83.28
9	2017/18	Spring	1577	1524.0	1558	103.48
10		Summer	A		1532.375	116.93
11		Autumn	1463		1526.875	96.00
12		Winter	1274		1525.125	83.60
13	2018/19	Spring	1600	1519.75	C	105.28
14		Summer	1745		1525.25	114.82
15		Autumn	1504			98.96
16		Winter	1230			80.93

a Calculate the value of A, B and C.   (3 marks)

b Copy and complete the table showing the seasonal index for each season.   (1 mark)

Summer	Autumn	Winter	Spring
1.1545		0.8268	1.0432

c Calculate the deseasonalised value for Winter 2017/18.   (2 marks)

d Comment on the effect the seasonal index had on the value found in part **c**.   (1 mark)

e The least-squares line using deseasonalised data is $R = -12.071n + 1681.25$. Use this line to predict the total number of rooms occupied during Spring 2020/21.   (2 marks)

When a prediction was made for Spring 2020/21, using the least-squares line based on the 4-point centred moving averages, the answer was 1481.

f Explain why this is different from the answer obtained in part **e**.   (1 mark)

The manager of the hotel attended a meeting with the owners of the hotel. She explained to the owners that the reduction in occupancy was due to the downturn in the Western Australian economy in recent years.

g Comment on the statement made by the hotel manager.   (2 marks)

h What practical advice, in the context of the question, would you give to the manager of the hotel?   (1 mark)

Reading the question

- Note the decimal places in the table: *A* will have no decimal place, *B* will have two decimal places and *C* will have three decimal places. Always answer the question based on the decimal places shown in the table.
- You must know the difference between predictions based on the least-squares line from the deseasonalised data and the moving averages.

Thinking about the question

- You need to calculate the original data from the seasonal mean, determine a 4-point centred moving average and calculate a seasonal mean.
- You need to calculate the deseasonalised value.

Worked solution ($\checkmark$ = 1 mark)

a Using the seasonal mean for 2017/18:

$\dfrac{1577 + A + 1463 + 1274}{4} = 1524$. Therefore, $A = \mathbf{1782}$ $\checkmark$

Calculating the percentage of seasonal mean for Summer 2015/16

$\dfrac{1904}{1660.5} \times 100 = B$. Therefore, $B = \mathbf{114.66}$ $\checkmark$

Calculating the 4-point centred moving average:

$\dfrac{\dfrac{1463}{2} + 1274 + 1600 + 1745 + \dfrac{1504}{2}}{4} = 4C$. Therefore, $C = \mathbf{1525.625}$ $\checkmark$

b The seasonal index for Autumn:

$1.1545 + \text{SI for Autumn} + 0.8268 + 1.0432 = 4$

Therefore, the seasonal index for Autumn is **0.9755** $\checkmark$

c The deseasonalised value for Winter 2017/18 $= \dfrac{1274}{0.8268}$ $\checkmark$

$= 1540.88$

$= \mathbf{1541}$ $\checkmark$ (to the nearest whole number)

d The value is **increased** due to the underlying trend. $\checkmark$

e Spring 2020/21 $\Rightarrow n = \mathbf{21}$ $\checkmark$

Therefore, $R = -12.071(21) + 1681.25 = 1427.8$

Since R is a deseasonalised value, we have to multiply it by the seasonal index for Spring.

$1427.8 \times 1.0432 = \mathbf{1489.4}$

The prediction is **1489** (to the nearest whole number) $\checkmark$

f **Two different methods were used for smoothing**. $\checkmark$

g **The comment was not appropriate.** $\checkmark$

The cause was not established. $\checkmark$

h **Reduce room rate, advertise etc**.

gives a valid reason $\checkmark$

Recap

Use the following information to answer Questions 1–3.

The seasonal indices for the first 11 months of the year for sales in a sporting equipment store are shown in the table.

Month	Jan	Feb	Mar	Apr	May	June	July	Aug	Sep	Oct	Nov	Dec
Seasonal index	1.23	0.96	1.12	1.08	0.89	0.98	0.86	0.76	0.76	0.95	1.12	

1 Determine the seasonal index for December.

2 The store sold $213 956 worth of sporting equipment in May. Determine the deseasonalised sales figure.

3 The deseasonalised sales for July was $230 000. Determine the actual sales for the month of July.

Mastery

4 [WORKED EXAMPLE 9] The following table lists the deseasonalised number of sales of a particular costume in a fancy-dress store for each quarter in 2022–2023 and the seasonal indices.

Quarter	1	2	3	4
Deseasonalised number of sales in 2022	33	33	25	28
Deseasonalised number of sales in 2023	22	21	11	7
Seasonal index	0.9	0.6	1.3	1.2

a Find the equation of the least-squares trend line for the deseasonalised time series data for 2022–2023. Round the slope and intercept to two decimal places.

b Plot the time series and draw the trend line for the deseasonalised data on the same axes. Comment on the trend by interpreting the slope of the trend line equation.

c Use the trend line equation to forecast the deseasonalised number of sales for Quarter 1 2024.

d Use the trend line equation to forecast the actual number of sales for Quarter 1 2024.

5 The following deseasonalised data represents the monthly sales, in dollars, of a market stall over a period of 2 years.

	Jan	Feb	Mar	Apr	May	Jun	Jul	Aug	Sep	Oct	Nov	Dec
2021 sales ($)	185	286	199	177	178	256	211	172	181	180	177	287
2022 sales ($)	194	288	198	192	197	295	200	195	183	191	212	195
Seasonal index	1.81	0.70	0.77	0.73	0.86	0.89	0.76	1.13	1.07	0.97	1.22	1.09

a Find the equation of the least-squares trend line for the deseasonalised time series data for 2021–2022. Round the slope and intercept to three decimal places.

b Use the trend line equation to forecast the deseasonalised sales for May of 2023, correct to the nearest dollar.

c Use the trend line equation to forecast the actual sales for May of 2023, correct to the nearest dollar.

6 © SCSA MA2020 Q14 (14 marks) The table below shows the number of sprinkler systems installed by a local reticulation business over the past four years.

Year	Season	n	Number of systems	Seasonal mean	Number of systems as a percentage of the seasonal mean	Seasonally adjusted figures
2017	Summer	1	A	14	71.4	10.4
	Autumn	2	18		B	15.7
	Winter	3	11		78.6	14.7
	Spring	4	17		121.4	14.7
2018	Summer	5	15	C	105.3	15.7
	Autumn	6	16		112.3	14.0
	Winter	7	11		77.2	14.7
	Spring	8	15		105.3	13.0
2019	Summer	9	13	11.75	110.6	13.6
	Autumn	10	12		102.1	10.5
	Winter	11	8		68.1	10.7
	Spring	12	14		119.1	12.1
2020	Summer	13	16	–	–	–
	Autumn	14	15		–	–

a Calculate the value of A, B and C. (3 marks)

b Copy and complete the table showing the seasonal index for each season. (2 marks)

Season	Summer	Autumn	Winter	Spring
Seasonal index	95.8	114.3		

c Show how the seasonally adjusted figure of 13.6 for Summer 2019 was calculated. (2 marks)

d During which season could more employees be given annual holidays with least disruption to sprinkler installations? Use mathematical evidence to support your answer. (2 marks)

e Determine the least-squares line using the seasonally adjusted figures. (1 mark)

f Using your line from part **e**, estimate the number of sprinkler systems that will be installed in Summer 2021. (2 marks)

g Comment on the long-term prospects of the business. (2 marks)

7 © SCSA MA2021 Q8 (11 marks) The graph below shows the number of bookings at a dog grooming salon over its first few weeks of business.

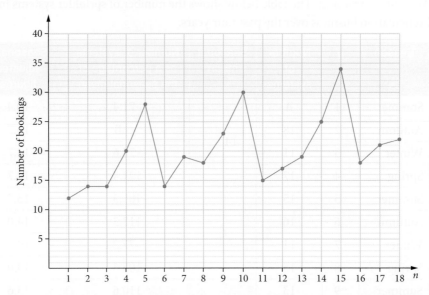

a Which simple moving average would be the **most** suitable for the data displayed in this graph? (1 mark)

A more detailed view of the same data is given in the table below.

Week	Day	n	Number of bookings	Seasonal mean	Number of bookings as a percentage of the seasonal mean	Seasonally adjusted figures
1	Tuesday	1	12	17.6	A	17.7
	Wednesday	2	14		79.55	16.9
	Thursday	3	14		79.55	16.6
	Friday	4	20		113.64	17.8
	Saturday	5	28		159.09	18.3
2	Tuesday	6	14	B	67.31	20.6
	Wednesday	7	19		91.35	23.0
	Thursday	8	18		86.54	21.4
	Friday	9	23		110.58	20.4
	Saturday	10	30		144.23	19.7
3	Tuesday	11	C	22	68.18	22.1
	Wednesday	12	17		77.27	20.6
	Thursday	13	19		86.36	22.6
	Friday	14	25		113.64	22.2
	Saturday	15	34		154.55	22.3
4	Tuesday	16	18	–	–	–
	Wednesday	17	21		–	–
	Thursday	18	22		–	–

b Calculate the value of A, B and C in the table. (3 marks)

c Calculate the seasonal index for Saturday. (1 mark)

d The equation of the least-squares line using the seasonally adjusted figures is $y = 0.40n + 16.94$. Copy the graph and draw this line. (2 marks)

e **i** Use the equation of the least-squares line given in part **d** to predict the number of bookings that will be made for the Saturday of Week 5. (2 marks)

   ii Comment on this prediction. (2 marks)

Time series plots

A **time series** involves data where the explanatory variable is time measured at equally spaced intervals such as hours, days, weeks, months, seasons and years.

A **time series plot** is a scatterplot of a time series where the data points are joined by straight lines.

Time series features

Trend

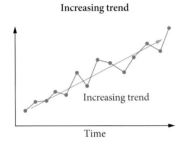

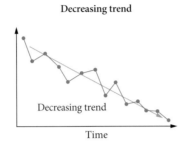

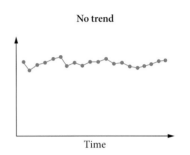

Seasonality

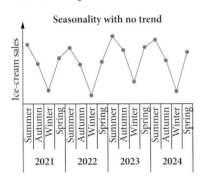

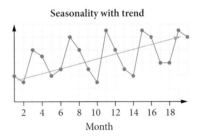

Irregular variations

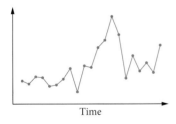

Outliers in time series

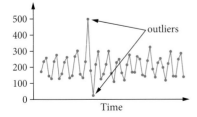

Smoothing

- **Smoothing** levels out fluctuations in time series to produce a smoother graph so we can see the underlying trends more clearly.
- **Moving averages** involves finding a series of means of the data points. An odd number of data points involves one-step smoothing, whereas an even number of data points involves a second step called **centring**.

Moving average smoothing with an odd number of points

3-point moving averages

Time	Original data	Smoothed data
1	*	
2	*	average → *
3	*	average → *
4	*	

5-point moving averages

Time	Original data	Smoothed data
1	*	
2	*	
3	*	average → *
4	*	average → *
5	*	
6	*	

Moving average smoothing with an even number of points

4-point moving averages with centring

Time	Original data	Centring	Smoothed data
1	*		
2	*		
		average → *	
3	*		average → *
		average → *	
4	*		
5	*		

Faster method for 4-point moving averages with centring

Time	Original data	Smoothed data
1	*	
2	*	
3	*	$\dfrac{0.5* + * + * + * + 0.5*}{4}$
4	*	
5	*	

Seasonal adjustment

- **Seasonal indices** are used to make **seasonal adjustments**.
- Seasonal indices compare each season to an average season.
- A season that is exactly average has a seasonal index of 1.
- An above average season has a seasonal index greater than 1.
- A below average season has a seasonal index less than 1.
- The mean of seasonal indices is 1.
- To interpret seasonal indices, convert them to percentages.
- The sum of the seasonal indices = the number of seasons.

Type of data	No. of seasons	Cycle	Sum of seasonal indices
Daily figures for data from Monday to Sunday	7	full week	7
Daily figures for data from Monday to Friday	5	working week	5
Monthly figures	12	year	12
Quarterly accounts	4	year	4

- Use the following versions of the same formula depending on what you are asked to find.
  - seasonal index:

    $$\text{seasonal index} = \frac{\text{actual figure}}{\text{deseasonalised figure}}$$

  - deseasonalise the data:

    $$\text{deseasonalised figure} = \frac{\text{actual figure}}{\text{seasonal index}}$$

  - re-seasonalise the data:

    $$\text{actual figure} = \text{deseasonalised figure} \times \text{seasonal index}$$

Fitting a least-squares line

- We can use a **least-squares line of best fit** to model time series trend as long as the data appears to be linear, but if there is seasonality then we usually need to deseasonalise the data first before fitting the least-squares line.
- The least-squares line can be used to forecast, but the result will give a deseasonalised figure. This value needs to be re-seasonalised to give the actual figure.

Total number of marks: 18   Reading time: 2 minutes   Working time: 18 minutes

1 © SCSA MA2020 Q4defg (6 marks) The table shows data comparing the gestation period (in days) with the birth weight (in grams) for ten Tasmanian possums.

Gestation period in days (d)	36	37	37	38	38	42	43	44	44	45
Birth weight in grams (g)	42	43	44	43	44	41	42	43	41	42

The correlation coefficient for these observations is approximately –0.6 and the least-squares line is $g = -0.17d + 49$.

a State the meaning of the coefficient of determination in the context of the question.   (1 mark)

b Use the least-squares line to predict the birth weight of a possum after 40 days gestation.   (1 mark)

c Comment on the validity of this prediction.   (2 marks)

d Is there any statistical evidence to support the research view that a higher birth weight will cause a shorter gestation period? Justify your answer.   (2 marks)

2 (4 marks) Consider the sequence generated by the recursive rule $T_{n+1} = T_n - 4$, $T_2 = 3$.

a What are the first three terms of the sequence?   (1 mark)

b Write the rule for the nth term of the sequence.   (1 mark)

c At what position of the sequence is the value –81?   (2 marks)

3 (8 marks) Consider the following graph.

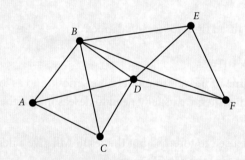

a Redraw the graph to show that it is planar.   (2 marks)

b State the number of faces in the graph and, hence, verify Euler's formula.   (2 marks)

c Determine the number of edges that need to be added to the graph in order to make it complete, listing those edges.   (2 marks)

d Determine the minimum number of edges that need to be removed from the graph in order to make it Eulerian. Justify your answer.   (2 marks)

Cumulative examination: Calculator-assumed

Total number of marks: 36   Reading time: 4 minutes   Working time: 36 minutes

1 ©SCSA MA2019 Q8abc (5 marks) Abdul has a lawnmowing business and is investigating if there is a relationship between the size of a lawn and the length of time it takes to cut the lawn. He takes a random sample of eight customers and measures the areas of their lawns and notes the times, in minutes, it takes to mow their lawns. The results are in the table below, where A is the area of the lawn in square metres and T is the time in minutes. (Note: some values are missing.)

Customer	A	B	C	D	E	F	G	H
A (m^2)		260		480	540	600	860	1180
T (min)	25	55	50	70	90	70	135	140

a Copy and complete the scatterplot below. (1 mark)

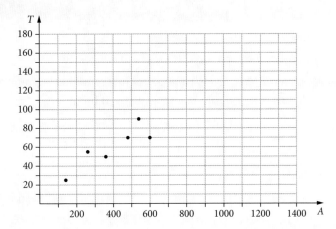

b From the information below, determine the equation of the least-squares line in terms of A and T and state the coefficient of determination for these data. (2 marks)

$$y = ax + b$$
$$a = 0.114691$$
$$b = 16.008241$$
$$r = 0.9510026$$
$$r^2 = 0.9044059$$

c Interpret the value of the gradient of the least-squares line in the context of the question. (2 marks)

2 (3 marks) A new team, the Recursive Roos, has joined the Peel Football League. From week to week, a certain percentage of the crowd does not return and 158 new people attend. The weekly attendance at games is modelled by the recursive rule:

$$A_{n+1} = 0.945A_n + 158$$

a What percentage of the crowd does not return from week to week? (1 mark)

b In the long run, how many people will attend the Recursive Roos games each week? Give your answer to the nearest whole number. (2 marks)

3 (10 marks) The weighted digraph below shows the number of minutes it takes to walk directly between seven stores in a shopping complex, whereby the vertex S represents the entrance to the complex.

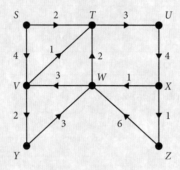

a Determine the shortest path from S to W, stating the corresponding amount of time it would take to complete this path without stopping at any of the stores. (2 marks)

b Find an open Hamiltonian path starting at S and, hence, state the amount of time it would take to complete this path without stopping at any of the stores. (2 marks)

c Copy and complete the adjacency matrix to represent the weighted digraph provided. (2 marks)

$$
A =
\begin{array}{c}
 \\ S \\ T \\ U \\ V \\ W \\ X \\ Y \\ Z
\end{array}
\begin{array}{c}
S\ T\ U\ V\ W\ X\ Y\ Z \\
\left[
\begin{array}{cccccccc}
0 & 1 &   &   &   &   &   &   \\
0 & 0 &   &   &   &   &   &   \\
0 & 0 & 0 &   &   &   &   &   \\
0 & 1 & 0 & 0 &   &   &   &   \\
0 & 1 & 0 & 1 & 0 &   &   &   \\
0 & 0 & 0 & 0 & 1 & 0 &   &   \\
0 & 0 & 0 & 0 & 1 & 0 & 0 &   \\
0 & 0 & 0 & 0 & 1 & 0 & 0 & 0 \\
\end{array}
\right]
\end{array}
$$

d Hence, identify which of the stores cannot be reached directly from store X. (1 mark)

e Evaluate a_{62}^3 and interpret the significance of the value in context of the question. (3 marks)

4 (18 marks) The following table shows the total seasonal rainfall (mm) for Perth from 2017/18 to 2019/20.

Year	Season	Period	Total rainfall (mm)	4-point centred moving averages
2017/18	Summer	1	120.6	
	Autumn	2	84.2	
	Winter	3	486.4	A
	Spring	4	80	160.525
2018/19	Summer	5	7.4	144.9
	Autumn	6	52.4	132.025
	Winter	7	393.2	132.85
	Spring	8	70.2	147.475
2019/20	Summer	9	23.8	145.25
	Autumn	10	153	144.55
	Winter	11	274.8	
	Spring	12	B	

Data: Australian Government, Bureau of Meteorology, www.bom.gov.au, (accessed 29 Dec 2022)

a Calculate the value of A and B. (2 marks)

b The data is smoothed by using the 4-point centred moving average. Give an explanation for why this is a suitable method to smooth the data. (1 mark)

c Determine the equation of the least-squares regression line for the 4-point centred moving averages (M). (2 marks)

d Comment on the trend of the rainfall data based on the equation of the least-squares regression line. (2 marks)

The data is now rearranged in the following table.

Year	Season	Period	Total rainfall (mm)	Yearly mean	Percentage of yearly mean	Deseasonalised rainfall
2017/18	Summer	1	120.6	C	62.55	
	Autumn	2	84.2		43.67	
	Winter	3	486.4		252.28	
	Spring	4	80		41.49	
2018/19	Summer	5	7.4	130.8	5.66	
	Autumn	6	D		40.06	
	Winter	7	393.2		300.61	
	Spring	8	70.2		53.67	
2019/20	Summer	9	E	F	15.00	
	Autumn	10	153		96.44	
	Winter	11	274.8		G	H
	Spring	12	183		115.35	

e Calculate the value of C, D, E, F and G. (5 marks)

f Determine the seasonal index for Winter, clearly showing your calculations. (2 marks)

g Hence or otherwise, determine the value of H. (1 mark)

h The equation of the least-squares regression for the deseasonalised rainfall, R, is $\hat{R} = -5.6754t + 202.00$.

What is the predicted rainfall for Summer 2021/22 to the nearest mm?
Comment on the reliability of this predictions. (3 marks)

CHAPTER

7

LOANS, INVESTMENTS AND ANNUITIES

TOPIC 4.2: LOANS, INVESTMENTS AND ANNUITIES

Compound interest loans and investments

4.2.1 use a recurrence relation to model a compound interest loan or investment and investigate (numerically or graphically) the effect of the interest rate and the number of compounding periods on the future value of the loan or investment

4.2.2 calculate the effective annual rate of interest and use the results to compare investment returns and cost of loans when interest is paid or charged daily, monthly, quarterly or six-monthly

4.2.3 with the aid of a calculator or computer-based financial software, solve problems involving compound interest loans, investments and depreciating assets

Reducing balance loans (compound interest loans with periodic repayments)

4.2.4 use a recurrence relation to model a reducing balance loan and investigate (numerically or graphically) the effect of the interest rate and repayment amount on the time taken to repay the loan

4.2.5 with the aid of a financial calculator or computer-based financial software, solve problems involving reducing balance loans

Annuities and perpetuities (compound interest investments with periodic payments made from the investment)

4.2.6 use a recurrence relation to model an annuity, and investigate (numerically or graphically) the effect of the amount invested, the interest rate, and the payment amount on the duration of the annuity

4.2.7 with the aid of a financial calculator or computer-based financial software, solve problems involving annuities (including perpetuities as a special case)

Mathematics Applications ATAR Course Year 12 syllabus p. 13 © SCSA

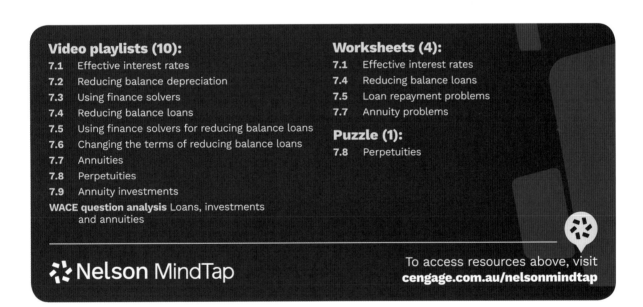

Video playlists (10):

7.1 Effective interest rates
7.2 Reducing balance depreciation
7.3 Using finance solvers
7.4 Reducing balance loans
7.5 Using finance solvers for reducing balance loans
7.6 Changing the terms of reducing balance loans
7.7 Annuities
7.8 Perpetuities
7.9 Annuity investments
WACE question analysis Loans, investments and annuities

Worksheets (4):

7.1 Effective interest rates
7.4 Reducing balance loans
7.5 Loan repayment problems
7.7 Annuity problems

Puzzle (1):

7.8 Perpetuities

Nelson MindTap

To access resources above, visit
cengage.com.au/nelsonmindtap

Video playlist
Effective
interest rates

Worksheet
Effective
interest rates

(7.1) Effective interest rates

Nominal vs effective interest rates

When deciding on an investment or loan, it's crucial to compare **interest** rates well. The **compound interest** rates we have discussed thus far are known as **nominal interest rates** (or **reducible interest rates**). These rates are provided as an annual rate coupled with a **compounding period**.

Comparing nominal interest rates can be challenging when the rates and compounding periods are different. To address this, we calculate **effective interest rates**, which take compounding into account.

- For an investment, the best option is the one with the highest effective interest rate.
- For a loan, the best option is the one with the lowest effective interest rate.

Effective interest rate formula

$$i_{\text{effective}} = \left(1 + \frac{i}{n}\right)^n - 1$$

where i = nominal interest rate *per year* as a fraction or decimal

n = number of compounding periods *per year*.

 Exam hack

The i in the effective interest rate formula is per year, *not* per compounding period.
The n in the effective interest rate formula is also per year, *not* per compounding period.
This formula appears on the formula sheet for the exam.

WORKED EXAMPLE 1 | Finding effective interest rates using the formula

Emma is looking to invest her money. She has done some research on interest rates and found the best offers from four different banks.

Bank 1: 8.45% p.a. compounding daily

Bank 2: 8.6% p.a. compounding monthly

Bank 3: 8.7% p.a. compounding six-monthly

Bank 4: 8.8% p.a. compounding annually

a Find the effective interest rate for each bank, rounding to two decimal places.

b Which bank should Emma choose if she wants to earn the most interest?

c Which bank would earn Emma the least interest?

d Why are the nominal and effective interest rates for Bank 4 the same?

Steps	Working
a For each option, substitute the known variables into the effective interest rate formula and solve, rounding to two decimal places. $$i_{\text{effective}} = \left(1 + \frac{i}{n}\right)^{n} - 1$$	**Bank 1** $i = 8.45\% = 0.0845$, $n = 365$ $$i_{\text{effective}} = \left(1 + \frac{0.0845}{365}\right)^{365} - 1 = 0.0882 = 8.82\% \text{ p.a.}$$ **Bank 2** $i = 8.6\% = 0.086$, $n = 12$ $$i_{\text{effective}} = \left(1 + \frac{0.086}{12}\right)^{12} - 1 = 0.0895 = 8.95\% \text{ p.a.}$$ **Bank 3** $i = 8.7\% = 0.087$, $n = 2$ $$i_{\text{effective}} = \left(1 + \frac{0.087}{2}\right)^{2} - 1 = 0.0889 = 8.89\% \text{ p.a.}$$ **Bank 4** $i = 8.8\% = 0.088$, $n = 1$ $$i_{\text{effective}} = \left(1 + \frac{0.088}{1}\right)^{1} - 1 = 0.088 = 8.8\% \text{ p.a.}$$

🔒 Exam hack

If a compound interest rate compounds annually, the effective interest rate is always the same as the nominal interest rate.

Steps	Working
b Compare the four results and choose the largest.	Emma should choose Bank 2 because it pays the higher effective rate of interest and will therefore pay more interest.
c Compare the four results and choose the smallest.	Bank 4 would earn Emma the least interest.
d Compare the nominal and effective interest rates.	The nominal and effective interest rates for Bank 4 are the same because the rate compounds annually.

USING CAS 1 | Finding effective interest rates

Determine the effective interest rate for an investment that offers 3.6% interest compounding fortnightly, giving your answer correct to two decimal places.

ClassPad

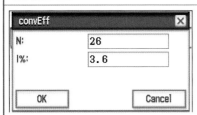

convEff	☒
N:	26
I%:	3.6

| OK | Cancel |

☼ Edit Action Interactive

convEff(26, 3.6)
　　　　　　　3.663003398

1 Tap **Interactive > Financial > Interest Conversion > ConvEff**.

2 In the **N:** field, enter **26**, which is the number of compounds per year.

3 In the **I%:** field, enter **3.6**, which is the nominal rate of interest.

4 Tap **OK**.

5 The effective rate of interest will be displayed.

The effective interest rate, correct to two decimal places, is 3.66%.

TI-Nspire

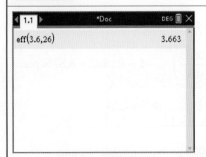

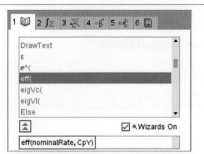

1 Press **menu** > **Finance** > **Interest Conversion** > **Effective Interest Rate**.

2 For the first parameter, enter **3.6**, which is the nominal interest rate.

3 For the second parameter, enter **26**, which is the number of compounds per year.

4 Press **enter**.

5 The effective rate of interest of **3.66**% will be displayed.

6 Alternatively, press **catalog** > **E** then scroll down to the **eff** function to display the parameters for the effective rate of interest function, shown in the red box above.

7 The two parameters are **nominalRate** and **CpY**.

8 Press **enter** to display the function in the **Calculator** page.

9 Enter the parameters and press **enter**.

The effective interest rate, correct to two decimal places, is 3.66%.

EXERCISE 7.1 Effective interest rates

ANSWERS p. 445

Mastery

1 **WORKED EXAMPLE 1** Georgio is looking to invest his money. He has done some research on interest rates and found the best offers from four different banks.

Bank 1: 7.4% p.a. compounding weekly

Bank 2: 7.5% p.a. compounding fortnightly

Bank 3: 7.6% p.a. compounding six-monthly

Bank 4: 7.7% p.a. compounding annually

a Find the effective interest rate for each bank, rounding to two decimal places.

b Which bank should Georgio choose if he wants to earn the most interest?

c Which bank would earn Georgio the least interest?

d What do you notice about the nominal and effective interest rates for Bank 4? Why is this the case?

2 **Using CAS 1** For each of the following investments determine the effective interest rate, giving your answer correct to two decimal places.

a 9% p.a. compounding monthly

b 11% p.a. compounding weekly

c 12% p.a. compounding six-monthly

d 6% p.a. compounding daily

Calculator-free

3 © SCSA | MA2017 Q4 | (5 marks) Ryan was keen to compare interest rates offered by different banks, so he decided to construct a table showing the effective annual rates of interest (%). Part of his table is shown below.

Compounding period	Rate of interest (p.a.)				
	4%	4.5%	5%	5.5%	6%
Quarterly	4.060	4.577	5.095	5.614	6.136
Monthly	4.074	4.594	5.116	5.641	6.168
Daily	4.081	4.602	5.127	5.654	6.183

a Ryan wants to borrow $5000 to purchase a second-hand car. A bank offers to lend him the money at the rate of 6% p.a. for one year. He plans to pay off the entire loan (including the interest) at the end of the year. Which compounding period should he sign up for? Justify your decision. (2 marks)

b Ryan is curious to know how much interest he would earn by investing $100 for a year, earning 4% p.a. with interest compounded quarterly. Determine the interest he would earn. (1 mark)

c Ryan's sister has $3000 to invest for a year. She has been offered a rate of 5% p.a., with interest compounded daily. Determine the value of her investment at the end of the year. (2 marks)

Calculator-assumed

4 (5 marks) An amount of money is deposited into an account that earns compound interest.

a Calculate the effective interest rates for each of the options below. (3 marks)

  A 3.7% per annum, compounding weekly

  B 3.7% per annum, compounding monthly

  C 3.7% per annum, compounding quarterly

b Comment on the effective interest rates for options A and B. From part a, which will be the best for a loan or an investment? (2 marks)

5 (3 marks) Eva has $1200 that she plans to invest for one year. One company offers to pay her interest at the rate of 6.75% per annum compounding daily.

a What is the effective interest rate for this investment? (1 mark)

b What will be the interest earned on this investment? (2 marks)

6 (4 marks) Consider the following nominal interest rates.

  I 7.9% p.a. compounding quarterly

  II 7.7% p.a. compounding fortnightly

  III 7.5% p.a. compounding monthly

  IV 7.45% p.a. compounding daily

  V 7.58% p.a. compounding six-monthly

a Which has the lowest effective interest rate? (3 marks)

b Which is the best rate for a loan? (1 mark)

7 (2 marks) Murray has chosen an investment with an interest rate of 5.4% p.a. compounded quarterly. What is the difference between the nominal interest rate and the effective interest rate?

8 © SCSA MA2016 Q13b (3 marks) Simon has $5000 that he wants to invest for a period of time without touching it. He is currently deciding between two options and wishes to compare them.

Option A: Invest the $5000 in an account earning compound interest at the rate of 5.5% per annum, with interest paid monthly.

Option B: Invest the $5000 in an account earning compound interest at the rate of 5.4% per annum, with interest paid daily.

Simon decides to calculate the effective annual rate of interest for each option, in order to compare the possible investments. He determines that Option A has an effective annual rate of interest of 5.64%, correct to two decimal places.

Calculate the effective annual rate of interest for Option B using the formula, correct to two decimal places, and hence decide on the better option for Simon.

9 © SCSA MA2018 Q16abc (4 marks) Natalia inherits a sum of money from her grandfather. She wishes to place it in a high-interest savings account.

She is considering the following two options:

Account A: Interest rate 4.40% per annum, compounded monthly

Account B: Interest rate 4.30% per annum, compounded daily.

a The effective annual interest rate for Account A is 4.49% (correct to two decimal places). Determine the effective annual interest rate for Account B. (1 mark)

Natalia's bank offers her another account, C, with an interest rate of 4.50% per annum.

b Under what circumstances will this interest rate and the effective annual interest rate be the same? (1 mark)

c Which account (A, B or C) should Natalia choose to maximise her savings? Explain your reasoning. (2 marks)

10 (8 marks) Jillian is choosing between the following investment options:

Bank of Western Australia: 7.8% p.a. compounding quarterly

Power Bank: 8% p.a. compounding yearly

Aussie Bank: 8% p.a. compounding half-yearly

a What nominal interest rate is the Bank of Western Australia offering? (1 mark)

b Does Power Bank or Aussie Bank offer the higher effective interest rate? Explain why you don't need to do a calculation to decide. (2 marks)

c What is the effective interest rate for Power Bank? Explain why you don't need to do a calculation. (2 marks)

d Calculate the effective interest rates for Bank of Western Australia and Aussie Bank, correct to two decimal places. Which should Jillian choose? (3 marks)

7.2 Reducing balance depreciation

Appreciation and depreciation

When things such as property, gold, antiques and collectibles increase in value over time, it's called **appreciation**. On the other hand, items purchased by businesses to help them function decrease in value over time. These items, such as computers and machines, are called **assets**. We use the term **depreciation** to describe this decrease in value. Depreciation occurs due to age, amount of use or lack of demand. The estimate of the value of an item at any point in time is called the **future value**.

Reducing balance depreciation

Reducing balance depreciation calculates the future value of an asset by reducing the value every year by a fixed percentage.

Reducing balance depreciation recurrence relation

The recurrence relation for the value of an asset V_n being depreciated using reducing balance depreciation is

V_0 = initial value of the asset, $\quad V_{n+1} = \left(1 - \dfrac{r}{100}\right) \times V_n$

where r = percentage depreciation rate per year

$\quad n$ = number of years

$\quad V_n$ = value of an asset after n years.

The graph of a reducing balance depreciation recurrence relation would look like this.

$V_0 - V_n$ = total amount of depreciation after n years

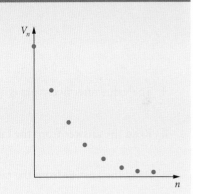

Video playlist
Reducing balance depreciation

🔓 Exam hack

The main difference between reducing balance depreciation and compound interest investment is we subtract $\dfrac{r}{100}$ rather than add it. Also, we always depreciate per year, so we don't have to worry about different compounding periods.

WORKED EXAMPLE 2 Using reducing balance depreciation recurrence relations

A business purchased a photocopier for $10 000. It is depreciated using reducing balance depreciation at a rate of 18% per annum. Give all answers to the nearest dollar.

a Copy and complete the table to find

  i the value of the photocopier after five years

  ii the amount of depreciation in the third year

  iii when the photocopier first depreciates to under $5000.

n	Depreciation	Value after n years ($)
0		10 000
1	$\dfrac{18}{100} \times 10\,000 = 1800$	$10\,000 - 1800 = 8200$
2	$\dfrac{18}{100} \times 8200 = 1476$	$8200 - 1476 = 6724$
3		
4		
5		

▶ **b** Write down a recurrence relation that gives the value of the photocopier after n years.

c What percentage of the previous value is each new value?

d Describe the sort of growth or decay modelled by the recurrence relation.

Steps	Working
a Calculate the percentage of successive values and subtract from the previous value. Use CAS's recursive computation where possible. Give all values to the nearest dollar, but don't round until after all the calculations have been done. (Note: Answers can vary slightly depending on when values are rounded.)	<table><tr><td>n</td><td>Depreciation</td><td>Value after n years (\$)</td></tr><tr><td>0</td><td></td><td>10 000</td></tr><tr><td>1</td><td>$\frac{18}{100} \times 10\,000 = 1800$</td><td>$10\,000 - 1800 = 8200$</td></tr><tr><td>2</td><td>$\frac{18}{100} \times 8200 = 1476$</td><td>$8200 - 1476 = 6724$</td></tr><tr><td>3</td><td>$\frac{18}{100} \times 6724 = 1210$</td><td>$6724 - 1210 = 5514$</td></tr><tr><td>4</td><td>$\frac{18}{100} \times 5514 = 993$</td><td>$5514 - 993 = 4521$</td></tr><tr><td>5</td><td>$\frac{18}{100} \times 4521 = 814$</td><td>$4521 - 814 = 3707$</td></tr></table>
i The value after five years is V_5.	The value of the photocopier after five years is \$3707.
ii Read the answer from the table.	The amount of depreciation in the third year is \$1210.
iii Read the answer from the table.	Value < 5000 when $n = 4$. The photocopier first depreciates to under \$5000 after four years.
b 1 Identify V_n, V_0 and r.	Let V_n equal the value of the photocopier after n years. $V_0 = 10\,000$, $r = 18$
2 Substitute the values into V_0 = initial value of the asset, $V_{n+1} = \left(1 - \dfrac{r}{100}\right)V_n$ and simplify.	$V_0 = 10\,000,\quad V_{n+1} = \left(1 - \dfrac{18}{100}\right)V_n$ $V_0 = 10\,000,\quad V_{n+1} = (1 - 0.18)V_n$ $V_0 = 10\,000,\quad V_{n+1} = 0.82V_n$
c Look at the decimal in front of V_n and convert it to a percentage.	The value of the photocopier in any year is 82% of its value the previous year.
d Is addition or subtraction involved? Is multiplication by a number greater than 1 or between 0 and 1 involved?	No addition or subtraction is involved. Multiplication by a number between 0 and 1 is involved. So, the recurrence relation models geometric decay.

Herman the Handyman depreciates his power tools using the reducing balance method. The value of the tools, in dollars, after n years, V_n, can be modelled by the recurrence relation

$$V_0 = 30\,000, \quad V_{n+1} = 0.7V_n$$

a Use recursion to show that the value of the tools after two years, V_2, is $14\,700.

b What is the annual percentage rate of depreciation used by Herman?

c If Herman plans to replace these tools when their value first falls below $2000, after how many years will Herman replace these tools?

Steps	Working
a Step out the recurrence relation working to find V_2.	$V_0 = 30\,000$ $\begin{aligned} V_1 &= 0.7V_0 \\ &= 0.7 \times 30\,000 \\ &= 21\,000 \end{aligned}$ $\qquad$ $\begin{aligned} V_2 &= 0.7V_1 \\ &= 0.7 \times 21\,000 \\ &= 14\,700 \end{aligned}$
b 1 Use the recurrence relation to find an equation for r.	Comparing: $V_{n+1} = \left(1 - \dfrac{r}{100}\right)V_n$ $V_{n+1} = 0.7V_n$ we can see that $1 - \dfrac{r}{100} = 0.7$
2 Solve for r, using CAS if necessary.	$\dfrac{r}{100} = 1 - 0.7$ $\dfrac{r}{100} = 0.3$ $r = 30$
3 Write the answer.	The annual percentage rate of depreciation is 30%.
c 1 Use CAS recursive to generate the sequence. The value of the tools after n years is V_n.	

ClassPad

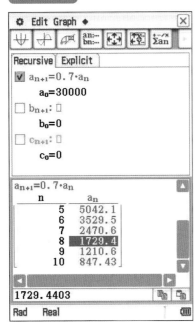

TI-Nspire

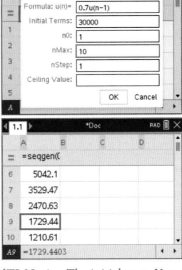

[TI-Nspire: The initial term V_0 occurs in row 1, so row 9 is V_8.]

2 Write the answer. $\qquad$ $V_8 = 1729.44$, $n = 8$

Herman will replace these tools after eight years.

Reducing balance depreciation general rule

Reducing balance depreciation general rule

The general rule for the value V_n of a depreciated asset using reducing balance depreciation is

$$V_n = \left(1 - \frac{r}{100}\right)^n \times V_0$$

where V_0 = initial value of the asset

r = percentage depreciation rate per year

n = number of years.

When solving for n, always round *up*, never down, to the nearest whole number.

$V_{n-1} - V_n$ = amount of depreciation in the nth year

WORKED EXAMPLE 4	Using the reducing balance depreciation rule

A truck bought for $250 000 is depreciated using the reducing balance method at a rate of 20% p.a.

a Write a rule that will calculate the value of the truck after n years.

b Use the rule to find the value of the truck after nine years to the nearest dollar.

c Use the rule to find how many years it would take for the truck to depreciate to under $20 000.

d How much is the truck depreciated by in the sixth year?

Steps	Working
a Substitute the values of V_0 and r into the reducing balance depreciation general rule and simplify.	$V_0 = 250\,000$, $r = 20$ $V_n = \left(1 - \dfrac{r}{100}\right)^n \times V_0$ $V_n = \left(1 - \dfrac{20}{100}\right)^n \times 250\,000$ $V_n = 0.8^n \times 250\,000$
b Substitute the value of n into your rule and solve.	$n = 9$ $V_9 = 0.8^9 \times 250\,000$ $\qquad = 33\,554.43\ldots$ The value of the truck after nine years is $33 554.
c 1 Identify what we know and what we need to find from the reducing balance depreciation rule.	$V_0 = 250\,000$, $r = 20$, $V_n = 20\,000$, $n = ?$ $V_n = \left(1 - \dfrac{r}{100}\right)^n \times V_0$
2 Substitute into the rule and solve using CAS.	$20\,000 = 0.8^n \times 250\,000$

ClassPad	TI-Nspire
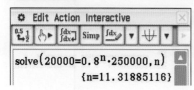	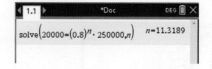

3 Write the answer, rounding *up* to the nearest year.	It will take the truck 12 years to depreciate to under $20 000.

d Amount of depreciation in the nth year
$= V_{n-1} - V_n$

$n = 6$

$V_5 = 0.8^5 \times 250\,000 = 81\,920$

$V_6 = 0.8^6 \times 250\,000 = 65\,536$

$V_5 - V_6 = 81\,920 - 65\,536 = 16\,384$

The amount of depreciation in the sixth year is $16\,384.

WORKED EXAMPLE 5 Working with the reducing balance depreciation rule

Find each of the following using the reducing balance depreciation rule.

a Saskia bought a car for $28\,000. After two years of being depreciated on a reducing balance basis, the car is now valued at $20\,230. Show that the annual rate of depreciation in the value of the car is 15%.

b Lloyd is using the reducing balance method to depreciate the carpet in his investment property. The annual rate of depreciation in the value of the carpet is 3%. If after 10 years the value of the carpet is $20\,647.88, what was the original price of the carpet to the nearest dollar?

Steps	Working
a 1 Identify what we know and what we need to find from the reducing balance depreciation rule.	$V_0 = 28\,000,\ n = 2,\ V_2 = 20\,230,\ r = ?$ $$V_n = \left(1 - \frac{r}{100}\right)^n \times V_0$$
2 Substitute into the rule and show the steps to solve for r. If this wasn't a 'show' question, we could use CAS.	$$V_2 = \left(1 - \frac{r}{100}\right)^2 \times 28\,000 = 20\,230$$ $$\left(1 - \frac{r}{100}\right)^2 = \frac{20\,230}{28\,000} = 0.7225$$ $$1 - \frac{r}{100} = \sqrt{0.7225} = 0.85$$ $$\frac{r}{100} = 1 - 0.85 = 0.15$$ The annual rate of depreciation is 15%.
b 1 Identify what we know and what we need to find from the reducing balance depreciation rule.	$r = 3,\ n = 10,\ V_{10} = 20\,647.88,\ V_0 = ?$ $$V_n = \left(1 - \frac{r}{100}\right)^n \times V_0$$
2 Substitute into the rule and solve using CAS.	$$V_{10} = \left(1 - \frac{3}{100}\right)^{10} \times V_0 = 20\,647.88$$
ClassPad 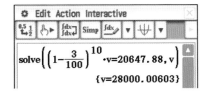	TI-Nspire 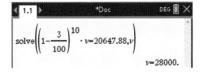
3 Write the rounded answer.	The original price of the carpet is $28\,000.

Recap

1 Shari wants to invest $13 000. Which of the following nominal interest rates will give the highest effective interest rate?

A 4.9% per annum, compounding quarterly   **B** 4.9% per annum, compounding weekly

C 4.9% per annum, compounding yearly   **D** 4.9% per annum, compounding monthly

E 4.9% per annum, compounding daily

2 A bank offers a compound interest rate of 8% p.a. compounding weekly. Which of the following is closest to the difference between this rate and the effective interest rate?

A 0.08%   **B** 0.15%   **C** 0.3%   **D** 5.2%   **E** 8.3%

Mastery

3 🔗 WORKED EXAMPLE 2   A business purchased a workstation for $12 000. It is depreciated using reducing balance depreciation at a rate of 16% per annum. Give all answers to the nearest dollar.

a Copy and complete the table to find

   i the value of the workstation after five years

   ii the amount of depreciation in the fourth year

   iii when the workstation first depreciates to less than $8000.

n	Depreciation ($)	Value after n years ($)
0		12 000
1	$\dfrac{16}{100} \times 12\,000 = 1920$	$12\,000 - 1920 = 10\,080$
2	$\dfrac{16}{100} \times 10\,080 = 1613$	$10\,080 - 1613 = 8467$
3		
4		
5		

b Write down a recurrence relation that gives the value of the workstation after n years.

c What percentage of the previous value is each new value?

d Describe the sort of growth or decay modelled by the recurrence relation.

e Use the graph to find the workstation's approximate value after 14 years.

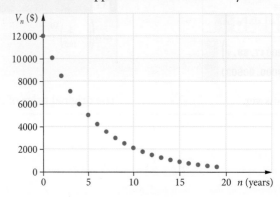

4 [WORKED EXAMPLE 3] Melanie runs a mowing service and she depreciates her ride-on mower using the reducing balance method. The value of the mower, in dollars, after n years, V_n, can be modelled by the recurrence relation

$$V_0 = 25\,000, \quad V_{n+1} = 0.6\,V_n$$

   a Use recursion to show that the value of the mower after two years, V_2, is $9000.

   b What is the annual percentage rate of depreciation used by Melanie?

   c If Melanie plans to replace the mower when its value first falls below $2000, after how many years will Melanie replace the mower?

5 [WORKED EXAMPLE 4] A truck was bought for $200\,000 and is being depreciated on a reducing balance basis rate of 25% per annum.

   a Write a rule that will calculate the value of the truck after n years.

   b Use the rule to find the value of the truck after eight years to the nearest dollar.

   c Use the rule to find how many years it would take for the truck to depreciate to under $40\,000.

   d How much is the truck depreciated by in the seventh year?

6 [WORKED EXAMPLE 5] Find each of the following using the reducing balance depreciation rule.

   a Spencer bought a sports car for $64\,000. After two years of depreciation on a reducing balance basis, the car is now valued at $40\,960. Show that the annual rate of depreciation in the value of the car is 20%.

   b Talia's Tree Lopping is using the reducing balance method to depreciate a mulcher. The annual rate of depreciation in the value of the mulcher is 16%. If after eight years the value of the mulcher is $1735.13, what was the original price of the mulcher to the nearest dollar?

Calculator-free

7 (4 marks) A computer purchased for $4000 is depreciated using reducing balance depreciation at a rate of 15% per annum. Copy and complete the depreciation table.

n	Depreciation ($)	Value after n years ($)
0		4000
1		
2		

8 (4 marks)

   a A car purchased for $32\,000 depreciates on a reducing balance basis at a rate of 11% per annum. Write a rule that will calculate the value of the car after n years. (2 marks)

   b A truck was purchased for $134\,000. Using the reducing balance method, the value of the truck is depreciated by 8.5% each year. Write the recurrence relation that could be used to determine the value of the truck after n years. (2 marks)

9 © SCSA │ MA2016 Q7 MODIFIED │ (9 marks) Julie buys a car with a purchase price of $13 000. However, she has been told to expect the car to depreciate in value. The value of the car after *n* years can be determined by using the recursive rule

$$T_{n+1} = 0.85T_n, \quad T_0 = 13\,000$$

a Copy and complete the table below to show the value of the car at the end of each year, to the nearest dollar. (2 marks)

n	0	1	2	3
Value of car after *n* years ($)	13 000			

b Use the information above to determine the rate of depreciation of Julie's car per year. (1 mark)

c Determine a rule for the *n*th term of the sequence of values found in part **a**. (2 marks)

d Determine the value of Julie's car after eight years, correct to the nearest dollar. (2 marks)

e Julie decides that she will sell her car at the end of the year in which its value drops to half of the purchase price. After how many years should she sell her car? (2 marks)

10 (3 marks) Phil is a builder who has purchased a large set of tools for $60 000. The value of Phil's tools is depreciated using the reducing balance method. The value of the tools, in dollars, after *n* years, V_n, can be modelled by the recurrence relation shown below.

$$V_0 = 60\,000, \quad V_{n+1} = 0.9V_n$$

a Use recursion to show that the value of the tools after two years, V_2, is $48 600. (1 mark)

b What is the annual percentage rate of depreciation used by Phil? (1 mark)

c Phil plans to replace these tools when their value first falls below $20 000. After how many years will Phil replace these tools? (1 mark)

11 (3 marks) Julie withdraws $14 000 from her account to purchase a car for her business. For tax purposes, she plans to depreciate the value of her car using the reducing balance method. The value of Julie's car, in dollars, after *n* years, C_n, can be modelled by the recurrence relation shown below.

$$C_0 = 14\,000, \quad C_{n+1} = R \times C_n$$

a For each of the first three years of reducing balance depreciation, the value of *R* is 0.85. What is the annual rate of depreciation in the value of the car during these three years? (1 mark)

b For the next five years of reducing balance depreciation, the annual rate of depreciation in the value of the car is changed to 8.6%. What is the value of the car eight years after it was purchased? Round your answer to the nearest cent. (2 marks)

 Exam hack

Watch out for the word 'next' in questions. This means there are two stages to look at.

12 (2 marks) A snooker table at a community centre was purchased for $3000. The value of the snooker table was depreciated using the reducing balance method of depreciation. After one year, the value of the snooker table is $2760. After two years, the value of the snooker table is $2539.20.

a Show that the annual rate of depreciation in the value of the snooker table is 8%. (1 mark)

b Let S_n be the value of the snooker table after *n* years. Write down a recurrence relation, in terms of S_{n+1} and S_n, that can be used to determine the value of the snooker table after *n* years using this reducing balance method. (1 mark)

7.3 Using finance solvers

Introducing CAS finance solvers

CAS **finance solvers** are a fast way of solving complex financial problems. We will be dealing with more complex loan and investment problems in this chapter, but first we will look at how to use CAS finance solvers for the sorts of situations that we have covered so far.

Follow these steps to access the finance solver on CAS.

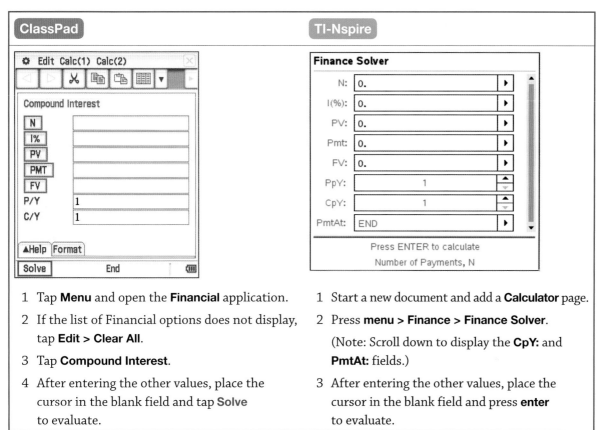

ClassPad

1 Tap **Menu** and open the **Financial** application.

2 If the list of Financial options does not display, tap **Edit > Clear All**.

3 Tap **Compound Interest**.

4 After entering the other values, place the cursor in the blank field and tap **Solve** to evaluate.

TI-Nspire

1 Start a new document and add a **Calculator** page.

2 Press **menu > Finance > Finance Solver**.

(Note: Scroll down to display the **CpY:** and **PmtAt:** fields.)

3 After entering the other values, place the cursor in the blank field and press **enter** to evaluate.

For the Using CAS examples in this chapter, the **ClassPad** and **TI-Nspire** screens will be combined as follows, with the value to be calculated highlighted in red:

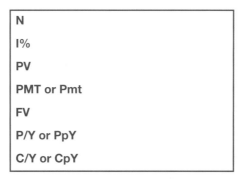

N
I%
PV
PMT or Pmt
FV
P/Y or PpY
C/Y or CpY

Video playlist
Using finance solvers

Values in a finance solver can be positive, negative or zero.

- Money coming *to the person* is positive.
- Money going away *from the person* is negative.

Compound interest investments with no payments

N	**Total number** of compounding periods
I%	**Interest rate** per year
PV	**Present value** for an investment is **negative** because the money is going away from the person to the bank.
PMT or Pmt	Regular **payments** for the investments we have looked at so far are **zero**.
FV	**Future value** has the opposite sign of the present value, so it will be **positive**.
P/Y or PpY	Number of **payments per year**. This will always take the same value as **C/Y** or **CpY**.
C/Y or CpY	Number of **compounding periods per year**

Reducing balance depreciation

N	**Total number** of compounding periods
I%	**Interest rate** per year will be **negative** because the asset is losing money.
PV	**Present value** for reducing balance depreciation is **negative** because the person has spent money to buy the asset so the money is going away from them.
PMT or Pmt	Regular **payments** for depreciation are **zero**.
FV	**Future value** has the opposite sign of the present value, so it will be **positive** (or zero).
P/Y or PpY	Number of **payments per year**. This will always take the same value as **C/Y** or **CpY**.
C/Y or CpY	Number of **compounding periods per year**

PV and FV always have opposite signs (except when FV is zero).

Compound interest and finance solvers

Finance solvers can be used to solve problems involving compound interest investments and reducing balance depreciation. The examples we have looked at so far have had no regular payments involved so **PMT** or **Pmt** is 0.

WORKED EXAMPLE 6	Using finance solvers for compound interest investments

Collette invested $36 000 in a term deposit earning 8.1% p.a. interest compounding monthly.
Give answers to the nearest cent.

a What is the value of Collette's investment after five years?

b How long will it take for her investment to grow to $46 000?

c i How long will it take for her investment to grow to $46 000 if the interest compounded quarterly?

  ii How much longer will it take for quarterly compounding to reach $46 000 compared to monthly compounding?

Steps	Working
The cell we are solving, in finance solver, is shown in red in each part of the question.	**ClassPad** ➕ **TI-Nspire**

a 1 This is an investment so money will go away from us at the start meaning PV is negative. The money will return to us at the end, so FV will be positive. Enter the values into the finance solver, then place the cursor in FV and solve.

Total number of compounding periods = 5 × 12 = 60 months	N	60
Annual interest rate	I%	8.1
Present value for an investment is negative.	PV	−36 000
Payment amount	PMT or Pmt	0
Future value has the opposite sign to present value	FV	53 901.493…
Same as C/Y or CpY	P/Y or PpY	12
Number of compounding periods per year	C/Y or CpY	12

2 Write the answer, rounding to nearest cent. | The value of Collette's investment after five years will be $53 901.49.

b 1 Enter the values into the finance solver, then place the cursor in N and solve.

Total number of compounding periods	N	36.436…
Annual interest rate	I%	8.1
Present value for an investment is negative.	PV	−36 000
Payment amount	PMT or Pmt	0
Future value has the opposite sign to present value.	FV	46 000
Same as C/Y or CpY	P/Y or PpY	12
Number of compounding periods per year	C/Y or CpY	12

2 Write the answer, rounding up to the next whole number. | It will take 37 months for the investment to grow to $46 000.

c 1 Change C/Y or CpY to compounding quarterly, place the cursor in N and solve.

Total number of compounding periods	N	12.2269…
Annual interest rate	I%	8.1
Present value for an investment is negative.	PV	−36 000
Payment amount	PMT or Pmt	0
Future value has the opposite sign to present value.	FV	46 000
Same as C/Y or CpY	P/Y or PpY	4
Number of compounding periods per year	C/Y or CpY	4

2 i Write the answer, rounding up to the next whole number. | It will take 13 quarters for the investment to grow to $46 000.

ii Use the fact that there are three months in a quarter to compare. | It takes 13 × 3 = 39 months to reach $46 000 compounding quarterly.

So, compounding quarterly takes two months longer than compounding monthly.

Exam hack

Finance solvers are often the quickest way to answer a question. You can use them if the question doesn't specifically say to use a different method.

WORKED EXAMPLE 7 | Using finance solvers for reducing balance depreciation

Fran has been depreciating her farm machinery on a reducing balance basis rate per annum.

a Fran bought a planter for $44 000 and after seven years its value is $22 390. What is the rate of depreciation, rounded to one decimal place?

b Fran has been depreciating her combine harvester at a rate of 12% p.a. After five years its value was $168 874. What was the original cost of the combine harvester to the nearest dollar?

Steps		Working
		ClassPad ╋ **TI-Nspire**

a 1 Enter the values into the finance solver, then place the cursor in I% and solve. As this is a reducing balance depreciation PV is negative and FV is positive. Find I from negative PV to positive FV.

Total number of compounding periods	N	7
Annual interest rate for depreciation is negative	I%	−9.199 98…
Present value for depreciation is negative.	PV	−44 000
Payment amount	PMT or Pmt	0
Future value has the opposite sign to present value.	FV	22 390
Same as C/Y or CpY	P/Y or PpY	1
Number of compounding periods per year	C/Y or CpY	1

2 Write the answer, rounding to one decimal place.   The rate of depreciation is 9.2% p.a.

b 1 The value of the harvester after 5 years is $168 874, so this is the value of FV. Enter all the values into the finance solver, then place the cursor in PV and solve.

Total number of compounding periods	N	5
Annual interest rate for depreciation is negative.	I%	−12
Present value for depreciation is negative.	PV	−319 999.595…
Payment amount	PMT or Pmt	0
Future value has the opposite sign to present value.	FV	168 874
Same as C/Y or CpY	P/Y or PpY	1
Number of compounding periods per year	C/Y or CpY	1

2 Write the answer, rounding to the nearest dollar.   The combine harvester was originally bought for $320 000.

	WORKED EXAMPLE 8	Using finance solvers for two-step compound interest investment problems

The balance of Maddie's investment, which compounds monthly, was $5416 after four months and $5604 after a year.

a What is the rate of compound interest per annum, correct to two decimal places?

b What was the amount of money that she initially invested to the nearest dollar?

Steps		Working
		ClassPad ✛ TI-Nspire

a 1 This is a compound interest investment, so PV is negative and FV is positive. The first step is to find the interest rate. Enter all the values into the finance solver, then place the cursor in I% and solve.

	N	8
Total number of compounding periods = 12 − 4 = 8 months		
Annual interest rate	I%	5.129395…
Present value for an investment is negative.	PV	−5416
Payment amount	PMT or Pmt	0
Future value has the opposite sign to present value.	FV	5604
Same as C/Y or CpY	P/Y or PpY	12
Number of compounding periods per year	C/Y or CpY	12

2 Write the answer, rounding to two decimal places.　The interest rate is 5.13%.

b 1 Use the unrounded answer for I% find the initial amount PV.

	N	4
Total number of compounding periods		
Annual interest rate from previous solver (unrounded)	I%	5.129395580514 1
Present value for an investment is negative.	PV	−5324.378489
Payment amount	PMT or Pmt	0
Future value has the opposite sign to present value.	FV	5416
Same as C/Y or CpY	P/Y or PpY	12
Number of compounding periods per year	C/Y or CpY	12

2 Write the answer, rounding to the nearest dollar.　The amount of money that Maddie initially invested was $5324.

Recap

1 A farmer purchased a tractor for $100 000. It is depreciated using reducing balance depreciation at a rate of 15% per annum. Give all answers to the nearest dollar. Copy and complete the table to find

   a   the value of the tractor after four years   **b**   the amount of depreciation in the first year

   c   the total depreciation after four years.

n	Depreciation	Value after n years ($)
0		100 000
1	$\dfrac{15}{100} \times 100\,000 = 15\,000$	$100\,000 - 15\,000 = 85\,000$
2	$\dfrac{15}{100} \times 85\,000 = 12\,750$	$85\,000 - 12\,750 = 72\,250$
3		
4		

Mastery

2   WORKED EXAMPLE 6   Sara invested $40 000 in a term deposit earning 6.3% p.a. compounding monthly.

   a   What is the value of her investment after seven years? Give answers to the nearest cent.

   b   How long will it take for her investment to grow to $50 000?

   c   **i**   How long will it take for her investment to grow to $50 000 if the interest is compounded quarterly?

         ii   How much longer will it take to reach $50 000 if compounding quarterly, rather than compounding monthly?

3   WORKED EXAMPLE 7   Eugen depreciates his printing machinery on a reducing balance basis rate per annum.

   a   He bought a paper guillotine for $33 000 and after eight years its value is $17 840. What is the rate of depreciation, rounded to one decimal place?

   b   He depreciates his printing press at a rate of 10%. After seven years its value is $64 570. What was the original cost of the printing press, rounded to the nearest dollar?

4 **a**   Kirily invested $45 700 in an account earning 2.9% p.a. compounding weekly. To find the balance after three years, what values would you enter for N and C/Y or CpY in a CAS finance solver?

   b   Fletcher invested $22 000 at 4.1% p.a. compounding six-monthly. To find the value of his investment after five years, what values would you enter for N and P/Y or PpY in a CAS finance solver?

   c   Sinjin's investment, which compounds monthly, has grown from $15 000 to $17 000 in the last four years. To find the interest rate of the account, what values would you enter for N and PV in a CAS finance solver?

   d   Heidi invested $4000 in an account earning 3.6% interest compounding fortnightly. To find out how long it will take to grow to $5000, what values would you enter for I%, PV and FV in a CAS finance solver?

   e   A company depreciates a machine bought for $70 000 on a reducing balance basis rate per annum. If its value after six years is $3200, what would you enter for N, PV and C/Y or CpY in a CAS finance solver to find the rate of depreciation?

   f   A business has been depreciating its laptops at a rate of 12.5% p.a. After four years their value was $18 600. To find the original value of the laptops, what would you enter for N, I% and FV in a CAS finance solver?

5 WORKED EXAMPLE 8 Solve the following two-step finance solver question.

The balance of Arthur's investment, which compounds monthly, was $6234 after eight months and $6357 after one year. What was the amount of money that he initially invested to the nearest dollar?

Calculator-assumed

6 © SCSA MA2016 Q13 (6 marks) Simon has $5000 that he wants to invest for a period of time without touching it.

 a If he chooses to invest this money in an account earning compound interest at the rate of 6.5% per annum, determine the

  i value of his investment after three years, if interest is paid annually (1 mark)

  ii time required for him to double his investment, if interest is paid monthly. (2 marks)

 b Simon is currently deciding between two options and wishes to compare them.

  Option A: Invest the $5000 in an account earning compound interest at the rate of 5.5% per annum, with interest paid monthly.

  Option B: Invest the $5000 in an account earning compound interest at the rate of 5.4% per annum, with interest paid daily.

  He decides to calculate the effective annual rate of interest for each option, in order to compare the possible investments. He determines that Option A has an effective annual rate of interest of 5.64%, correct to two decimal places.

  Calculate the effective annual rate of interest for Option B using the CAS effective interest function (ClassPad only), correct to two decimal places, and hence decide on the better option for Simon. (3 marks)

7 © SCSA MA2021 Q12 (6 marks) Virat purchases a new motor vehicle for $24 500. For the first two years the vehicle depreciates at a rate of 13% per year and for the third year it depreciates at a lower rate of 9.5% per year.

 a Calculate the value of the vehicle after one year. (1 mark)

 b Calculate the value of the vehicle after the first three years. (2 marks)

 For the next three years the rate of depreciation is constant at r% per year. The average rate of depreciation for the first six years is 11% per year.

 c Calculate the value of r as a percentage. (3 marks)

8 (2 marks) Answer the following questions.

 a Ardy invests $150 000 for six years at an interest rate of 3.5% per annum, compounding annually. Find the value of the investment after six years. (1 mark)

 b A file server costs $30 000. The file server depreciates by 20% of its value each year. Find the value of the server after three years. (1 mark)

9 (4 marks) Millie invested $20 000 in an account at her bank with interest compounding monthly. After one year, the balance of Millie's account was $20 732. Find

 a the rate of interest per annum used by the bank, correct to 2 decimal places (2 marks)

 b the difference between the rate of interest per annum used by her bank and the effective annual rate of interest for Millie's investment. (2 marks)

10 (4 marks) A computer originally purchased for $6000 is depreciated each year using the reducing balance method. If the computer is valued at $2000 after four years, find

 a the annual rate of depreciation, correct to two decimal places (2 marks)

 b the value of the computer after 6 years. (2 marks)

11 (3 marks) Richard is selling his stereo system to help pay for a holiday. The stereo system was originally purchased for $8500. He is using a reducing balance depreciation method, with an annual depreciation rate of 8%.

    a Using this depreciation method, what is the value of the stereo system four years after it was purchased? Round your answer to the nearest cent.   (1 mark)

    b Four years after it was purchased, Richard sold his stereo system for $4500. Assuming a reducing balance depreciation method was used, what annual percentage rate of depreciation did this represent? Round your answer to one decimal place.   (2 marks)

> 🔓 **Exam hack**
>
> When answering extended answer questions with finance solvers, show your working by listing the values your entered for N, I%, PV etc.

(7.4) Reducing balance loans

Reducing balance loan recurrence relations

A **reducing balance loan** is a compound interest loan with regular payments. Home loans and personal loans are examples of reducing balance loans. For each compounding period, the balance of the loan changes in two ways. The balance:

- increases, because compound interest is being charged by the bank
- decreases, because a fixed amount is paid into the account.

The change in the amount owed is a combination of geometric growth and linear decay.

> **Reducing balance loan recurrence relation**
>
> The recurrence relation for the value V_n of a reducing balance loan is
>
> $$V_0 = \text{principal}, \quad V_{n+1} = \left(1 + \frac{r}{100}\right)V_n - d$$
>
> where r = percentage interest rate per compounding period
>
> n = number of compounding periods
>
> d = payment made per compounding period.
>
> The graph of a reducing balance loan recurrence relation would look like this.
>
>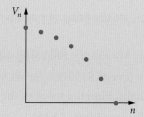

> 🔓 **Exam hack**
>
> Don't confuse reducing balance depreciation and reducing balance loans. Reducing balance loans must have a payment.

WORKED EXAMPLE 9	Finding reducing balance loan recurrence relations

Freya has taken out a loan of \$17 000 with an interest rate of 12.3% per annum compounding monthly, and she makes regular monthly payments of \$540.

a Write a recurrence relation for the balance.

b Describe the sort of growth or decay modelled by the recurrence relation.

c Sketch the shape of the graph of the recurrence relation.

Steps	Working
a 1 Find the number of compounding periods per year.	There are 12 compounding periods per year.
2 Identify V_n, V_0, r and d.	Let V_n = the balance after n compounding periods. $$V_0 = 17\,000, r = \frac{12.3}{12} = 1.025, d = 540$$
3 Substitute the values into these formulas and simplify. V_0 = principal $$V_{n+1} = \left(1 + \frac{r}{100}\right)V_n - d$$	$$V_{n+1} = \left(1 + \frac{1.025}{100}\right)V_n - 540$$ $$= (1 + 0.010\,25)V_n - 540$$ $$= 1.010\,25V_n - 540$$ $$V_0 = 17\,000, \quad V_{n+1} = 1.010\,25V_n - 540$$
b Is there addition or subtraction involved?	There is subtraction involved.
Is there multiplication involved by a number greater than 1 or between 0 and 1?	There is multiplication involved by a number greater than 1. So, the recurrence relation models a combination of linear decay *and* geometric growth.
c Show the points forming the shape of the curve.	

Exam hack

Questions don't always state that the loan is a reducing balance one. You know it's a reducing balance loan because it involves compound interest and a regular payment.

WORKED EXAMPLE 10	Working with reducing balance loan recurrence relations

Gary has taken out a reducing balance loan, compounding quarterly and with regular quarterly payments, according to the recurrence relation

$$V_0 = 14\,000, \quad V_{n+1} = 1.0215V_n - 655,$$

where V_n is the value of the loan after n compounding periods.

a How much money did Gary borrow?

b How much are the regular quarterly payments?

c Use recursion to write down calculations that show that the amount owing on his loan after three quarters will be \$12 915 when rounded to the nearest dollar.

d What is the annual percentage compound interest rate for this loan?

e After how many quarters will the balance of Gary's loan fall below \$12 000?

Steps	Working
a Identify V_0.	Gary borrowed $14 000.
b Identify d.	The regular quarterly payments are $655.
c Step out the recurrence relation calculations to find V_3, giving answers to the nearest dollar.	$V_0 = 14 000$ $V_1 = 1.0215V_0 - 655$ $\quad = 1.0215 \times 14 000 - 655$ $\quad = 13 646.00$ $V_2 = 1.0215V_1 - 655$ $\quad = 1.0215 \times 13 646.00 - 655$ $\quad = 13 284.39$ $V_3 = 1.0215V_2 - 655$ $\quad = 1.0215 \times 13 284.39 - 655$ $\quad = 12 915$
d 1 Use the recurrence relation to find an equation for r, the percentage interest rate per compounding period.	Comparing $V_{n+1} = \left(1 + \dfrac{r}{100}\right)V_n - d$ $V_{n+1} = 1.0215V_n - 655$ we can see that $1 + \dfrac{r}{100} = 1.0215$
2 Solve for r, using CAS if necessary.	$\dfrac{r}{100} = 1.0215 - 1$ $\dfrac{r}{100} = 0.0215$ $r = 2.15$ The quarterly interest rate is 2.15%.
3 Multiply r by the compounding period to find the annual percentage compound interest rate.	The interest compounds quarterly, so the annual percentage compound interest rate is $2.15\% \times 4 = 8.6\%$
e Use CAS sequence to enter the recurrence relation.	$V_6 = 11 758.51$, $n = 6$ The balance of Gary's loan will fall below $12 000 after six quarters.

ClassPad

TI-Nspire

Reducing balance loan graphs

If a payment is missed in a reducing balance loan, then extra interest needs to be paid on the next balance. This means the loan will take longer to repay even if the next payment is doubled to try to make up for the missed payment. The graph of a reducing balance loan where a payment was missed and the next payment was doubled would look like this.

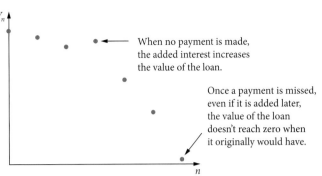

When no payment is made, the added interest increases the value of the loan.

Once a payment is missed, even if it is added later, the value of the loan doesn't reach zero when it originally would have.

For a reducing balance loan, the regular payment is part interest and part principal.

- The amount of interest ($) paid **decreases** with each compounding period. Towards the end of the life of the loan, the interest becomes a small part of the regular payment.
- The amount paid off the principal ($) **increases** with each compounding period. Towards the end of the life of the loan, the amount paid off the principal becomes a large part of the regular payment.

Reducing balance loan payment graphs

Graph shape of total interest paid ($)

Graph shape of total amount paid off principal ($)

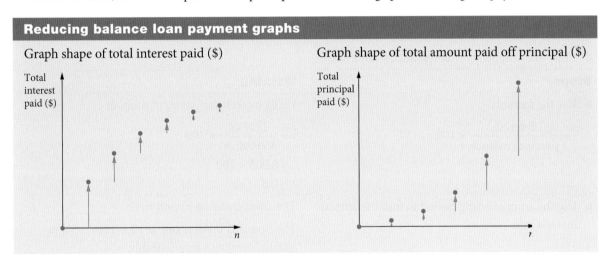

Reducing balance loan amortisation tables

Amortisation tables help us track the interest paid, the opening closing balance and the balance of the loan for each compounding period.

Reducing balance loan amortisation table

$$r = \frac{\text{interest}}{\text{previous balance}} \times 100$$

Period number	Opening balance	Interest	Repayment	Closing balance
1	V_0	$V_0 \times \dfrac{r}{100}$	d	$V_1 = V_0 + \left(V_0 \times \dfrac{r}{100}\right) - d$
2	V_1	$V_1 \times \dfrac{r}{100}$	d	$V_2 = V_1 + \left(V_1 \times \dfrac{r}{100}\right) - d$

where n = period number

d = payment made per compounding period

r = percentage interest rate per compounding period

V_0 = principal

V_n = closing balance of nth period.

Exam hack

When entering values into an amortisation table, always round to the nearest cent. The usual rule of not rounding until the very end of your calculations doesn't apply.

WORKED EXAMPLE 11 | **Using reducing balance loan amortisation tables**

Reggie has taken out a housing loan of $450 000 with interest compounding monthly and is paying monthly instalments of $3500. The amortisation table below shows the first few months' calculations for his loan.

Month	Opening balance	Interest	Repayment	Closing balance
1	450 000.00	2250.00	3500.00	448 750.00
2				
3				

a Use the table to find the percentage interest rate per month, r, and show the calculation in your answer.

b What is the nominal interest rate for the loan?

c Copy and complete the amortisation table above.

Steps	Working
a Use the formula $$r = \frac{interest}{previous\ balance} \times 100$$	Using interest for payment number 1: $$r = \frac{2250.00}{450000.00} \times 100$$ $$= 0.005 \times 100$$ $$= 0.5$$
b Use the compounding period to find the annual interest rate.	The compounding period is 12. The nominal interest rate is $12 \times 0.5\% = 6\%$ per annum compounding monthly.
c Complete the table using $$interest = \frac{r}{100} \times opening\ balance$$ closing balance = opening balance + interest − repayment Give all values to the nearest cent.	

Month	Opening balance	Interest	Repayment	Closing balance
1	450 000.00	2250.00	3500.00	448 750.00
2	448 750.00	448 750 × 0.005 = 2243.75	3500.00	448 750.00 + 2243.75 − 3500 = 447 493.75
3	447 493.75	2237.47	3500.00	446 231.22

 Exam hack

When calculating r from an amortisation table, you can use any of the interest values. Choose the option with the easiest calculations, which is nearly always the interest value for payment 1.

9780170476959

WORKED EXAMPLE12 Analysing reducing balance loan amortisation tables

Karl has taken out a short-term personal loan of $20 000 for a gap year trip to Europe. The interest rate is 18% p.a. compounding quarterly and it needs to be repaid by making four quarterly payments of $5570.

a Find the values of A, B, C and D.

b How much of the principal has been repaid by the end of the 3rd quarter?

Quarter	Opening balance	Interest	Repayment	Closing balance
1	20 000.00	900.00	5570.00	15 330.00
2	15 330.00	689.85	5570.00	A
3	B	C	5570.00	D

Steps	Working
a 1 Closing balance = opening balance + interest − repayment	The balance after two quarters $A = 15\,330.00 + 689.85 - 5570$ $= 10\,649.85$
2 Find B. Closing balance at quarter 2 = opening balance at quarter 3	$A = B$ $B = 10\,649.85$
3 Find r, the percentage interest rate per compounding period.	$r = \dfrac{18}{4} = 4.5$
4 Find C. Interest $= \dfrac{r}{100} \times$ opening balance	$C =$ interest for quarter 3 $= \dfrac{4.5}{100} \times 10\,649.85$ $= 0.045 \times 10\,649.85$ $= 479.24$
5 Find D. Closing balance = opening balance + interest − repayment	The closing balance after three quarters $D = 10\,649.85 + 479.24 - 5570$ $D = \$5559.09$
b Amount repaid at the end of quarter 3 = principal − closing balance quarter 3	Amount repaid at the end of quarter 3 $= 20\,000 - 5559.09$ $= \$14\,440.91$

Interest-only loan recurrence relations and formula

For reducing balance loans, the regular payments are greater than the interest at each compounding period causing the balance to decrease each period. If the payments were less than the interest, then the balance of the loan would keep growing, and the borrower would owe an ever-increasing amount.

An **interest-only loan** is a reducing balance loan where the payments are *exactly* equal to the interest for each compounding period. This means the balance doesn't change. This sort of loan is used when the borrower expects what they've bought with the loan to increase in value so that it can be sold and the loan repaid in full from the sale.

Interest-only loan

For an interest-only loan with value $V_n = V_0$ after n compounding periods

$$d = \frac{r}{100} \times V_0$$

where d = payment made per compounding period

r = percentage interest rate per compounding period

V_0 = principal.

The graph of an interest-only loan would look like this.

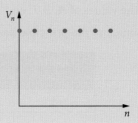

WORKED EXAMPLE 13 | Working with interest-only loans

a Elsa has taken out a loan with a recurrence relation for V_n, the value of the loan after n compounding periods, of

$$V_0 = 14\,500, \quad V_{n+1} = 1.008V_n - 116$$

Find V_1 and V_2 and explain why this shows that this models an interest-only loan.

b Lewis has taken out an interest-only loan with an interest rate of 4.8% per annum compounding monthly with regular monthly payments of $1200. How much did Lewis borrow?

c Jessie has taken out a $500\,000 interest-only loan. The interest rate is 3.5% per annum compounding quarterly. What are Jessie's regular quarterly repayments?

Steps	Working
a 1 Find V_1.	$V_1 = 1.008V_0 - 116$ $\quad = 1.008 \times 14\,500 - 116$ $\quad = 14\,500$
2 Find V_2.	$V_2 = 1.008V_1 - 116$ $\quad = 1.008 \times 14\,500 - 116$ $\quad = 14\,500$
3 For interest-only loans, $V_n = V_0$.	The value of the loan stays at the principal value $14\,500 for all compounding periods.
b 1 Identify what we know and what we need to find from the interest-only loan formula.	$r = \dfrac{4.8}{12} = 0.4, d = 1200, V_0 = ?$ $d = \dfrac{r}{100} \times V_0$
2 Substitute into the formula and solve, using CAS if necessary.	$1200 = \dfrac{0.4}{100} \times V_0$ $V_0 = \dfrac{1200 \times 100}{0.4}$ $\quad = 300\,000$
3 Write the answer.	Lewis borrowed $300\,000.
c 1 Identify what we know and what we need to find from the interest-only loan formula.	$r = \dfrac{3.5}{4} = 0.875, V_0 = 500\,000, d = ?$ $d = \dfrac{r}{100} \times V_0$
2 Substitute into the formula and solve.	$d = \dfrac{0.875}{100} \times 500\,000$ $\quad = 4375$
3 Write the answer.	Jessie's regular quarterly repayment for this loan is $4375.

Recap

1 An investment is made at 4% interest per annum compounding fortnightly. The value of the investment at the end of the tenth year is $20 879.13. How much money was originally invested?

   A $7395   **B** $14 000   **C** $18 821   **D** $20 560   **E** $20 561

2 A workstation originally purchased for $8000 is depreciated each year using the reducing balance method. If the computer is valued at $3000 after five years, then the annual rate of depreciation is closest to

   A −89%   **B** 17%   **C** 18%   **D** 22%   **E** 23%

Mastery

3   WORKED EXAMPLE 9   Aisling has taken out a loan of $30 000 with an interest rate of 15.4% per annum compounding quarterly and is making regular quarterly payments of $2300.

   a Write a recurrence relation for the balance.

   b Describe the sort of growth or decay modelled by the recurrence relation.

   c Sketch the shape of the graph of the recurrence relation.

4   WORKED EXAMPLE 10   Ishmael has taken out a reducing balance loan, compounding monthly and with regular monthly payments, according to the recurrence relation

$$V_0 = 15\,000, \quad V_{n+1} = 1.006V_n - 465$$

where V_n is the value of the loan after n compounding periods.

   a How much money did Ishmael borrow?

   b How much are the regular monthly payments?

   c Use recursion to write down calculations that show that the amount owing on his loan after three months will be $13 868, rounded to the nearest dollar.

   d What is the annual percentage compound interest rate for this loan?

   e After how many months will the balance of Ishmael's loan fall below $12 500?

5   WORKED EXAMPLE 11   Praveen has taken out a housing loan of $380 000 with interest compounding monthly and is paying monthly instalments of $3854. The amortisation table below shows the first few calculations for his loan.

Month	Opening balance	Interest	Repayment	Closing balance
1	380 000.00	2850.00	3854.00	378 996.00
2	378 996.00	2842.47	3854.00	377 984.47
3				

   a Use the table to find the percentage interest rate per month, r, and show the calculation in your answer.

   b What is the nominal interest rate for the loan?

   c Copy and complete the amortisation table above.

6 🔗 WORKED EXAMPLE 12 | Chester has taken out a short-term personal loan of $10 000 for a car. The interest rate is 16% p.a. compounding quarterly and it needs to be repaid in four quarterly payments of $2750. Answer the following questions using the amortisation table of the loan given below, which has some missing entries.

Quarter	Opening balance	Interest	Repayment	Closing balance
1	10 000.00	400.00	2750.00	7650.00
2	7 650.00	306.00	2750.00	5206.00
3	5 206.00	208.24	2750.00	A
4	B	C	2750.00	D

a Find the values of A, B, C and D.

b How much of the principal has been repaid by the end of the 4th quarter?

7 🔗 WORKED EXAMPLE 13

a Venice has taken out a loan with a recurrence relation for V_n, the value of the loan after n compounding periods, of
$$V_0 = 80 000, \quad V_{n+1} = 1.0036 V_n - 288$$
Find V_1 and V_2 and explain why this shows that this models an interest-only loan.

b Jim has taken out an interest-only loan with an interest rate of 5.2% per annum, compounding fortnightly with regular fortnightly payments of $400. How much did Jim borrow?

c Myrtle has taken out a $650 000 interest-only loan. The interest rate is 4.2% per annum compounded quarterly. What are Myrtle's regular quarterly repayments?

Calculator-assumed

8 (3 marks) Shirley would like to purchase a new home. She will establish a loan for $225 000 with interest charged at the rate of 3.6% per annum, compounding monthly. Each month, Shirley will pay only the interest charged for that month.

a Find the amount Shirley will owe after three years. (1 mark)

b Let V_n be the value of Shirley's loan, in dollars, at the start of month n. Find the recurrence relation that models the value of V_n. (2 marks)

9780170476959

9 ©SCSA | MA2021 Q10 | (14 marks) Wendy moved into an apartment and organised a loan of \$16 000 to purchase new furniture. To pay off the loan Wendy makes repayments of \$600 at the end of each month. The spreadsheet below shows the progress of her loan.

	A	B	C	D	E
1	Month	Opening balance	Interest	Repayment	Closing balance
2	1	16,000.00	98.67	600.00	15,498.67
3	2	15,498.67	95.58	600.00	14,994.24
4	3	14,994.24	92.46	600.00	14,486.71
5					

a Write a calculation to show that the yearly interest rate is approximately 7.4%. (2 marks)

b Copy and complete the last row of the above spreadsheet. (3 marks)

c Write a recursive rule to determine the closing balance of the loan at the end of each month. (2 marks)

d Determine how many months it will take Wendy to pay off the loan. (1 mark)

e Calculate how much interest is paid over the duration of the loan. (3 marks)

On reflection, Wendy realised she could have repaid \$800 each month.

f Determine the maximum amount Wendy would have been able to borrow, if all other details of the loan and repayment time remained the same. (3 marks)

10 ©SCSA | MA2016 Q12 | (11 marks) Thomas has borrowed \$16 000 from a bank at a reducible interest rate of 18% per annum with interest accrued and repayments made monthly. Standard repayments are set at \$500 per month.

The table below shows the progress of the loan for the first six months. All values have been rounded to the nearest cent.

Month	Amount owing at beginning of month	Interest for the month	Repayment	Amount owing at end of month
1	16 000.00	240.00	500.00	15 740.00
2	15 740.00	236.10	500.00	15 476.10
3	15 476.10	232.14	500.00	15 208.24
4	15 208.24	228.13	500.00	14 936.37
5	14 936.37	224.04	500.00	14 660.41
6	14 660.41	A	500.00	B

a What is the monthly interest rate? (1 mark)

b Determine the values of A and B. (2 marks)

c Determine the length of time it will take Thomas to pay off the loan. (1 mark)

d Determine the total amount Thomas pays over the duration of the loan. (3 marks)

e The bank suggests that Thomas need only make repayments of \$240 per month. Describe how this would affect the length of time and total amount he pays over the duration of the loan. (2 marks)

f After listening to advice, Thomas decides that he wants to pay off the loan completely in two years, making equal payments each month over that time. Determine the amount of each repayment he will need to make in order to make this happen, correct to the nearest cent. (2 marks)

▶ **11** © SCSA MA2020 Q12 (10 marks) Jessica wants to borrow $15 000 from her parents to purchase a car. They will be charging her compound interest at the rate of 4% per annum, with interest added yearly.

a Jessica is currently studying so she will not want to be making any regular repayments.

   i Copy and complete the table below to show the amount she will owe her parents at the end of each year. (2 marks)

Number of years (n)	0	1	2	3
Amount owing ($)	15 000			

   ii Write a recursive rule to calculate the amount owing at the end of each year. (2 marks)

Jessica's parents are encouraging her to get a part-time job so that she can make repayments along the way. Jessica estimates that she will be able to earn enough money to pay off $2400 each year.

b If interest is charged yearly and she repays the $2400 at the end of each year, write a recursive rule to calculate the amount owing at the end of each year. (1 mark)

c If interest is charged monthly and she makes equal monthly repayments

   i write a recursive rule to calculate the amount owing at the end of each month (1 mark)

   ii calculate how many months it will take to repay the loan (1 mark)

   iii calculate the total amount Jessica would pay over the duration of the loan. (3 marks)

12 (3 marks) Samuel has a reducing balance loan. The first five lines of the amortisation table for Samuel's loan are shown below.

Payment number	Payment	Interest	Principal reduction	Balance of loan
0	0.00	0.00	0.00	320 000.00
1	1600.00	960.00	640.00	319 360.00
2	1600.00	958.08	641.92	318 718.08
3	1600.00	956.15		318 074.23
4	1600.00			

Interest is calculated monthly and Samuel makes monthly payments of $1600. Interest is charged on this loan at the rate of 3.6% per annum.

a Using the values in the amortisation table, calculate the

   i principal reduction associated with payment number 3 (1 mark)

   ii balance of the loan after payment number 4 is made. Round your answer to the nearest cent. (1 mark)

b Let S_n be the balance of Samuel's loan after n months. Write down a recurrence relation, in terms of S_0, S_{n+1} and S_n, that could be used to model the month-to-month balance of the loan. (1 mark)

7.5 Using finance solvers for reducing balance loans

Reducing balance loans and finance solvers

At the start of this chapter we used CAS finance solvers to answer questions about compound interest investments and reducing balance depreciation, which don't involve regular payments. We will now look at how to use finance solvers for financial questions where payments are involved, starting with reducing balance loans.

Using finance solvers for reducing balance loans

N	Total **number** of compounding periods
I%	**Interest** rate per year
PV	**Present value** for a loan is **positive** because the money has come to the person.
PMT or Pmt	Regular **payments** for a loan are **negative** because the money is going away from the person.
FV	**Future value** has the opposite sign of the present value so it will be **negative**. A future value of **zero** means the loan has been fully repaid.
P/Y or PpY	Number of **payments per year**.
C/Y or CpY	Number of **compounding** periods per year

Video playlist
Using finance solvers for reducing balance loans

Worksheet
Loan repayment problems

When solving for N, always round *up*, never down, to the nearest whole number.

Ignore the negative sign in values when using the following formulas:

total interest paid = N × PMT − (PV − FV)

total loan cost = N × PMT

$$\text{percentage decrease in loan balance} = \frac{\text{PV} - \text{FV}}{\text{PV}} \times 100\%$$

WORKED EXAMPLE 14 | Using finance solvers for annual loan interest rates

Shari's reducing balance loan of $200 000 is to be fully repaid over 25 years with quarterly repayments of $3985.14. Find the interest rate per annum correct to one decimal place and explain why you need to enter the payment as a negative.

Steps	Working	
	ClassPad ＋ TI-Nspire	
1 Enter the values into the finance solver, then place the cursor in I% and solve.		
Total number of compounding periods	N	100
There are 25 × 4 = 100 quarters.		
Annual interest rate	I%	6.299 992…
Present value for a loan is positive.	PV	200 000
Money moving away from a person is negative.	PMT or Pmt	−3985.14
Future value is zero when a loan is fully repaid.	FV	0
Same as C/Y or CpY	P/Y or PpY	4
Number of compounding periods per year	C/Y or CpY	4

2 After entering all the other values, the I% field displays the annual interest rate. Round to the correct number of decimal places.

The annual interest rate is **6.3%**.

3 Is the money moving to the person or away from the person?

The value is negative because Shari is making payments to the bank, so the money is moving *away* from her.

WORKED EXAMPLE 15 Using finance solvers for loan repayments, total loan cost and total interest paid

Dougal borrowed $312 000 at 5.5% per annum for 25 years with monthly repayments and interest compounding monthly. Find, to the nearest cent, the

a repayment required to repay the loan in full

b total cost of the loan

c total interest paid on the loan.

Steps	Working

	ClassPad **+** TI-Nspire

a 1 Enter the values into the finance solver, then place the cursor in PMT and solve.

Total number of compounding periods	N	300
There are 25 × 12 = 300 months.		
Annual interest rate	I%	5.5
Present value for a loan is positive.	PV	312 000
Money moving away from a person is negative.	PMT or Pmt	−1915.952...
Future value is zero when a loan is fully repaid.	FV	0
Same as C/Y or CpY	P/Y or PpY	12
Number of compounding periods per year	C/Y or CpY	12

2 Write the answer correct to the nearest cent.

Dougal's monthly repayment is $1915.95.

b Total loan cost = N × PMT

Ignore negative sign in the PMT value.
Round to the nearest cent.

Total loan amount repaid
= N × PMT
= 300 × $1915.95
= $574 785.00

c Total interest paid = N × PMT − (PV − FV)
Ignore the negative signs in values.
Round to the nearest cent.

Amount of interest paid
= N × PMT − (PV − FV)
= 300 × $1915.95 − (312 000 − 0)
= 574 785.00 − 312 000
= $262 785.00

WORKED EXAMPLE 16 Using finance solvers for amount owed and number of loan repayments

Jordan takes out a loan of $30 000 at a rate of 8.5% per annum, compounding monthly, with repayments of $650 per month.

a How much is still owing on Jordan's loan after two years to the nearest cent?

b By what percentage has Jordan reduced the balance of his loan? Round to the nearest percentage.

c How many repayments are required in order for him to repay the loan in full?

d How much is still owing on Jordan's loan after 56 repayments to the nearest cent?

e Jordan decides to adjust the value of the 56th repayment so that the loan is fully repaid. What will the adjusted 56th payment be to the nearest cent?

Steps	Working
	ClassPad ✛ **TI-Nspire**
a **1** Find FV.	
Total number of compounding periods = 2 × 12 = 24 months	N 24
Annual interest rate	I% 8.5
Present value for a loan is positive.	PV 30 000
Money moving away from a person is negative.	PMT or Pmt −650
Future value has the opposite sign to present value.	FV −18 598.558…
Same as C/Y or CpY	P/Y or PpY 12
Number of compounding periods per year	C/Y or CpY 12
2 Round to the nearest cent.	After two years, Jordan still owes $18 598.56.
b Percentage decrease in loan balance = $\dfrac{\text{PV} - \text{FV}}{\text{PV}} \times 100\%$.	$\dfrac{30\,000 - 18\,598.56}{30\,000} \times 100\%$
Ignore the negative signs in values. Round to the nearest percentage.	= 0.380 048 × 100% = 38%
c **1** Find N when FV = 0.	
Total number of compounding periods	N 56.0888…
Annual interest rate	I% 8.5
Present value for a loan is positive.	PV 30 000
Money moving away from a person is negative.	PMT or Pmt −650
Future value is zero when a loan is fully repaid.	FV 0
Same as C/Y or CpY	P/Y or PpY 12
Number of compounding periods per year	C/Y or CpY 12
2 When solving for N, always round *up*, never down, to the nearest whole number.	Jordan will need to make 57 repayments. So **56** repayments won't be enough. The 57th repayment will be less than the usual $650.
d **1** Find FV after 56 months.	
Total number of compounding periods	N 56
Annual interest rate	I% 8.5
Present value for a loan is positive.	PV 30 000
Money moving away from a person is negative.	PMT or Pmt −650
Future value has the opposite sign to present value.	FV −57.541 594…
Same as C/Y or CpY	P/Y or PpY 12
Number of compounding periods per year	C/Y or CpY 12
2 After entering all the other values, the **FV** field displays the amount still owing. Round to the nearest cent.	Jordan still owes $57.54.
e For the last payment, add the amount still owing on the loan to the regular payment, rounding to the nearest cent.	$650 + $57.54 = $707.54 To repay the loan in full the final payment will be $707.54.

Interest-only loans and finance solvers

Finance solvers can also be used to answer questions about **interest-only loans** where the balance of the loan doesn't change because the payments are exactly equal to the interest for each compounding period.

Using finance solvers for interest-only loans

For an interest-only loan:

- all compounding periods are the same so N = 1
- FV and PV have the same value but opposite signs so FV = −PV.

WORKED EXAMPLE 17 Using finance solvers for interest-only loans

Use finance solvers to find each of the following.

a Savaratnam has a $100 000 interest-only loan compounding monthly with monthly payments of $348. What is the annual interest rate correct to one decimal place?

b Voula has taken out a $300 000 interest-only loan. The interest rate is 4.3% per annum compounding weekly. What are Voula's regular weekly repayments to the nearest cent?

Steps		Working	
		ClassPad ➕ **TI-Nspire**	
a 1	Find I% when the value of the loan stays the same.		
	All compounding periods are the same for interest-only loans.	N	1
	Annual interest rate	I%	4.176
	Present value for a loan is positive.	PV	100 000
	Money moving away from a person is negative.	PMT or Pmt	−348
	FV = −PV for interest-only loans	FV	−100 000
	Same as C/Y or CpY	P/Y or PpY	12
	Number of compounding periods per year	C/Y or CpY	12
2	Round your answer to one decimal place.	The annual interest rate is 4.2%.	
b 1	Find Pmt when the value of the loan stays the same.		
	All compounding periods are the same for interest-only loans.	N	1
	Annual interest rate	I%	4.3
	Present value for a loan is positive.	PV	300 000
	Money moving away from a person is negative.	PMT or Pmt	−248.076...
	FV = −PV for interest-only loans	FV	−300 000
	Same as C/Y or CpY	P/Y or PpY	52
	Number of compounding periods per year	C/Y or CpY	52
2	Round your answer to the nearest cent.	Voula's weekly repayments are $248.08.	

9780170476959

Recap

1 Marina borrows $200\,000$ at a rate of 6% per annum, calculated monthly on the reducing balance. Each month, she repays $1700. If A_n is the amount owing after the nth repayment, a recurrence relation that models the amount owing on this loan is

A $A_0 = 200\,000$, $\quad A_{n+1} = 0.06A_n - 1700$

B $A_0 = 1700$, $\quad A_{n+1} = 0.06A_n - 200\,000$

C $A_0 = 200\,000$, $\quad A_{n+1} = 0.005A_n - 1700$

D $A_0 = 200\,000$, $\quad A_{n+1} = 1.005A_n - 1700$

E $A_0 = 1700$, $\quad A_{n+1} = 1.005A_n - 200\,000$

2 The first two lines of an amortisation table for a reducing balance loan are shown below.

Month	Opening balance	Interest	Repayment	Closing balance
1	2500	50	100	2450
2	2450	49	100	2399
3				

Find the

a monthly rate of interest

b closing balance in the third month.

Mastery

3 [WORKED EXAMPLE 14] Find the interest rate per annum for each of the following reducing balance loans. Answer correct to one decimal place.

a Arthur's loan of $500\,000$ is fully repaid over 20 years with monthly repayments of $3355.27.

b Celeste's loan of $150\,000$ is fully repaid over 12 years with monthly repayments of $1300.34.

c Florence's loan of $250\,000$ is fully repaid over 20 years with quarterly repayments of $5642.56.

d Yaya's loan of $500\,000$ is fully repaid over 25 years with fortnightly repayments of $4045.55.

e Why do you need to enter the payments into the finance solver as a negative?

4 [WORKED EXAMPLE 15] For each of the following, find to the nearest cent, the

   i repayment required to repay the loan in full

   ii total cost of the loan

   iii total interest paid on the loan.

a Fidel borrowed $8000 at 6% per annum for five years with monthly repayments and interest compounding monthly.

b Kim borrowed $300\,000$ at 7.5% per annum for 30 years with monthly repayments and interest compounding monthly.

c Hector borrowed $35\,000$ at 12% per annum for 10 years with quarterly repayments and interest compounding quarterly.

d Marisol borrowed $75\,000$ at 9.25% per annum for 10 years with fortnightly repayments and interest compounding fortnightly.

▶ **5** 🔗 WORKED EXAMPLE 16

    a Chantelle takes out a loan of $58 000 at a rate of 9.5% per annum, compounding monthly, with repayments of $835 per month.

        i How much is still owing on Chantelle's loan after four years, to the nearest cent?

        ii By what percentage has Chantelle reduced the balance of her loan, correct to the nearest percentage?

        iii How many repayments are required in order for her to repay the loan in full?

        iv How much is still owing on Chantelle's loan after 101 repayments, to the nearest cent?

        v Chantelle decides to adjust the value of the 101st repayment so that the loan is fully repaid. What will be the adjusted 101st payment, correct to the nearest cent?

    b Briony takes out a loan of $40 000 at a rate of 11.2% per annum, compounding fortnightly, with repayments of $250 per fortnight.

        i How much is still owing on Briony's loan after three years, correct to the nearest cent?

        ii By what percentage has Briony reduced the balance of her loan, correct to the nearest percentage?

        iii How many repayments are required in order for her to repay the loan in full?

        iv How much is still owing on Briony's loan after 271 repayments, correct to the nearest cent?

        v Briony decides to adjust the value of the 271st repayment so that the loan is fully repaid. What will be the adjusted 271st payment, correct to the nearest cent?

6 🔗 WORKED EXAMPLE 17   Use finance solvers to find each of the following.

    a Senad has a $92 000 interest-only loan compounding weekly with weekly payments of $80. What is the annual interest rate, correct to one decimal place?

    b Cedric has taken out a $268 000 interest-only loan. The interest rate is 5.25% per annum compounding quarterly. What are Cedric's regular quarterly repayments, correct to the nearest cent?

    c Taiko has a $173 000 interest-only loan compounding fortnightly with fortnightly payments of $210. What is the annual interest rate, correct to one decimal place?

    d Luisa has taken out a $500 000 interest-only loan. The interest rate is 3.7% per annum compounding six-monthly. What are Luisa's regular six-monthly repayments, correct to the nearest cent?

Calculator-assumed

7 (3 marks)

    a A loan of $17 500 is to be paid back over four years at an interest rate of 6.25% per annum on a reducing monthly balance. Find the monthly repayment, correct to the nearest cent.   (1 mark)

    b A loan of $300 000 is to be repaid over a period of 20 years. Interest is charged at the rate of 7.25% per annum compounding quarterly. Find the quarterly repayment to the nearest cent.   (1 mark)

    c Swee borrowed $150 000 at 6.2% per annum compounding monthly. The repayments are $1100 per month. Find the balance of the loan at the end of five years, correct to the nearest cent.   (1 mark)

 Exam hack

This loan has compounding and regular repayments, so you should know it's a reducing balance loan even though this isn't mentioned.

▶

8 (4 marks)

   a Joseph borrowed $50 000 to buy a new car. Interest on this loan is charged at the rate of 7.5% per annum, compounding monthly. Joseph will fully repay this loan with 60 monthly repayments over five years. Immediately after the 59th repayment is made, Joseph still owes $995.49. Find the value of his final repayment, to the nearest cent.   (2 marks)

   b Adam has a home loan with a present value of $175 260.56. The interest rate for Adam's loan is 3.72% per annum, compounding monthly. His monthly repayment is $3200. The loan is to be fully repaid after five years. Adam knows that the loan cannot be exactly repaid with 60 repayments of $3200. To solve this problem, Adam will make 59 repayments of $3200. He will then adjust the value of the final repayment so that the loan is fully repaid with the 60th repayment. Find the value of the 60th repayment.   (2 marks)

9 (4 marks) An area of a club needs to be refurbished. $40 000 is borrowed at an interest rate of 7.8% per annum. Interest on the unpaid balance is charged to the loan account monthly. Suppose the $40 000 loan is to be fully repaid in equal monthly instalments over five years.

   a Determine the monthly payment, correct to the nearest cent.   (1 mark)

   b If, instead, the monthly payment was $1000, how many months will it take to fully repay the $40 000?   (1 mark)

   c Suppose no payments are made on the loan in the first 12 months.

     i What will be the balance of the loan account after the first 12 months, correct to the nearest dollar?   (1 mark)

     ii After the first 12 months, only the interest on the loan is paid each month. Determine the monthly interest payment, correct to the nearest cent.   (1 mark)

10 (2 marks) Samuel took out a reducing balance loan. The interest rate for this loan was 4.1% per annum, compounding monthly. The balance of the loan after four years of monthly repayments was $329 587.25. The balance of the loan after seven years of monthly repayments was $280 875.15. Samuel will continue to make the same monthly repayment. To ensure the loan is fully repaid, to the nearest cent, the required final repayment will be lower. In the first seven years, Samuel made 84 monthly repayments. From this point on, how many more monthly repayments will Samuel make to repay the loan in full?

7.6 Changing the terms of reducing balance loans

The terms of a reducing balance loan can change mid-loan. Not all interest rates are fixed. Many types of loans have interest rates that increase or decrease due to economic factors. A borrower can also negotiate the frequency of their repayments or the amount of the repayment. Changes to terms have an effect on the length of a loan, the total interest paid, and the total cost of the loan. Finance solvers can be used to deal with these sorts of changes.

WORKED EXAMPLE 18	Changing the length of the loan

Allison wants to buy a house and has borrowed $225 000 at an interest rate of 4.23% per annum, fixed for 10 years. Interest is calculated monthly and monthly repayments are set at $1279. After 10 years, Allison renegotiates the conditions for the balance of her loan. The new interest rate will be 4.05% per annum. She will pay $1620 per month. How many years will it take her to repay the loan in full?

Steps	Working
	ClassPad + **TI-Nspire**
1 Enter values into finance solver and find FV.	
Total number of compounding periods = 10 × 12 = 120 months for first part of loan.	N 120
Annual interest rate for first part of loan.	I% 4.23
Present value for a loan is positive.	PV 225 000
Money moving away from a person is negative.	PMT or Pmt −1279
Future value has the opposite sign to present value.	FV −152 580.600…
Same as C/Y or CpY	P/Y or PpY 12
Number of compounding periods per year	C/Y or CpY 12
2 Use the unrounded FV value as the positive PV value to find N for second part of loan.	
Total number of compounding periods	N 113.537 59…
Annual interest rate for second part of loan.	I% 4.05
Present value for a loan is positive.	PV 152 580.600 443 51
Money moving away from a person is negative.	PMT or Pmt −1620
Future value is zero when a loan is fully repaid.	FV 0
Same as C/Y or CpY	P/Y or PpY 12
Number of compounding periods per year	C/Y or CpY 12
3 Always round up your answer for N. Convert to years.	The length of the renegotiated part of the loan is 114 months. $$\frac{114}{12} = 9.5 \text{ years}$$
4 Add the lengths of the two parts of the loan.	10 + 9.5 = 19.5 It will take $19\frac{1}{2}$ years for Allison to repay the loan in full.

WORKED EXAMPLE 19 | Changing the interest rate

Yuki borrows $100 000 at 8% per annum compounding monthly, to be repaid over 10 years. The repayments for the first five years are $1213.28 each month. After five years, the interest rate is reduced to 7.5% per annum.

a What is the new repayment, to the nearest cent, required for Yuki to repay the loan?

b How much total interest, to the nearest cent, has Yuki paid over the 10 years?

Steps	Working
	ClassPad ➕ **TI-Nspire**
a 1 Enter values into finance solver and find FV.	
Total number of compounding periods = 5 × 12 = 60 months for first part of loan	N 60
Annual interest rate	I% 8
Present value for a loan is positive.	PV 100 000
Money moving away from a person is negative.	PMT or Pmt –1213.28
Future value has the opposite sign to present value.	FV –59 836.570...
Same as C/Y or CpY	P/Y or PpY 12
Number of compounding periods per year	C/Y or CpY 12
2 Use the unrounded FV as the positive PV for the second part of the loan and find PMT.	
Total number of compounding periods	N 60
Annual interest rate	I% 7.5
Present value for a loan is positive.	PV 59 836.570 684 934
Money moving away from a person is negative.	PMT or Pmt –1199.00
Future value is zero when a loan is fully repaid.	FV 0
Same as C/Y or CpY	P/Y or PpY 12
Number of compounding periods per year	C/Y or CpY 12
3 Write your answer to the nearest cent.	Over the last five years, the monthly repayments required to repay the loan are $1199.00.
b 1 Use the formula for both parts of the loan: Total interest paid = N × PMT – (PV – FV) Ignore the negative sign in values.	Total interest paid for first five years = 60 × 1213.28 – (100 000 – 59 836.57) = 72 796.80 – 40 163.43 = 32 633.37
	Total interest paid for second five years = 60 × 1199.00 – (59 836.57 – 0) = 71 940.00 – 59 836.57 = 12 103.43
2 Add the two totals.	Total interest over ten years = 32 633.37 + 12 103.43 = $44 736.80

WORKED EXAMPLE 20 | Changing the repayments

Ryan has taken out a loan of $15 000 at the rate of 10.4% per annum, compounding monthly and with regular monthly payments, to pay for an overseas trip.

a Ryan will make interest-only repayments for the first two years of this loan. How much is each interest-only repayment to the nearest dollar?

b For the next two years, Ryan will increase his monthly repayments so that the balance of the loan is $8320. What are Ryan's repayments, to the nearest cent, each month during these two years?

c Ryan will fully repay the outstanding balance of $8320 over the next three years. The first 35 monthly repayments will each be $270. The 36th repayment will have a different value to ensure the loan is repaid exactly to the nearest cent. What is the value of the 36th repayment, rounded to the nearest cent?

Steps		Working	
		ClassPad + **TI-Nspire**	
a 1 Find PMT when the value of the loan stays the same.			
	All compounding periods are the same for interest-only loans.	**N**	1
	Annual interest rate	**I%**	10.4
	Present value for a loan is positive.	**PV**	15 000
	Money moving away from a person is negative.	**PMT or Pmt**	−130
	FV = −PV for interest-only loans	**FV**	−15 000
	Same as C/Y or CpY	**P/Y or PpY**	12
	Number of compounding periods per year	**C/Y or CpY**	12
2	Write your answer to the nearest dollar.	Each interest-only repayment is $130.	
b 1 Find PMT.			
	Total number of compounding periods = 2 × 12 = 24 months	**N**	24
	Annual interest rate	**I%**	10.4
	Present value for a loan is positive.	**PV**	15 000
	Money moving away from a person is negative.	**PMT or Pmt**	−381.5894…
	Future value has the opposite sign to present value.	**FV**	−8320
	Same as C/Y or CpY	**P/Y or PpY**	12
	Number of compounding periods per year	**C/Y or CpY**	12
2	Write your answer to the nearest cent.	The repayments are $381.59.	
c 1 Find FV.			
	Total number of compounding periods = 3 × 12 = 36 months	**N**	36
	Annual interest rate	**I%**	10.4
	Present value for a loan is positive.	**PV**	8320
	Money moving away from a person is negative.	**PMT or Pmt**	−270
	Future value has the opposite sign to present value.	**FV**	−1.184 847…
	Same as C/Y or CpY	**P/Y or PpY**	12
	Number of compounding periods per year	**C/Y or CpY**	12
2	For the last payment, add the amount still owing on the loan to the regular payment, rounding to the nearest cent.	$270 + $1.18 = $271.18 The adjusted value of the 36th repayment to exactly pay the loan is $271.18.	

Recap

1 A loan of $300 000 is taken out to finance a new business venture. The loan is to be repaid fully over 20 years with quarterly payments of $6727.80. Interest is calculated quarterly on the reducing balance. The annual interest rate for this loan is closest to

A 4.1%   **B** 6.5%   **C** 7.3%   **D** 19.5%   **E** 26.7%

2 Rho takes a 20-year loan of $172 000 at 6% per annum, compounding monthly and with monthly repayments. To repay the loan in full in 20 years, the amount he must repay each month is

A $716.67   **B** $1216.54   **C** $1232.26   **D** $9058.63   **E** $10 320.00

Mastery

3 WORKED EXAMPLE 18 Indira has borrowed $190 000 to buy a house and she makes monthly repayments of $1470. The interest rate of 4.1% per annum is fixed for eight years and is calculated monthly. After eight years, Indira renegotiates the conditions for the balance of her loan. The new interest rate will be 3.9% per annum and she will pay $1530 per month. How many years will it take her to repay the loan in full?

4 WORKED EXAMPLE 19 Jodie borrows $130 000 at 6% per annum compounding monthly, to be repaid over 14 years. The repayments for the first seven years are $1352.73 each month. After seven years, the interest rate is reduced to 5.5% per annum.

  a What are the new monthly repayments, to the nearest cent, required for Jodie to pay out the loan?

  b How much total interest, to the nearest cent, has Jodie paid over the fourteen years?

5 WORKED EXAMPLE 20 Saverio has taken out a loan of $19 000 at the rate of 10.2% per annum, compounding monthly and with regular monthly payments, to pay for an overseas trip.

  a Saverio will make interest-only repayments for the first two years of this loan. How much is each interest-only repayment to the nearest cent?

  b For the next three years, Saverio will increase his monthly repayments so that the balance of the loan is $10 200. What are Saverio's repayments, to the nearest cent, each month during these three years?

  c Saverio will fully repay the outstanding balance of $10 200 over the next two years. The first 23 monthly repayments will each be $462. The 24th repayment will have a different value to ensure the loan is repaid exactly to the nearest cent. What is the value of the 24th repayment, rounded to the nearest cent?

Calculator-assumed

6 (6 marks) Xavier borrowed $245 000 to pay for a house. For the first 10 years of the loan, the interest rate was 4.35% per annum, compounding monthly. Xavier made monthly repayments of $1800. After 10 years, the interest rate changed. If Xavier now makes monthly repayments of $2000, he could repay the loan in a further five years. Find the new annual interest rate for Xavier's loan, correct to one decimal place.

7 (6 marks) To purchase a house Sam has borrowed $250 000 at an interest rate of 4.45% per annum, fixed for 10 years. Interest is calculated monthly on the reducing balance of the loan. Monthly repayments are set at $1382.50. After 10 years, Sam renegotiates the conditions for the balance of his loan. The new interest rate will be 4.25% per annum. He will pay $1750 per month. Find the total time it will take him to pay out the loan fully.

8 © SCSA MA2020 Q11 (9 marks) Shari requires a loan of $325 000 for the purchase of a new house. She wishes to make two equally-spaced repayments of $700 each month.

Shari is offered a choice of two loan options for the first three years, both of which have interest calculated daily.

Option 1: An introductory compound interest rate of 2.55% per annum for the first year which changes to 2.99% per annum for the next two years.

Option 2: A compound interest rate of 2.85% per annum fixed for the first three years.

a Describe briefly the benefit of making two repayments of $700 each month instead of one repayment of $1400 at the end of each month. (1 mark)

b For Option 1, calculate

  i the loan balance at the end of the first year (3 marks)

  ii the loan balance at the end of the third year. (2 marks)

c Determine which option gives the best result for Shari after three years and by how much. (3 marks)

9 (3 marks) Khan decides to extend his home office and borrows $30 000 for building costs. Interest is charged on the loan at a rate of 9% per annum compounding monthly. Assume Khan will pay only the interest on the loan at the end of each month.

a Calculate the amount of interest he will pay each month. (1 mark)

Suppose the interest rate remains at 9% per annum compounding monthly and Khan pays $400 each month for five years.

b Determine the amount of the loan that is outstanding at the end of five years. Write your answer correct to the nearest dollar. (1 mark)

Khan decides to repay the $30 000 loan fully in equal monthly instalments over five years. The interest rate is 9% per annum compounding monthly.

c Determine the amount of each monthly instalment. Write your answer correct to the nearest cent. (1 mark)

10 (3 marks) Andrew borrowed $10 000 to pay for a holiday and other expenses. Interest on this loan will be charged at the rate of 12.9% per annum, compounding monthly. Immediately after the interest has been calculated and charged each month, Andrew will make a repayment.

a For the first year of this loan, Andrew will make interest-only repayments each month. What is the value of each interest-only repayment? (1 mark)

b For the next three years of this loan, Andrew will make equal monthly repayments. After these three years, the balance of Andrew's loan will be $3776.15. What amount, in dollars, will Andrew repay each month during these three years? (1 mark)

c Andrew will fully repay the outstanding balance of $3776.15 with a further 12 monthly repayments. The first 11 repayments will each be $330. The twelfth repayment will have a different value to ensure the loan is repaid exactly to the nearest cent. What is the value of the twelfth repayment? Round your answer to the nearest cent. (1 mark)

11 (3 marks) Tania takes out a reducing balance loan of $265 000 to pay for her house. Her monthly repayments will be $1980. Interest on the loan will be calculated and paid monthly at the rate of 7.62% per annum.

a **i** How many monthly repayments are required to repay the loan? Write your answer to the nearest month. (1 mark)

  ii Determine the amount that is paid off the principal of this loan in the first year. Write your answer to the nearest cent. (1 mark)

Immediately after Tania made her twelfth payment, the interest rate on her loan increased to 8.2% per annum, compounding monthly. Tania decided to increase her monthly repayment so that the loan would be repaid in a further nineteen years.

b Determine the new monthly repayment. Write your answer to the nearest cent. (1 mark)

12 (4 marks) Ken has borrowed \$70 000 to buy a new caravan. He will be charged interest at the rate of 6.9% per annum, compounding monthly.

a For the first year (12 months), Ken will make monthly repayments of \$800.

   i Find the amount that Ken will owe on his loan after he has made 12 repayments. (1 mark)

   ii What is the total interest that Ken will have paid after 12 repayments? (1 mark)

b After another two years, Ken will make a lump sum payment of \$L in order to reduce the balance of his loan. This lump sum payment will ensure that Ken's loan is fully repaid in a further three years. Ken's repayment amount remains at \$800 per month and the interest rate remains at 6.9% per annum, compounding monthly. What is the value of Ken's lump sum payment, \$L? Round your answer to the nearest dollar. (2 marks)

(7.7) Annuities

Annuity recurrence relations

An **annuity** is a compound interest investment with regular payments to the investor called withdrawals. It is exactly the same as a reducing balance loan except the bank is borrowing from the person rather than the person borrowing from the bank.

The recurrence relation for an annuity is the same as the recurrence relation for a reducing balance loan.

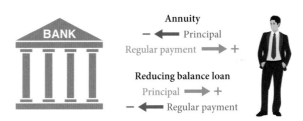

Video playlist
Annuities

Worksheet
Annuity problems

Annuity recurrence relation

The recurrence relation for the value V_{n+1} of an annuity

$$V_0 = \text{principal}, \quad V_{n+1} = \left(1 + \frac{r}{100}\right)V_n - d$$

where

   r = percentage interest rate per compounding period

   n = number of compounding periods

   d = payment made per compounding period

   V_n = closing balance of the annuity after n compounding periods.

The graph of an annuity recurrence relation will look like this.

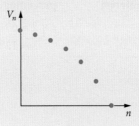

Serena plans to invest $40 000 in an annuity at a rate of 8.4% per annum compounding quarterly where she makes regular quarterly withdrawals of $300.

a Write a recurrence relation for the balance.

Serena changes her mind and decides to invest the $40 000 in an annuity compounding six-monthly with six-monthly withdrawals that has the recurrence relation

$$V_0 = 40\,000, \quad V_{n+1} = 1.06V_n - 300$$

b What is the annual interest rate of this investment?

c How much has the investment increased between the third and fourth six-monthly periods, to the nearest cent?

Steps	Working
a 1 Find the number of compounding periods per year.	There are four compounding periods per year.
2 Identify V_n, V_0, r and d.	Let V_n = the balance after n compounding periods. $$V_0 = 40\,000, r = \frac{8.4}{4} = 2.1, d = 300$$
3 Substitute the values into these formulas and simplify. V_0 = principal $$V_{n+1} = \left(1 + \frac{r}{100}\right)V_n - d$$	$$V_{n+1} = \left(1 + \frac{2.1}{100}\right)V_n - 300$$ $$= (1 + 0.021)V_n - 300$$ $$= 1.021V_n - 300$$ $$V_0 = 40\,000, \quad V_{n+1} = 1.021V_n - 300$$
b 1 Use the recurrence relation to find an equation for r, the percentage interest rate per compounding period.	Comparing $$V_{n+1} = \left(1 + \frac{r}{100}\right)V_n - d$$ $$V_{n+1} = 1.06V_n - 300$$ we can see that $$1 + \frac{r}{100} = 1.06$$
2 Solve for r, using CAS if necessary.	$$\frac{r}{100} = 1.06 - 1$$ $$\frac{r}{100} = 0.06$$ $$r = 6\% \text{ per six months.}$$
3 Multiply r by the compounding period to find the annual percentage compound interest rate.	The interest compounds six-monthly, so the annual percentage compound interest rate is $$r \times 2 = 6\% \times 2$$ $$= 12\%$$

c Use CAS recursive computation to find the balances, and subtract to find the difference, rounding to the nearest cent.

$$V_4 - V_3 = 49\,186.6936 - 46\,685.56$$
$$= 2501.1336$$

The investment increased by \$2501.13.

ClassPad

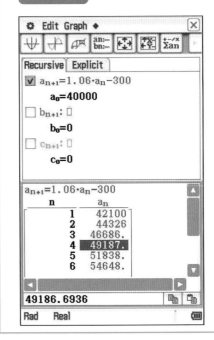

TI-Nspire

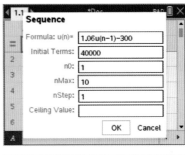

Annuity tables

Annuity tables work the same way for annuities as for reducing balance loans.

Annuity table

$$r = \frac{\text{interest}}{\text{previous balance}} \times 100$$

Period number	Opening balance	Investment gain	Payment	Closing balance
1	V_0	$V_0 \times \dfrac{r}{100}$	d	$V_1 = V_0 + V_0 \times \dfrac{r}{100} - d$
2	V_1	$V_1 \times \dfrac{r}{100}$	d	$V_2 = V_1 + V_1 \times \dfrac{r}{100} - d$

where n = payment number

d = payment made per compounding period

r = percentage interest rate per compounding period

V_0 = principal

V_n = closing balance of the annuity after n compounding periods.

Analysing annuity amortisation tables

The following amortisation table shows the first three payments of an annuity with monthly compounding interest and monthly withdrawals.

Month	Opening balance	Interest	Payment	Closing balance
1	450 000.00	2250.00	3797.00	448 453.00
2	448 453.00	2242.27	3797.00	446 898.27
3	446 898.27	2234.49	3797.00	445 335.76

Use the table to find the

a amount invested

b regular monthly withdrawal

c interest rate per compounding period

d interest rate per annum

e interest paid in the third month

f amount by which the principal has been reduced in the third month

g balance after one withdrawal.

Steps	Working
a Read the principal from the table.	amount invested = $450 000.00
b Read the regular payment from the table.	regular monthly withdrawal = $3797.00
c $r = \dfrac{\text{interest}}{\text{previous balance}} \times 100$ Choose the option with the easiest calculations.	$r = \dfrac{2250.00}{450000.00} \times 100 = 0.5$ Interest rate per compounding period is 0.5%
d Multiply r by the compounding period.	interest rate per annum = 12 × 0.5% = 6%
e Read the interest from the table.	interest for month 3 = $2234.49
f Principal reduction = opening balance − closing balance	principal reduction in the 3rd month = 446 898.27 − 445 335.76 = $1562.51
g Read the closing balance from the table.	balance after one withdrawal = $448 453

 9780170476959

Annuities and finance solvers

Annuities can be analysed using finance solvers in a similar way to reducing balance loans. The difference is:

- an annuity is an investment, not a loan, so the PV is always negative
- PMT is always positive because the payments are coming to the person.

7.7

Using finance solvers for annuities

N	**Total number** of compounding periods
I%	**Interest** rate per year
PV	**Present value** for an annuity is **negative** because the money is going away from the person.
PMT or Pmt	Regular **payments** for an annuity are **positive** because the money is coming to the person.
FV	**Future value** has the opposite sign of the present value so it will be **positive**. A future value of **zero** means the investment has been fully paid out.
P/Y or PpY	Number of **payments per year**.
C/Y or CpY	Number of **compounding** periods per year

When solving for N, always round *up*, never down, to the nearest whole number.

Ignore the negative sign in values when using the following formula:

total interest paid = N × PMT − (PV − FV)

WORKED EXAMPLE 23 Calculating how long an annuity will last

Roger purchases a $400 000 annuity. Interest is paid at 8% per annum compounding monthly. If he receives monthly payments of $3500, how many years will the annuity last?

Steps	Working
	ClassPad ✛ **TI-Nspire**
1 Find N when FV = 0.	
Total number of compounding periods	N 215.9794...
Annual interest rate	I% 8
Present value for an investment is negative.	PV −400 000
Money moving to a person is positive.	PMT or Pmt 3500
Future value is zero when an investment is fully paid out.	FV 0
Same as C/Y or CpY	P/Y or PpY 12
Number of compounding periods per year	C/Y or CpY 12
2 Always round N *up* to the next whole number.	The annuity will last for **216** months.
3 Convert to years and answer the question.	216 months = $\dfrac{216}{12}$ = 18 years The annuity will last 18 years.

WORKED EXAMPLE 24 Calculating how much to withdraw from an annuity

Caroline invests $350 000 in an annuity with interest paid at 7.5% per annum compounding monthly. She receives monthly payments from this investment. What monthly payment will she receive, to the nearest cent, if she wishes to receive payments for 15 years?

Steps	Working	
	ClassPad ╋ **TI-Nspire**	
1 Find PMT, given		
Total number of compounding periods = 15 × 12 = 180 months	**N**	180
Annual interest rate	**I%**	7.5
Present value for an investment is negative.	**PV**	−350 000
Money moving to a person is positive.	**PMT or Pmt**	3244.543…
Future value is zero when an investment is fully paid out.	**FV**	0
Same as C/Y or CpY	**P/Y or PpY**	12
Number of compounding periods per year	**C/Y or CpY**	12
2 Round your answer to the nearest cent.	The amount of each monthly payment is $3244.54.	

EXERCISE 7.7 Annuities

ANSWERS p. 448

Recap

1 Indira borrowed $29 000 to buy a car and was charged interest at the rate of 12.5% per annum, compounding monthly. For the first year of the loan, Indira made monthly repayments of $425. For the second year of the loan, Indira made monthly repayments of $500. The total amount of interest that Indira paid over this two-year period is closest to

    A $2500   **B** $4300   **C** $5900   **D** $6800   **E** $7700

2 Xavier borrows $45 000 from the bank to buy a car. He is offered a reducing balance loan for three years with an interest rate of 9.75% per annum, compounding monthly. He can repay this loan by making 36 equal monthly payments. Instead, Xavier decides to repay the loan in 18 equal monthly payments. If there are no penalties for repaying the loan early, the amount he will save is closest to

    A $2697   **B** $3530   **C** $3553   **D** $6581   **E** $7083

Mastery

3   WORKED EXAMPLE 21   Ashleigh plans to invest $45 000 in an annuity at a rate of 13% per annum compounding fortnightly where she makes regular fortnightly withdrawals of $250.

    a Write a recurrence relation for the balance.

Ashleigh changes her mind and decides to invest the $45 000 in an annuity compounding weekly with weekly withdrawals that has the recurrence relation

$$V_0 = 45\,000, \quad V_{n+1} = 1.002V_n - 25$$

    b What is the annual interest rate of this investment?

    c How much has the investment increased between the third and fourth week, correct to the nearest cent?

▶

4 ⟨⟨⟩⟩ WORKED EXAMPLE 22 The following table shows the first three payments of an annuity with monthly compounding interest and monthly withdrawals.

Month	Opening balance	Interest	Payment	Closing balance
1	380 000.00	2850.00	3854.00	378 996.00
2	378 996.00	2842.47	3854.00	377 984.47
3	377 984.47	2834.88	3854.00	376 965.35

Use the table to find the

a amount invested

b regular monthly withdrawal

c interest rate per compounding period

d interest rate per annum

e interest paid in the third month

f amount by which the principal has been reduced in the first month

g balance after three withdrawals.

5 ⟨⟨⟩⟩ WORKED EXAMPLE 23 Ashton purchases a $500 000 annuity. Interest is paid at 7% per annum compounding monthly. If he receives monthly payments of $4200, for how many years will the annuity last?

6 ⟨⟨⟩⟩ WORKED EXAMPLE 24 Anton invests $450 000 in an annuity with interest paid at 6.5% per annum compounding monthly. He receives monthly payments from this investment. What will be the monthly payment, to the nearest cent, if he wishes to receive payments for 20 years?

Calculator-assumed

7 (4 marks) Find the following.

a Cheryl invested $175 000 in an annuity. This investment earns interest at the rate of 4.8% per annum, compounding quarterly. Immediately after the interest has been added to the account each quarter, Cheryl withdraws a payment of $3500. Find a recurrence relation that can be used to determine the value of Cheryl's investment after n quarters. (2 marks)

b The value of an annuity, V_n, after n monthly payments of $555 have been made, can be determined using the recurrence relation

$$V_0 = 100\,000, \quad V_{n+1} = 1.0025V_n - 555.$$

Find the value of the annuity after five payments have been made. (2 marks)

8 (6 marks) Kim invests $400 000 in an annuity paying 3.2% interest per annum. The annuity is designed to give her an annual payment of $47 372 for 10 years. The table for this annuity is shown below. Some of the information is missing.

Year	Opening balance	Interest	Payment	Closing balance
1	400 000.00	12 800.00	47 372.00	
2		11 693.70	47 372.00	329 749.70
3	329 749.70	10 551.99	47 372.00	292 929.69
4	292 929.69	9373.75	47 372.00	254 931.44
5	254 931.44	8157.81	47 372.00	215 717.24
6	215 717.24	6902.95	47 372.00	175 248.19
7	175 248.19	5607.94	47 372.00	133 484.14
8	133 484.14			90 383.63
9	90 383.63	2 892.28	47 372.00	45 903.90
10	45 903.90	1 468.92	47 372.00	0.83

a Find the balance of the annuity after one payment has been made. (2 marks)

b Find the reduction in the principal of the annuity in year 5. (2 marks)

c Find the amount of payment number 8 that is the interest earned. (2 marks)

9 (6 marks) Ekaterina invests $200 000 at a rate of 9% p.a. compounding monthly. Each month, after interest is paid, she withdraws an income of $2000. The value of her investment, V_n, at the end of month n is given by a recurrence relation of the form

$$V_0 = a, \quad V_{n+1} = RV_n + d$$

a Find the values of a, R and d. (1 mark)

b For how many months will Ekaterina receive an income from this investment? (1 mark)

c If she withdrew $3000 a month

   i how long would it last? (1 mark)

   ii what is the value of the final payment? (1 mark)

d If the interest rate was 8% and her monthly withdrawals were $2000

   i how long would it last? (1 mark)

   ii what is the value of the final payment? (1 mark)

10 (3 marks) A record producer gave the band $50 000 to write and record an album of songs. This $50 000 was invested in an annuity that provides a monthly payment to the band. The annuity pays interest at the rate of 3.12% per annum, compounding monthly. After six months of writing and recording, the band has $32 667.68 remaining in the annuity.

a What is the value, in dollars, of the monthly payment to the band? (1 mark)

b After six months of writing and recording, the band decided that it needs more time to finish the album. To extend the time that the annuity will last, the band will work for three more months without withdrawing a payment. After this, the band will receive monthly payments of $3800 for as long as possible. The annuity will end with one final monthly payment that will be smaller than all of the others. Calculate the **total** number of months that this annuity will last. (2 marks)

9780170476959

7.8 Perpetuities

A **perpetuity** (or **perpetuity investment**) is a type of annuity where the payments are *exactly* equal to the interest for each compounding period. This means the balance doesn't change and investment provides regular payments that continue forever. Often scholarship funds are set up as perpetuities. Perpetuities are investment versions of interest-only loans.

Perpetuity recurrence relations and formula

The perpetuity recurrence relations and formula work the same way as the ones for interest-only loans.

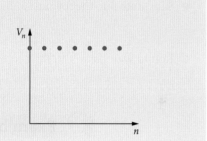

Video playlist
Perpetuities

Puzzle
Perpetuities

Perpetuity

For a perpetuity with value $V_n = V_0$ after n compounding periods:

$$d = \frac{r}{100} \times V_0$$

where

d = payment made per compounding period

r = percentage interest rate per compounding period

V_0 = principal.

The graph of perpetuity will look like this.

WORKED EXAMPLE 25 | Working with perpetuities

a Marina has an investment whose recurrence relation for V_n, the value of the investment after n compounding periods, is

$$V_0 = 24\,000, \quad V_{n+1} = 1.007V_n - 168$$

Find V_1 and V_2 and explain why this shows that this is a perpetuity.

b Lucy wishes to set up a scholarship fund in her name so that each year an amount of $5000 is awarded to a promising cross-country runner at the school she attended. If interest on the investment is 4% per year, compounded annually, how much should she invest in this perpetuity?

c Jeremiah has $500\,000 to set up a perpetuity for his granddaughter Sophie. He invests the money in bonds that return 3.5% per annum compounding quarterly. Use the perpetuity formula to find how much Sophie will receive each quarter from this investment, rounding to the nearest cent.

Steps	Working
a 1 Find V_1.	$V_1 = 1.007V_0 - 168$ $= 1.007 \times 24\,000 - 168$ $= 24\,000$
2 Find V_2.	$V_2 = 1.007V_1 - 168$ $= 1.007 \times 24\,000 - 168$ $= 24\,000$
3 For perpetuities $V_n = V_0$.	The value of the investment stays at the principal value $24\,000 for all compounding periods.

b 1 Identify what we know and what we need to find from the perpetuity formula.

$$r = \frac{4}{1} = 4, d = 5000, V_0 = ?$$

$$d = \frac{r}{100} \times V_0$$

2 Substitute into the formula and solve, using CAS if necessary.

$$5000 = \frac{4}{100} \times V_0$$

$$V_0 = \frac{5000 \times 100}{4}$$

$$= 125\,000$$

3 Write the answer.

Lucy should invest $125 000.

c 1 Identify what we know and what we need to find from the perpetuity formula.

$$r = \frac{3.5}{4} = 0.875, V_0 = 500\,000, d = ?$$

$$d = \frac{r}{100} \times V_0$$

2 Substitute into the formula and solve.

$$d = \frac{0.875}{100} \times 500\,000$$

$$= 4375$$

3 Round your answer to the nearest cent.

Sophie will receive $4375.00 quarterly from this investment.

> **🔒 Exam hack**
>
> Sometimes using the formula is quicker than using a finance solver.

Perpetuities and finance solvers

We can use finance solvers to answer questions about perpetuities in a similar way to how we used them for interest-only loans.

> **Using finance solvers for perpetuities**
>
> For a perpetuity
> - all compounding periods are the same so N = 1
> - FV and PV have the same value but opposite signs so FV = −PV.

WORKED EXAMPLE 26 | Using finance solvers for perpetuities

Use finance solvers to find each of the following.

a A magazine owner has $350 000 to invest in a perpetuity. The interest earned from this perpetuity will provide an annual literary prize of $10 000. What annual interest rate, correct to one decimal place, would be required for this investment?

b Jeremiah has $500 000 to set up a perpetuity for his granddaughter Sophie. He invests the money in bonds that return 3.5% per annum compounding quarterly. Use a finance solver to find how much Sophie will receive each quarter from this investment, rounding to the nearest cent.

Steps	Working
	ClassPad + **TI-Nspire**

a 1 Find I when the value of the investment stays the same.

All compounding periods are the same for perpetuities.	N	1
Annual interest rate	I%	2.857...
Present value for an investment is negative.	PV	−350 000
Money moving to a person is positive.	PMT or Pmt	10 000
FV = −PV for perpetuities	FV	350 000
Same as C/Y or CpY	P/Y or PpY	1
Number of compounding periods per year	C/Y or CpY	1

  2 Round your answer to one decimal place.   The annual interest rate is 2.9%.

b 1 Find PMT when the value of the investment stays the same.

All compounding periods are the same for perpetuities.	N	1
Annual interest rate	I%	3.5
Present value for an investment is negative.	PV	−500 000
Money moving to a person is positive.	PMT or Pmt	4375
FV = −PV for perpetuities	FV	500 000
Same as C/Y or CpY	P/Y or PpY	4
Number of compounding periods per year	C/Y or CpY	4

  2 Round your answer to the nearest cent.   Sophie will receive $4375.00 quarterly from this investment.

EXERCISE 7.8 Perpetuities

Recap

1 The following amortisation table shows the first payment of an annuity with quarterly compounding interest and quarterly withdrawals.

Quarter	Opening balance	Interest	Payment	Closing balance
1	60 000.00	720.00	1630.00	59 090.00
2				

Which of the following is closest to the annual interest rate of the annuity?

A 1.2%   **B** 4.8%   **C** 0.012%   **D** 0.048%   **E** 0.44%

2 Vusa has invested $420 000 in an annuity that pays interest at the rate of 3.6% per annum, compounding monthly. After the interest has been added each month, Vusa immediately receives a payment from the annuity. The value of Vusa's investment is $372 934.71 after three years. The monthly payment that Vusa receives from the annuity is closest to

A $1260   **B** $1310   **C** $2500   **D** $15 120   **E** $16 900

3 🔗 WORKED EXAMPLE 25

a The recurrence relation for an investment, where V_n is the value of the investment after n compounding periods, is

$$V_0 = 31\,000, \quad V_{n+1} = 1.006V_n - 186$$

Find V_1 and V_2 and explain why this shows that the investment is a perpetuity.

b Emil sets up a scholarship fund in his name so that each year an amount of $3000 is awarded to the top mathematics student in their final year at the school he attended. If interest on the investment is 5.6% per annum, compounded yearly, how much should he invest in this perpetuity, rounded to the nearest cent?

c Fleur invests $650 000 in a perpetuity that returns 4.2% per annum compounding six-monthly. How much will Fleur receive each six months from this investment, rounded to the nearest dollar?

4 🔗 WORKED EXAMPLE 26 Use finance solvers to find each of the following.

a A performing arts school has $400 000 to invest in a perpetuity. The interest earned from this perpetuity will provide an annual scholarship of $25 000. What annual interest rate, correct to one decimal place, would be required for this investment?

b Felicity has $330 000 in a perpetuity for her grandchildren, with an interest rate of 4.91% per annum compounding half yearly. Using a finance solver, find how much the grandchildren receive every six months from this investment, rounded to the nearest dollar.

c A billionaire has decided to use $2 000 000 to set up a perpetuity so that her only child will receive $14 500 a month for life. What annual interest rate, correct to one decimal place, would be required for this investment?

d A college awards a prize each year to the top student. A donation from a former student of $866 000 is used as a perpetuity to fund the prize. The investment has an interest rate of 5.23% per annum, compounding yearly. How much is the prize, rounded to the nearest dollar?

Calculator-assumed

5 (6 marks)

a Amir invested some money in a perpetuity from which he receives a monthly payment of $525. The perpetuity pays interest at an annual rate of 4.2%, paid monthly. How much money did Amir invest in the perpetuity? (2 marks)

b A music school has $80 000 to invest in a perpetuity. The interest earned from this perpetuity will provide an annual prize of $3000 to a talented musician from the school. What annual interest rate would be required for this investment? (2 marks)

c Jane invests in an ordinary perpetuity to provide her with a weekly payment of $500. The interest rate for the investment is 5.9% per annum. Assuming there are 52 weeks per year, find the amount, correct to the nearest dollar, that Jane needs to invest in the perpetuity. (2 marks)

6 (2 marks) A sponsor of a cricket club has invested $20 000 in a perpetuity. The annual interest from this perpetuity is $750. The interest from the perpetuity is given to the best player in the club every year, for a period of 10 years.

    a What is the annual rate of interest for this perpetuity investment? (1 mark)

    b After 10 years, how much money is still invested in the perpetuity? (1 mark)

7 © SCSA | MA2017 Q8 | (6 marks) Ming, a former high school student and now a successful business owner, wishes to set up a perpetuity of $6000 per year to be paid to a deserving student from her school. The perpetuity is to be paid at the start of the year in one single payment.

    a A financial institution has agreed to maintain an account for this perpetuity paying a fixed rate of 5.9% p.a. compounded monthly.

       Show that an amount of $98 974, to the nearest dollar, is required to maintain this perpetuity. (3 marks)

    b Ming allows herself five years to accumulate the required $98 974 by making regular quarterly payments into an account paying 5.4% p.a. compounded monthly.

       Determine the quarterly payment needed to reach the required amount after five years if Ming starts the account with an initial deposit of $1000. (3 marks)

> 🔒 **Exam hack**
>
> When using the finance solver in questions **7 b** and **9 b**, P/Y is not equal to C/Y.

8 (3 marks) Arthur invested $80 000 in a perpetuity that returns $1260 per quarter. Interest is calculated quarterly.

    a Calculate the annual interest rate of Arthur's investment. (1 mark)

    b After Arthur has received 20 quarterly payments, how much money remains invested in the perpetuity? (1 mark)

    c Arthur's wife, Martha, invested a sum of money at an interest rate of 9.4% per annum, compounding quarterly. She will be paid $1260 per quarter from her investment. After 10 years, the balance of Martha's investment will have reduced to $7000. Determine the initial sum of money Martha invested. Write your answer correct to the nearest dollar. (1 mark)

9 © SCSA | MA2021 Q14 | (11 marks) Patrick has retired and invested his lump sum superannuation payout of $717 850 at a rate of 5.7% per annum compounded monthly. He begins the investment strategy from 1 January.

    a Patrick will receive $4500 at the end of each month for general living expenses and will also receive a further $4000 at the end of each year for an annual holiday.

       i Identify this type of investment account. (1 mark)

       ii Determine the balance in the account at the end of the first year. (4 marks)

       iii Determine the balance in the account at the end of the second year. (3 marks)

    b When Patrick retired, he also considered the option of setting up a perpetuity with his superannuation payout still at 5.7% per annum compounded monthly. Calculate the quarterly payments Patrick would have received with this perpetuity in place. (3 marks)

10 © SCSA | MA2020 Q9 | (11 marks) Giuseppe wishes to set up an annuity. He is told that an annuity with quarterly investment returns and quarterly payments is modelled by the recursive rule

$$A_{n+1} = A_n \times 1.019 - P, \ A_0 = Q$$

with the values of P and Q consistent with the spreadsheet below.

	A	B	C	D	E
1	Quarter	Opening balance	Investment gain	Payment	Closing balance
2	1	$648 000	$12 312	$15 000	X
3	2		Y	$15 000	
4	3				

a Determine the values of P, Q, X and Y. Use these values to copy and complete the following table. (4 marks)

P	Q	X	Y

b What is the annual compound interest rate for this investment? (1 mark)

When the balance in the annuity first falls below $300 000, Giuseppe converts the payment to a perpetuity so that his children are left with some inherited benefits. The interest rate remains the same as that calculated in part **b**.

c Determine the number of years the annuity operates before the perpetuity starts. (2 marks)

d What are the quarterly payments under this perpetuity? (2 marks)

e Giuseppe believes that his investment returns are at an effective interest rate of 7.93% p.a. Use a clear calculation to comment on the accuracy of this belief. (2 marks)

Video playlist
Annuity
investments

(7.9) **Annuity investments**

Annuity investment recurrence relations

An **annuity investment** is a compound interest investment where regular payments are made into the account. This is different to an annuity where the regular payments are taken *out of* the account. People often use annuity investments when saving for retirement or for the deposit of a house.

> **Exam hack**
>
> Don't confuse annuities and annuity investments. Both are investments, but with annuities the regular payments are withdrawals and with annuity investments the regular payments are added.

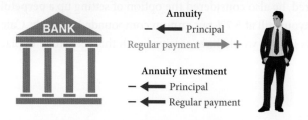

The recurrence relation for an annuity investment is the same as the recurrence relation for an annuity except the payment is added rather than subtracted.

Annuity investment recurrence relation

The recurrence relation for the value V_n of an annuity investment

$$V_0 = \text{principal}, \quad V_{n+1} = \left(1 + \frac{r}{100}\right)V_n + d$$

where r = percentage interest rate per compounding period

$\quad\quad n$ = number of compounding periods

$\quad\quad d$ = payment made per compounding period.

The graph of an annuity recurrence relation will look the same as a compound interest graph, except it will increase faster.

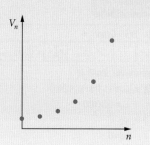

| **WORKED EXAMPLE 27** | Working with annuity investment recurrence relations |

Teofilo plans to invest \$55 000 in an annuity investment at a rate of 3.6% per annum compounding monthly, where he makes regular monthly additions of \$260.

a Write a recurrence relation for the balance.

Teofilo then changes his mind and decides to invest the \$55 000 in a different annuity investment compounding quarterly with quarterly additions that has the recurrence relation

$$V_0 = 55\,000, \quad V_{n+1} = 1.01V_n + 260$$

b What is the annual interest rate of this investment?

c How much has the investment increased between the third and fourth quarter, to the nearest cent?

Steps	Working
a 1 Find the number of compounding periods per year.	There are 12 compounding periods per year.
2 Identify V_n, V_0, r and d.	Let V_n = the balance after n compounding periods. $V_0 = 55\,000, r = \dfrac{3.6}{12} = 0.3, d = 260$
3 Substitute the values into $V_0 = \text{principal}, \quad V_{n+1} = \left(1 + \dfrac{r}{100}\right)V_n + d$ and simplify.	$V_{n+1} = \left(1 + \dfrac{0.3}{100}\right)V_n + 260$ $\quad\quad = (1 + 0.003)V_n + 260$ $\quad\quad = 1.003V_n + 260$ $V_0 = 55\,000, \quad V_{n+1} = 1.003V_n + 260$
b 1 Use the recurrence relation to find an equation for r, the percentage interest rate per compounding period.	Comparing: $V_{n+1} = \left(1 + \dfrac{r}{100}\right)V_n + d$ $V_{n+1} = 1.01V_n + 260$ we can see that $1 + \dfrac{r}{100} = 1.01$
2 Solve for r, using CAS if necessary.	$\dfrac{r}{100} = 1.01 - 1$ $\dfrac{r}{100} = 0.01$ $r = 1$ The rate is 1% per quarter.
3 Multiply r by the compounding period to find the annual percentage compound interest rate.	The interest compounds quarterly, so the annual percentage compound interest rate is $1\% \times 4 = 4\%$

c Use CAS sequence to enter the recurrence relation.

$n = 4$, $V_4 = 58\,288.924\,81$

$n = 3$, $V_3 = 57\,454.381$

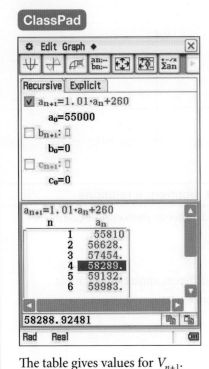

The table gives values for V_{n+1}.

$$V_4 - V_3 = 58\,288.924\,81 - 57\,454.381$$
$$= 834.543\,81$$

The investment increased by $834.54.

Annuity investment tables

Annuity investment tables work in a similar way to annuity tables. The difference is annuity investments involve a principal *addition* rather than a principal reduction.

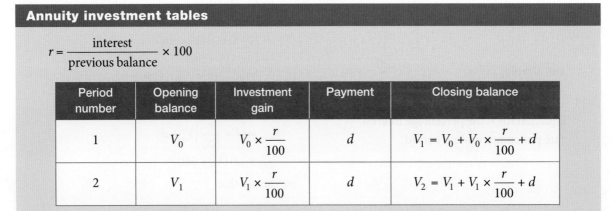

Annuity investment tables

$$r = \frac{\text{interest}}{\text{previous balance}} \times 100$$

Period number	Opening balance	Investment gain	Payment	Closing balance
1	V_0	$V_0 \times \dfrac{r}{100}$	d	$V_1 = V_0 + V_0 \times \dfrac{r}{100} + d$
2	V_1	$V_1 \times \dfrac{r}{100}$	d	$V_2 = V_1 + V_1 \times \dfrac{r}{100} + d$

where n = payment number

d = payment made per compounding period

r = percentage interest rate per compounding period

V_0 = principal

V_n = closing balance of the annuity after n compounding periods.

WORKED EXAMPLE 28	Using annuity investment tables

Vinh invests \$4000 in an annuity investment earning interest compounding annually. He deposits an extra \$1200 into the account each year after the initial deposit. The table below shows the first few calculations for his investment.

Year	Opening balance	Interest	Payment	Closing balance
1	4000.00	320.00	1200.00	5520.00
2	5520.00	441.60	1200.00	7161.60
3				

a Use the table to show a calculation that will give r, the percentage interest rate per compounding period.

b What is the nominal interest rate for the loan?

c Complete the investment table, for the third year of the annuity investment, giving all values to the nearest cent.

Steps	Working
a Use $r = \dfrac{\text{interest}}{\text{previous balance}} \times 100$.	Using interest for payment number 1 $$r = \frac{320.00}{4000.00} \times 100$$ $$= 0.08 \times 100$$ $$= 8$$
b Use the compounding period to find the annual interest rate.	The compounding period is 1. The nominal interest rate $= 1 \times 8\%$ $= 8\%$ per annum compounding annually

c Complete the table using

interest $= \dfrac{r}{100} \times$ previous balance

opening balance V_{n-1}

closing balance V_n

closing balance = opening balance + interest + payment

Give all values to the nearest cent.

Year	Opening balance	Interest	Payment	Closing balance
1	4000.00	320.00	1200.00	5520.00
2	5520.00	441.60	1200.00	7161.60
3	7161.60	7161.60 × 0.08 = 572.93	1200.00	7161.60 + 572.93 + 1200 = 8934.53

Annuity investments and finance solvers

Annuity investments can be analysed using finance solvers in a similar way to annuities. The difference is

- PMT is always negative because the payments are going away from the person
- FV will continue to increase and will never reach zero.

Using finance solvers for annuity investments

N	**Total number** of compounding periods
I%	**Interest rate** per year
PV	**Present value** for an annuity investment is **negative** because the money is going away from the person.
PMT or Pmt	**Regular payments** for an annuity investment are **negative** because the money is going away from the person.
FV	**Future value** has the opposite sign of the present value so it will be **positive**.
P/Y or PpY	Number of **payments per year**. This will always take the same value as **C/Y** or **CpY**.
C/Y or CpY	Number of **compounding** periods per year

When solving for N, always round *up*, never down, to the nearest whole number.

Ignore the negative sign in values when using the formula

$$\text{total interest paid} = N \times PMT - (PV - FV)$$

WORKED EXAMPLE 29 | Using finance solvers for annuity investments

Ruben's investment earns interest at the rate of 4.7% per annum, compounding monthly. Ruben initially invested $110 000 and adds monthly payments of $2000.

a After how many months will the value of this investment first exceed $130 000?

b Ruben wants to reach a target of $250 000 in four years. After one year, he increased his payments so that he would reach his target. What did Ruben increase his payments to? Give your answer to the nearest cent.

Steps	Working
	ClassPad + **TI-Nspire**
a 1 Enter the amount as FV and find N.	
Total number of compounding periods	**N** 8.113...
Annual interest rate	**I%** 4.7
Present value for an investment is negative.	**PV** −110 000
Money moving away from a person is negative.	**PMT or Pmt** −2000
Future value has the opposite sign to present value.	**FV** 130 000
Same as C/Y or CpY	**P/Y or PpY** 12
Number of compounding periods per year	**C/Y or CpY** 12
2 Round N *up* to the nearest whole number.	The value of the investment will first exceed $130 000 after nine months.

b 1 Find FV after 12 months.

Total number of compounding periods	N	12
Annual interest rate	I%	4.7
Present value for an investment is negative.	PV	−110 000
Money moving away from a person is negative.	PMT or Pmt	−2000
Future value has the opposite sign to present value.	FV	139 806.646…
Same as C/Y or CpY	P/Y or PpY	12
Number of compounding periods per year	C/Y or CpY	12

2 Use FV after one year as PV for the next three years. Find PMT.

Total number of compounding periods	N	36
Annual interest rate	I%	4.7
Present value for an investment is negative.	PV	−139 806.646 892
Money moving away from a person is negative.	PMT or Pmt	−2308.605…
Future value has the opposite sign to present value.	FV	250 000
Same as C/Y or CpY	P/Y or PpY	12
Number of compounding periods per year	C/Y or CpY	12

3 Write your answer to the nearest cent.

Ruben increased his payments to $2308.61.

Exam hack

Questions don't always state what type of compound interest investment you're dealing with. Look at the type of payment involved to decide.

- No payment means it's an ordinary compound interest investment.
- Withdrawals from the account mean it's an annuity.
- Additions to the account mean it's an annuity investment.

WORKED EXAMPLE 30 Using finance solvers for different types of compound interest investments

Antionette started her savings by investing $20 000 for 10 years in an account earning 4.8% p.a. interest compounding monthly. After the 10 years she put the balance into a superannuation account for her retirement earning 5.6% p.a. interest compounding monthly and made regular monthly payments of $600 for 30 years until she retired. Since Antionette retired, she has been taking out regular monthly payments of $4000 from this account. How much money, to the nearest cent, is in her account after she has been retired for five years?

Steps	Working
	ClassPad ╋ TI-Nspire

1 The first account is an ordinary compound interest investment. Find FV after 10 × 12 = 120 months.

Total number of compounding periods	N	120
Annual interest rate	I%	4.8
Present value for an investment is negative.	PV	−20 000
Payment amount	PMT or Pmt	0
Future value has the opposite sign to present value.	FV	32 290.556…
Same as C/Y or CpY	P/Y or PpY	12
Number of compounding periods per year	C/Y or CpY	12

2 The superannuation account is an annuity investment. Enter the unrounded FV from the first account as the new PV. Find FV.

Total number of compounding periods = 30 × 12 = 360 months	N	360
Annual interest rate	I%	5.6
Present value for an investment is negative.	PV	–32 290.556 720 832
Money moving away from a person is negative.	PMT or Pmt	–600
Future value has the opposite sign to present value.	FV	731 176.221...
Same as C/Y or CpY	P/Y or PpY	12
Number of compounding periods per year	C/Y or CpY	12

3 The superannuation account is now an annuity. Enter the unrounded FV from the annuity account as the new PV. Find FV after 5 × 12 = 60 months.

Total number of compounding periods	N	60
Annual interest rate	I%	5.6
Present value for an investment is negative.	PV	–731 176.221 612 1
Money moving to a person is positive.	PMT or Pmt	4000
Future value has the opposite sign to present value.	FV	690 581.164...
Same as C/Y or CpY	P/Y or PpY	12
Number of compounding periods per year	C/Y or CpY	12

4 Write your answer to the nearest cent.

Antionette has $690 581.16 in her account after she has been retired for five years.

Video
WACE question analysis: Loans, investments and annuities

WACE QUESTION ANALYSIS

© SCSA   MA2019 Q13   Calculator-assumed (10 marks)

Mehmet has saved $3600 from wages received at a part-time job. He is keen to invest this money in an account which earns 3.65% per annum, compounded monthly.

Over the next three years, Mehmet plans to continue working part-time and is aiming to make deposits of $250 at the end of each month.

a Write a recursive relation to give the value of the investment at the end of each month. (2 marks)

b Mehmet hopes that this investment will double his initial savings in one year. Justify whether this is possible. (2 marks)

c Determine the total amount of interest Mehmet would receive after three years. (3 marks)

Unfortunately, after two years, Mehmet's working hours are reduced and he is only able to deposit $120 at the end of each month.

d By how much would this reduce the value of his investment by the end of the three years? (3 marks)

Reading the question

- This question requires an understanding of annuity investments. These problems can be done using recursion.
- Highlight whether regular payments and interest rate compounding are made daily, weekly, fortnightly, monthly or yearly.
- Highlight the type of answer required in each part. This may be a recursion equation, an account balance or an amount of interest.
- Part **b** requires a justification of your answer, so make sure enough working is shown to demonstrate why this answer was obtained.

Thinking about the question

- This question requires an understanding of investment annuities.
- Annuities can be solved using recursion, finance solver on CAS and annuity tables. If a method is not indicated, you should choose the easiest method.
- You will need to be able to calculate the interest earned on an annuity.
- You will also need to solve a two-stage problem using a finance solver.

Worked solution ($\checkmark$ = 1 mark)

a Use the annuity formula,

$$V_0 = \text{principal}, \quad V_{n+1} = \left(1 + \frac{r}{100}\right)V_n + d$$

$$V_0 = 3600, \ r = \frac{3.65}{12}, \ d = 250$$

Recursion equation: $V_0 = 3600, \ V_{n+1} = \left(1 + \frac{0.0365}{12}\right)V_n + 250$

states recursive part of rule $\checkmark$

states correct starting value $\checkmark$

b The value of the annuity at the end of one year is **\$6784.32**, therefore Mehmet does **not double** his money in a year.

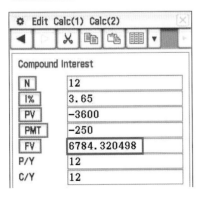

correctly states that Mehmet does not save \$7200 in one year $\checkmark$

correctly justifies this decision $\checkmark$

c Use a finance solver to find FV for the annuity, after 36 months.

Total number of compounding periods	**N**	**36**
Annual interest rate	**I%**	**3.65**
Present value for an investment is negative.	**PV**	**−3600**
Money moving away from a person is negative.	**PMT or Pmt**	**−250**
Future value has the opposite sign to present value.	**FV**	**13511.92**
Same as C/Y or CpY	**P/Y or PpY**	**12**
Number of compounding periods per year	**C/Y or CpY**	**12**

The value of the annuity at the end of three years is **$13511.92**.

Deposits = 3600 + 36 × $250 = **$12600**

Interest earned = $13511.92 − $3600 − 36 × $250 = **$911.92**

correctly determines final value ✓

correctly determines total of all deposits ✓

correctly determines interest received ✓

d The value of the annuity after 24 months = **$10086.83** ✓

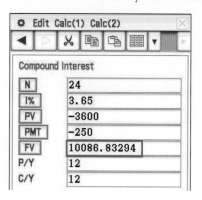
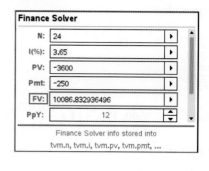

Reduce the payments to $120 for the next 12 months.

The value of the annuity after 24 months with monthly deposits of $250 and 12 months with monthly deposits of $120 is **$11925.56**. ✓

Total number of compounding periods	**N**	**12**
Annual interest rate	**I%**	**3.65**
Present value for an investment is negative.	**PV**	**−10086.83294**
Money moving away from a person is negative.	**PMT or Pmt**	**−120**
Future value has the opposite sign to present value.	**FV**	**11925.56...**
Same as C/Y or CpY	**P/Y or PpY**	**12**
Number of compounding periods per year	**C/Y or CpY**	**12**

This would reduce the value of his investment by the end of the three years by

$13511.92 − $11925.56 = **$1586.36** ✓

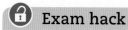 **Exam hack**

When answering extended questions with finance solvers, show your working by listing the values you entered for N, I%, PV etc.

9780170476959

Recap

1 Which one of the following recurrence relations could **not** be used to model the value of a perpetuity investment, P_n, after n months?

A $P_0 = 30\,000$, $P_{n+1} = 1.0063 \times P_n - 189$

B $P_0 = 50\,000$, $P_{n+1} = 1.0074 \times P_n - 370$

C $P_0 = 20\,000$, $P_{n+1} = 1.0017 \times P_n - 64$

D $P_0 = 60\,000$, $P_{n+1} = 1.0049 \times P_n - 294$

E $P_0 = 10\,000$, $P_{n+1} = 1.0058 \times P_n - 58$

2 Pia invests \$800\,000 in an ordinary perpetuity to provide an ongoing fortnightly pension for her retirement. The interest rate for this investment is 5.8% per annum. Assuming there are 26 fortnights per year, the amount she will receive at the end of each fortnight is closest to

A \$464   **B** \$892   **C** \$1422   **D** \$1785   **E** \$3867

Mastery

3 [WORKED EXAMPLE 27] Shane plans to invest \$27\,000 in an annuity investment, which has an interest rate of 7.2% per annum, compounding monthly. He makes regular monthly additional payments of \$310.

a Write a recurrence relation for the balance in terms of V_n.

Shane then changes his mind and decides to invest the \$27\,000 in a different annuity investment, compounding quarterly with quarterly additions, according to the recurrence relation

$$V_0 = 27\,000, \quad V_{n+1} = 1.018V_n + 310$$

b What is the annual interest rate of this investment?

c How much has the investment increased between the third and fourth quarter, to the nearest cent?

4 [WORKED EXAMPLE 28] Megan invests \$500 in an annuity investment earning interest compounded annually. She deposits an extra \$500 into the account each year after the initial deposit. The annuity investment table below shows the first few calculations for her investment.

Year	Opening balance	Interest	Payment	Closing balance
1	500.00	40.00	500.00	1040.00
2	1040.00	83.20	500.00	1623.20
3				

a Use the table to show two calculations that will give r, the percentage interest rate per compounding period.

b What is the nominal interest rate for the loan?

c Copy and complete the investment table, for the third year of the annuity investment. Give all values correct to the nearest cent.

5 [WORKED EXAMPLE 29] Phillip's investment earns interest at the rate of 4.2% per annum, compounding monthly. Phillip initially invested \$8000 and adds monthly payments of \$500.

a After how many months will the value of this investment first exceed \$12\,000?

b Phillip wants to reach a target of \$45\,000 in five years. After one year, Phillip increased his payments so that he would reach his target. What did Phillip increase his payments to? Give your answer to the nearest cent.

6 🔗 WORKED EXAMPLE 30 Mei started her savings by investing $25 000 for 12 years in an account earning 4.6% p.a. interest compounding monthly. After the 12 years she put the balance into a superannuation account for her retirement earning 5.9% p.a. interest compounding monthly and made regular monthly payments of $500 for 35 years until she retired. Since Mei retired, she has been taking out regular monthly payments of $5500 from this account. How much money, to the nearest cent, is in her account after she has been retired for four years?

Calculator-assumed

7 (3 marks) The opening balance of an annuity investment, in dollars, after n years, V_n, can be modelled by the recurrence relation

$$V_0 = 46\,000, \quad V_{n+1} = 1.0034V_n + 500$$

a What is the value of the regular payment added to the principal of this annuity investment? (1 mark)

b Find the increase in the value of this investment between the second and third years. (2 marks)

8 (4 marks) Find the following.

a An annuity investment earns interest at the rate of 3.8% per annum, compounding monthly. Cho initially invested $85 000 and will add monthly payments of $1500. Find the number of months when the value of this investment first exceeds $95 000. (2 marks)

b Sarah invests $5000 in a savings account that pays interest at the rate of 3.9% per annum compounding quarterly. At the end of each quarter, immediately after the interest has been paid, she adds $200 to her investment. Find the value of her investment after two years, to the nearest dollar. (2 marks)

9 (7 marks) Mariska plans to retire from work 10 years from now. Her retirement goal is to have a balance of $600 000 in an annuity investment at that time. The present value of this annuity investment is $265 298.48, on which she earns interest at the rate of 3.24% per annum, compounding monthly. To make this investment grow faster, Mariska will add a $1000 payment at the end of every month. Two years from now, she expects the interest rate of this investment to fall to 3.20% per annum, compounding monthly. It is expected to remain at this rate until Mariska retires. When the interest rate drops, she must increase her monthly payment if she is to reach her retirement goal. Find the

a value of the annuity after 2 years (2 marks)

b interest earned in the first 2 years (2 marks)

c value of this new monthly payment. (3 marks)

10 © SCSA MA2018 Q16 (11 marks) Natalia inherits a sum of money from her grandfather. She wishes to place it in a high-interest savings account.

She is considering the following two options:

Account A: Interest rate 4.40% per annum, compounded monthly

Account B: Interest rate 4.30% per annum, compounded daily

a The effective annual interest rate for Account A is 4.49%, correct to two decimal places. Determine the effective annual interest rate for Account B. (1 mark)

Natalia's bank offers her another account, C, with an interest rate of 4.50% per annum.

b Under what circumstances will this interest rate and the effective annual interest rate be the same? (1 mark)

c Which account (A, B or C) should Natalia choose to maximise her savings? Explain your reasoning. (2 marks)

Natalia's sister, Elena, has inherited $25 000 from her grandfather. She decides to invest this money in a high-interest savings account, with interest compounded monthly. Elena also chooses to deposit an additional $250 into this account at the end of each month.

The table below shows Elena's account balance over the first three months.

Month	Account balance at start of month	Interest earned	Deposit	Account balance at end of month
1	$25 000.00	$125.00	$250.00	$25 375.00
2	$25 375.00	$126.88	$250.00	$25 751.88
3	$25 751.88	$128.76	$250.00	$26 130.64

d Show that the annual interest rate that applies to Elena's account is 6%. (1 mark)

e The amount in Elena's account, A_n at the end of month n, can be expressed as a recursive rule, $A_{n+1} = cA_n + d$, $A_0 = 25 000$. Determine the values of c and d. (2 marks)

f After two years, Elena wishes to use the money she has saved as a deposit for a house. An amount of $35 000 will be required. Unfortunately, Elena has realised that by depositing $250 each month she will not reach her savings goal.

   i If she only deposits $250 each month, by how much will she be short of the required deposit? (2 marks)

   ii What increase in the monthly deposit is required for Elena to save the $35 000 in two years? (2 marks)

11 (3 marks) A community centre has received a donation of $5000. The donation is deposited into a savings account. This savings account pays interest compounding monthly. Immediately after the interest has been added each month, the community centre deposits a further $100 into the savings account. After five years, the community centre would like to have a total of $14 000 in the savings account.

a What is the annual interest rate, compounding monthly, that is required to achieve this goal? Write your answer correct to two decimal places. (1 mark)

b The interest rate for this savings account is actually 6.2% per annum, compounding monthly. After 36 deposits, the community centre stopped making the additional monthly deposits of $100. How much money will be in the savings account five years after it was opened? (2 marks)

12 (3 marks) Alex sold his mechanics' business for $360 000 and invested this amount in a perpetuity. The perpetuity earns interest at the rate of 5.2% per annum. Interest is calculated and paid monthly.

a What monthly payment will Alex receive from this investment? (1 mark)

b Later, Alex converts the perpetuity to an annuity investment. This annuity investment earns interest at the rate of 3.8% per annum, compounding monthly. For the first four years Alex makes a further payment each month of $500 to his investment. This monthly payment is made immediately after the interest is added. After four years of these regular monthly payments, Alex increases the monthly payment. This new monthly payment gives Alex a balance of $500 000 in his annuity after a further two years. What is the value of Alex's new monthly payment? Round your answer to the nearest cent. (2 marks)

13 (4 marks) Tisha plays drums in the same band as Marlon. She would like to buy a new drum kit and has saved $2500.

    a Tisha could invest this money in an account that pays interest compounding monthly. The balance of this investment after n months, T_n, could be determined using the recurrence relation.

$$T_0 = 2500, \quad T_{n+1} = 1.0036 \times T_n$$

    Calculate the total interest that would be earned by Tisha's investment in the first five months. Round your answer to the nearest cent.   (2 marks)

Tisha could invest the $2500 in a different account that pays interest at the rate of 4.08% per annum, compounding monthly. She would make a payment of $150 into this account every month.

    b Let V_n be the value of Tisha's investment after n months.

    Write down a recurrence relation, in terms of V_0, V_n and V_{n+1}, that would model the change in the value of this investment.   (1 mark)

    c Tisha would like to have a balance of $4500, to the nearest dollar, after 12 months. What annual interest rate would Tisha require? Round your answer to two decimal places.   (1 mark)

14 © SCSA   MA2019 Q17   (8 marks) Joel has set up a special investment fund that has a current balance of $350 000. He contributes 7.5% of his monthly income to the investment and has an overseas pension which contributes a further $355 per month. The investment fund has an interest rate of 6.5% per annum, compounded monthly. Joel's annual salary is $101 000 and he has just turned 60 years of age.

    a Calculate Joel's total monthly contribution to the fund.   (2 marks)

    b Calculate the lump sum that he could receive if he retires on his 67th birthday.   (2 marks)

Joel retires at 67 and wants to use his lump sum payment to set up a regular income. He decides to look at two options that offer monthly payments.

Option 1: A reducing balance annuity at 7% per annum, compounded monthly.

Option 2: A perpetuity at 7.5% per annum, compounded monthly.

    c Calculate his maximum monthly income for the next 20 years using Option 1.   (2 marks)

    d Calculate his monthly income using Option 2.   (2 marks)

Finance solvers

Values in a finance solver can be positive, negative or zero.

- Money coming *to the person* is positive.
- Money going away *from the person* is negative.
- PV and FV always have opposite signs (except when FV is zero).

N	**Total number** of compounding periods
I%	**Annual** interest rate
PV	**Present value**
PMT or Pmt	Regular **payments**
FV	**Future value** has the opposite sign of Present value (or can be zero).
P/Y or PpY	Number of **payments per year**.
C/Y or CpY	Number of **compounding** periods per year

Compound interest and finance solvers

	Compound interest investment	Reducing balance depreciation
I%	positive	negative
PV	negative	negative
PMT or Pmt	zero	zero
FV	positive	positive or zero

Loan and investment summary

	Reducing balance loan	Annuity	Annuity investment
Recurrence relation	$V_0 = \text{principal},$ $V_{n+1} = \left(1 + \dfrac{r}{100}\right)V_n - d$	$V_0 = \text{principal},$ $V_{n+1} = \left(1 + \dfrac{r}{100}\right)V_n - d$	$V_0 = \text{principal},$ $V_{n+1} = \left(1 + \dfrac{r}{100}\right)V_n + d$
Type	multiply by a number greater than 1 subtract an amount	multiply by a number greater than 1 subtract an amount	multiply by a number greater than 1 add an amount
Growth/decay	geometric growth linear decay	geometric growth linear decay	geometric growth linear growth
Graph			
Principal/Payment			
PV	positive	negative	negative
PMT or Pmt	negative	positive	negative
FV	negative or zero	positive or zero	positive

r = percentage interest rate per compounding period

n = number of compounding periods

d = payment made per compounding period

Working with finance solvers

When solving for N, always round *up*, never down, to the nearest whole number.

Ignore the negative sign in values when using the formulas

Total interest paid = N × PMT − (PV − FV)

Total loan cost = N × PMT

Percentage decrease in loan balance = $\dfrac{\text{PV} - \text{FV}}{\text{PV}} \times 100\%$

9780170476959

Reducing balance loan payment graphs

For a reducing balance loan, the regular payment is part interest and part principal.

- The amount of **interest** ($) **decreases** with each compounding period. Towards the end of the life of the loan, the interest becomes a small part of the regular payment.
- The amount paid off the **principal** ($) **increases** with each compounding period. Towards the end of the life of the loan, the amount paid off the principal becomes a large part of the regular payment.

Amortisation tables

$$r = \frac{\text{interest}}{\text{previous balance}} \times 100$$

Reducing balance loan amortisation table

Period number	Opening balance	Interest	Repayment	Closing balance
1	V_0	$V_0 \times \dfrac{r}{100}$	d	$V_1 = V_0 + \left(V_0 \times \dfrac{r}{100}\right) - d$
2	V_1	$V_1 \times \dfrac{r}{100}$	d	$V_2 = V_1 + \left(V_1 \times \dfrac{r}{100}\right) - d$

Annuity table

Payment number	Opening balance	Investment gain	Payment	Closing balance
1	V_0	$V_0 \times \dfrac{r}{100}$	d	$V_1 = V_0 + V_0 \times \dfrac{r}{100} - d$
2	V_1	$V_1 \times \dfrac{r}{100}$	d	$V_2 = V_1 + V_1 \times \dfrac{r}{100} - d$

Annuity investment table

Period number	Opening balance	Investment gain	Payment	Closing balance
1	V_0	$V_0 \times \dfrac{r}{100}$	d	$V_1 = V_0 + V_0 \times \dfrac{r}{100} + d$
2	V_1	$V_1 \times \dfrac{r}{100}$	d	$V_2 = V_1 + V_1 \times \dfrac{r}{100} + d$

n = payment number

d = payment made per compounding period

r = percentage interest rate per compounding period

V_0 = principal

Interest-only loans and perpetuities

	Interest-only loan	Perpetuity
Type	Reducing balance loan with $V_n = V_0$	Annuity with $V_n = V_0$
Formula	$d = \dfrac{r}{100} \times V_0$	$d = \dfrac{r}{100} \times V_0$
Graph		
Principal / payment		
N	1	1
PV	positive	negative
PMT or Pmt	negative	positive
FV	Same as PV but with opposite sign	Same as PV but with opposite sign

d = payment made per compounding period

r = percentage interest rate per compounding period

V_0 = principal

Total number of marks: 10   Reading time: 1 minute   Working time: 10 minutes

1 (4 marks) Consider the two graphs, graph G and graph H, below.

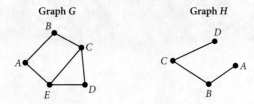

Graph G

Graph H

a Justify whether graph G is Eulerian, semi-Eulerian or neither.   (2 marks)

b Explain why graph H cannot contain a Hamiltonian cycle.   (2 marks)

2 (3 marks) A loan of \$20 000 is taken out to finance a new business venture. The loan is to be repaid fully over 5 years with quarterly payments of \$1223.13. Interest is calculated quarterly at 8% p.a on the reducing balance. Find the recurrence relation.

3 (3 marks) Mario has an investment whose recurrence relation for V_n, the value of the opening balance of an investment after n compounding periods, is

$$V_0 = 2000, \quad V_{n+1} = 1.01 V_n - 20$$

Find V_1 and explain why this shows that this is a perpetuity.

Total number of marks: 57   Reading time: 6 minutes   Working time: 57 minutes

1 (3 marks) The heights (in cm) and ages (in months) of a random sample of 15 boys have been plotted in the scatterplot. The least-squares line has been fitted to the data.

The equation of the least-squares line is

$height = 0.53 \times age + 75.4$

The correlation coefficient is $r = 0.7541$.

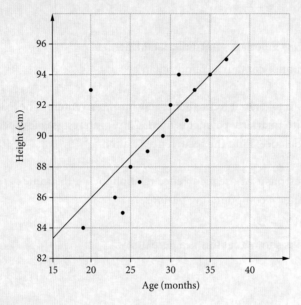

a Copy and complete the following sentence.

On average, the height of a boy increases by _____ cm for each one-month increase in age. (1 mark)

b **i** Evaluate the coefficient of determination.

Write your answer, as a percentage, correct to one decimal place. (1 mark)

ii Interpret the coefficient of determination in terms of the variables height and age. (1 mark)

2 (4 marks) Consider the sequence generated by the recursive rule

$B_{n+1} = 0.7B_n + b, \quad F_1 = 113$

a If $b = 4$, find the steady state to the nearest whole number. (2 marks)

b Find the value of b that would produce a steady state of 50. (2 marks)

3 (6 marks) Water will be pumped from a dam to eight locations on a farm. The pump and the eight locations (including the house) are shown as vertices in the network diagram. The weights on the edges give the shortest distances, in metres, between locations.

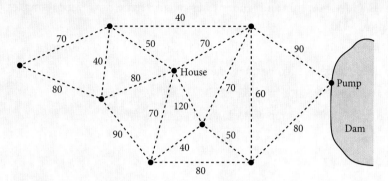

a State the shortest distance between the pump and the house. (1 mark)

b Explain why a closed Eulerian trail does not exists for this network. (2 marks)

c The total length of all edges in the network is 1180 m. If the farmer was to start from the house, walk around his farm such that he walks along every edge exactly once, and then finish at the house, state the shortest distance he could walk. Justify your answer. (2 marks)

d A Hamiltonian cycle, starting and ending at the house, can be found for this network. State the number of edges that this path involves. (1 mark)

4 (12 marks) Zen Supermarket is planning to acquire Chaos Grocers. To ascertain the long-term viability of Chaos Grocers, quarterly sales data over 12 years from first quarter 2010 to fourth quarter 2021 (inclusive) of Chaos Grocers were analysed by two analysts, Allan and Brianna from Zen Supermarket. Allan used a five-point moving average method to model the sales data while Brianna fitted a line of best fit through the deseasonalised data.

a State with justification which model is more appropriate in modelling the sales data. (2 marks)

Brianna's line of best fit through the deseasonalised data has the following equation

$$s = -1.9585t + 105$$

where s is the sales data in thousands ($000s) and $t = 1$ represents first quarter of 2010. The seasonal indices are shown in the table below.

Quarter	1	2	3	4
Seasonal index	125	92	x	78

b Determine the value of x. (1 mark)

c State with justification which quarter(s) had sales above the seasonal mean. (2 marks)

d i Predict the deseasonalised quarterly sales for fourth quarter of 2022. (3 marks)

　ii Comment on the reliability of this data with appropriate reason. (2 marks)

e Based on Brianna's analysis, would you recommend Zen Supermarket acquire Chaos Grocers? Provide a reason for your recommendation. (2 marks)

5 © SCSA MA2017 Q14 (13 marks) Andrew takes out a $14 999 loan to purchase his first car after paying a $1200 deposit. The car dealer offered the loan at an introductory interest rate of 1.80% p.a. for the first year and then the rate becomes 3.24% p.a. for the remaining time of the loan. Interest is added monthly and Andrew has calculated he can afford to make monthly repayments of $420.

a **i** Express the loan repayment process for the first year as a recursive formula. (2 marks)

ii How much does Andrew still owe after one year? (1 mark)

b How much does Andrew owe after two years? (3 marks)

c How long does it take Andrew to repay the loan? (2 marks)

d Determine the amount of the final repayment. (2 marks)

e Calculate the total cost of the car. (3 marks)

6 © SCSA MA2018 Q8 (7 marks) Anthon and Bryan each invest $4500 in accounts earning compound interest for a period of four years.

a Anthony places his money in an account earning interest at the rate of 3.24% per annum, compounded quarterly.

i Copy and complete the table below, showing the value of Anthony's investment at the end of the second and third quarters. (2 marks)

Number of quarters money is invested	1	2	3	...	16
Value of investment ($)	4536.45			...	5120.00

ii State the recursive rule for Anthony's investment, which gives the values shown in the table above. (2 marks)

b Bryan places his money in an account earning interest daily. After four years, the value of both Anthony's and Bryan's investments is the same.

Explain how the change to the compounding period has affected the annual rate of interest required for the value of Bryan's investment to be the same as that of Anthony. Include calculations to support your answer. (3 marks)

7 © SCSA MA2018 Q14 (12 marks) Marco is a plumber. Three years ago, he purchased a vehicle costing $48 000 for his business. He paid a deposit of $5000 and acquired a personal loan for the remainder from a financial institution, at a reducible interest rate of 22.5% per annum, compounded monthly. He agreed to make repayments of $1000 at the end of each month.

a **i** Use a recurrence relation to determine the amount Marco currently owes on the loan. (3 marks)

ii Determine how much longer it will take him to completely pay off the loan. (2 marks)

b After three years, Marco finds that his vehicle is only worth $27 150. Determine the average rate of depreciation of his vehicle, expressed as a percentage. (2 marks)

c When Marco originally took out a personal loan for the purchase of his vehicle, he was given two options by the financial institution. These were:

- increasing his monthly repayment by $200, or

- taking an option of reducing the interest rate to 18.5% and maintaining repayments of $1000 per month.

In terms of time taken to pay off the loan and total paid for his vehicle, which should he have chosen and why? (5 marks)

CHAPTER 8

TREES, FLOW AND ASSIGNMENT

(8.1) Minimum spanning trees

Trees and spanning trees

A **tree** is a connected graph with no loops, multiple edges between any two vertices, or cycles.

This graph is an example of a tree:

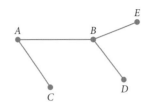

These are *not* trees:

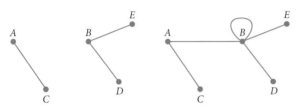

not a connected graph   contains a loop   contains multiple edges (BD)   contains a cycle

Video playlist
Minimum
spanning trees

Worksheet
Minimum
spanning trees

A **spanning tree** is a tree subgraph that connects all the vertices of the original graph. Every connected graph has at least one spanning tree.

The three spanning trees for

are shown:

Trees

- A tree is a connected graph with no loops, multiple edges between any two vertices, or cycles.
- The number of edges in a tree is always one less than the number of vertices.

Spanning trees

- A spanning tree is a tree subgraph that connects all the vertices of the original graph.
- Every connected graph has at least one spanning tree.

Which of the following graphs are a spanning tree of the graph shown on the right? For those that are spanning trees, verify that the number of edges is one less than the number of vertices. For those that aren't spanning trees, give a reason.

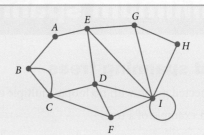

a

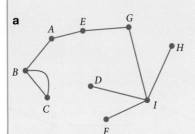

b

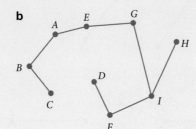

c

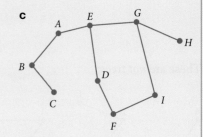

d

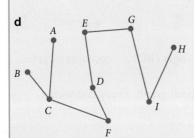

e

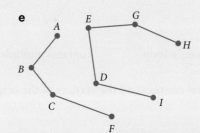

f

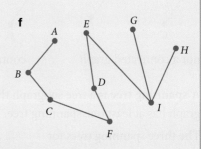

Steps	Working
Is it connected?	**a** This graph is not a spanning tree.
Does it have all the vertices of the original graph?	It has two edges between *B* and *C*.
Does it have no loops?	**b** This graph is a spanning tree.
Does it have no multiple edges between any two vertices?	It has 9 vertices and 8 edges.
Does it have no cycles?	**c** This graph is not a spanning tree.
	It has a cycle.
	d This graph is not a spanning tree.
	It has an edge (*AC*) that isn't in the original graph.
	e This graph is not a spanning tree.
	It is not connected.
	f This is a spanning tree.
	It has 9 vertices and 8 edges.

Minimum connector problems

A tree is a connected graph that contains the minimum number of edges. A **minimum spanning tree** is the tree with the smallest weight. Minimum spanning trees are an example of a **minimum connector** problem.

Spanning trees are used to solve minimum connector problems. These problems, like the shortest path problems, involve weights on the edges of the graph that represent quantities such as distance and cost. A minimum connector is the minimum weight path connecting *all* the vertices in a graph. This is different from the shortest path, which is the minimum weight path between *two* particular vertices and doesn't have to contain all the vertices in a graph.

To solve a minimum connector problem we need to find the minimum spanning tree. Minimum spanning trees are used when connecting broadband networks like the NBN, constructing water pipe systems and in real-time face recognition software.

WORKED EXAMPLE 2	Finding minimum spanning trees by inspection

Find all the spanning trees for the network shown and, hence, find the total weight of the minimum spanning tree.

Steps	Working
1 Work out how many edges need to be removed from the graph to create a spanning tree. In a spanning tree: • the number of edges is always one less than the number of vertices • the number of vertices is the same as the original graph. **2** Remove each edge in turn to see which options result in a spanning tree. Calculate the total weight of each spanning tree and find the one with the smallest total weight.	A spanning tree has 5 vertices and 4 edges. The original graph has 5 edges, so we need to remove one edge to create a spanning tree. Total weight of spanning tree = 20 Total weight of spanning tree = 21 Total weight of spanning tree = 18 Total weight of spanning tree = 20 This isn't a connected graph, so it's not a tree. The total weight of the minimum spanning tree is 18.

Prim's algorithm

Prim's algorithm is a series of steps to find
a minimum spanning tree for a graph.

Exam hack

Questions don't always say to use Prim's
algorithm, but you can use it for any
minimum spanning tree problem.

Prim's algorithm for finding a minimum spanning tree

1 Start at any vertex and choose the edge with the lowest weight connected to this vertex.

2 Look at *all* the edges connecting to the vertices you've chosen so far (*not just the last vertex
connected*) and choose the edge with the lowest weight that doesn't connect to a vertex already
in the tree. If there are edges with equal lowest weights, choose one of them.

3 Repeat step **2** until all the vertices in the graph are included in the tree.

WORKED EXAMPLE 3	Finding minimum spanning trees using Prim's algorithm

Use Prim's algorithm to find the minimum spanning tree for
the weighted graph shown and, hence, find the total weight
of the minimum spanning tree.

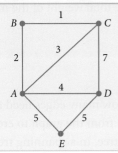

Steps

1 Start at any vertex and choose the edge with the lowest weight connected to this vertex.

2 Look at *all* the edges connecting to the vertices you've chosen so far (*not just the last vertex connected*)
and choose the edge with the lowest weight that doesn't connect to a vertex already in the tree. If there
are edges with equal lowest weights, choose one of them.

3 Repeat step **2** until all the vertices in the graph are included in the tree.

Working

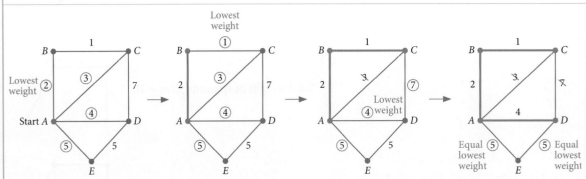

With the next step, all the vertices have
been included:

This gives us two minimum spanning trees
with the same total weight:

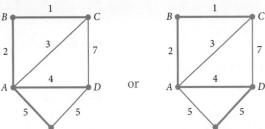

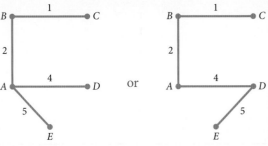

The total weight of the minimum spanning tree is $1 + 2 + 4 + 5 = 12$.

9780170476959

Prim's algorithm from a table

Prim's algorithm for finding a minimum spanning tree using a table

1 Pick any column vertex to start the spanning tree, highlight the column header, and strike out the row with the same letter.

2 Select and circle the lowest weight in the column.

3 If the lowest weight is in column *A* row *B*, highlight column header *B* and strike out row *B*.

4 Locate and circle the next lowest weight edge, selecting only from highlighted columns. You cannot select from the rows that have been eliminated by a strike-out line.

5 If there is more than one edge with the lowest weight, pick any of them.

6 Identify the row letter of the circled weight, highlight the column header and strike out the row with the same letter.

7 Continue until all the columns are highlighted and all the rows have been crossed out. The selected weights identify the edges in the minimum spanning tree.

WORKED EXAMPLE 4 | Prim's algorithm using a table

Use the table method of Prim's algorithm to find the minimum spanning tree for the weighted graph shown and, hence, find the total weight of the minimum spanning tree.

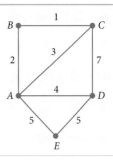

Steps	**Working**

1 Draw a table with five rows and five columns labelled *A*, *B*, *C*, *D* and *E*.

	A	*B*	*C*	*D*	*E*
A					
B					
C					
D					
E					

2 Enter the weights in the table to show the edges in the graph.

Edge *AB* has a weight of 2. This weight is shown in red in the table in cell *AB* (row *A* column *B*) and in cell *BA* (row *B* column *A*). Repeat this process for all the edges in the graph.

	A	*B*	*C*	*D*	*E*
A		2	3	4	5
B	2		1		
C	3	1		7	
D	4		7		5
E	5			5	

3 Like in Worked example 3, we start from vertex *A*, so highlight the column header of column *A* and strike out row *A*, then select the lowest weight in column *A*. This corresponds to 2 from column *A* in row *B*.

Strike out row *B* to ensure no cycles are created, then highlight the column *B* header.

	A	*B*	*C*	*D*	*E*
A		2	3	4	5
B	②		1		
C	3	1		7	
D	4		7		5
E	5			5	

4 Select the lowest weight, from the highlighted columns A and B, that has not been eliminated by a strike-out line. The lowest weight from column B is 1 in row C.

Highlight the column C header and strike out row C.

5 Select the lowest weight, from the highlighted columns A, B and C, that has not been eliminated by a strike-out line. The lowest weight from column A is 4 in row D.

Highlight the column D header and strike out row D.

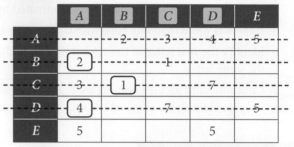

6 Select the lowest weight from the highlighted columns A, B, C and D. There are two possible lowest values, so either will produce the same minimum weight. The lowest weight chosen from column D is 5 in row E.

Highlight the column E header and strike out row E.

The process is now complete as all column headers are highlighted and all rows have a strike-out line.

7 Draw a tree with the edges of selected weights and add the weights to find the total weight of the minimum spanning tree.

The minimum spanning tree identified has edges AB, AD, BC and DE.

The weight of the minimum spanning tree is

$2 + 1 + 4 + 5 = 12$

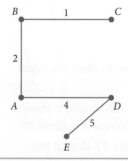

9780170476959

Mastery

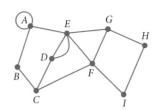

1 [WORKED EXAMPLE 1] Which of the following graphs are a spanning
tree of the graph shown on the right? For those that are spanning trees,
verify that the number of edges is one less than the number of vertices.
For those that aren't spanning trees, give a reason.

a 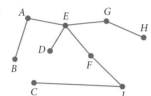 **b** 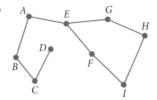 **c**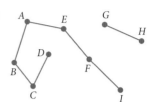

d 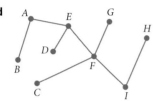 **e** 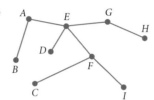 **f**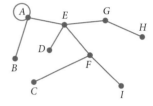

2 [WORKED EXAMPLE 2] Find all the spanning trees for the network
shown and, hence, find the total weight of the minimum spanning tree.

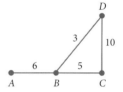

3 [WORKED EXAMPLE 3] Use Prim's algorithm to find the minimum
spanning tree for the weighted graph shown and, hence, find the
total weight of the minimum spanning tree.

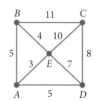

4 [WORKED EXAMPLE 3] A company is laying broadband cable
in a neighbourhood. The weighted graph shows the distances,
in metres, from the node N to each of the eight houses. Use
Prim's algorithm to find the minimum spanning tree and,
hence, find the minimum length of broadband cable needed
to connect to all eight houses.

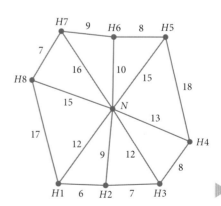

5 🔗 WORKED EXAMPLE 4   Use the table method of Prim's algorithm to find the minimum spanning tree and its weight.

a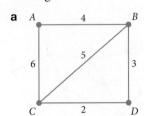

b

c

	A	B	C	D	E	F
A		10		4	5	
B	10		7		9	9
C		7		7	8	8
D	4		7		2	
E	5	9	8	2		
F		9	8			

d

	A	B	C	D	E	F	G
A		9				5	3
B	9		7				4
C		7		10			
D			10		9		6
E				9		8	7
F	5				8		
G	3	4		6	7		

Calculator-free

6 (2 marks) Consider the graph with five isolated vertices shown.

  a To form a tree, what is the minimum number of edges that must be added to the graph?

  b A simple, connected graph has 14 edges and 10 vertices. How many edges must be removed to form a tree?

7 (2 marks) Which of the following graphs is a tree? Justify your response.

A 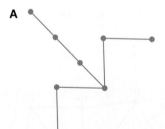   **B** 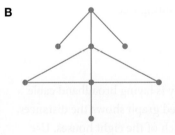   **C**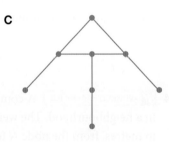

D 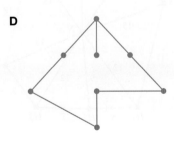   **E**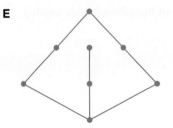

8 ©SCSA MA2016 Q1 (7 marks) Joe wishes to upgrade his sprinkler system using the least possible length of piping. The weighted graph below shows the existing system. The numbers on the edges indicate the length of each pipe, in metres, between sprinklers A, B, C, D, E, F, G and H.

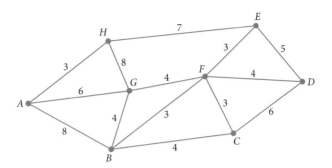

a Copy and complete the table below showing connections between each sprinkler. (2 marks)

	A	B	C	D	E	F	G	H
A	–	8						3
B	8	–						
C			–	6				
D			6	–				
E					–	3		7
F					3	–		
G							–	
H					7			–

b Copy the diagram below. Show the use of Prim's algorithm to establish a minimum spanning tree for the least length of piping required and draw this tree on your diagram. (5 marks)

9 © SCSA MA2018 Q1 (4 marks) The weighted network below represents an orienteering map where the vertices represent the various stations and the edges represent bush tracks joining the stations. The distances on the edges are in metres. The organisers wish to install freshwater fountains at each station using the minimum length of piping necessary to connect the stations along the bush tracks.

a Copy the diagram and highlight the bush tracks where the pipes should be installed. (2 marks)

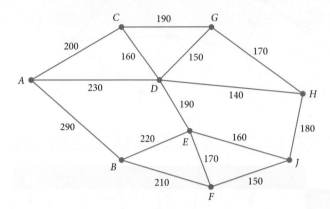

b Calculate the minimum length of piping required. (2 marks)

10 © SCSA MA2019 Q4c(i, ii) (6 marks) A marine park has attractions with paths connecting them. The vertices on the graph represent the attractions. Drinking water is already being supplied at E. The manager has recently received funding to establish drinking fountains at each attraction. For this to happen, pipelines will need to be laid along the paths to each attraction. He has drawn up a table to show the distances between attractions.

	E	P	Q	R	S	T	U	V
E	–	0.4	–	–	–	0.7	–	–
P	0.4	–	0.3	–	–	0.5	–	–
Q	–	0.3	–	0.2	–	0.5	–	–
R	–	–	0.2	–	0.4	0.6	–	0.8
S	–	–	–	0.4	–	0.2	–	0.5
T	0.7	0.5	0.5	0.6	0.2	–	0.3	–
U	–	–	–	–	–	0.3	–	1.2
V	–	–	–	0.8	0.5	–	1.2	–

a Use Prim's algorithm, or otherwise, to determine the minimum total length of pipelines. Copy the diagram and highlight the required pipelines. (4 marks)

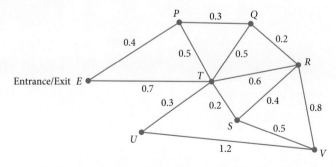

b The manager has been told that a pipeline of length 0.2 km could be laid from S to U. How, if at all, will this affect the total length of pipelines that should be laid in order to maintain a minimum length? (2 marks)

11 (3 marks) The network below shows the distances, in metres, between camp sites at a camping ground that has electricity. The vertices A to I represent the camp sites.

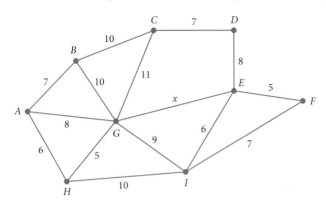

If the minimum length of cable required to connect all the camp sites is 53 m then EG must be at least x metres. Calculate the value of x.

12 (3 marks) The minimum spanning tree for the network shown includes the edge with weight labelled k.

The total weight of all edges for the minimum spanning tree is 33. Calculate the value of k.

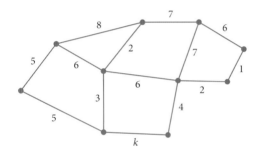

Calculator-assumed

13 (5 marks) A factory requires seven computer servers to communicate with each other through a connected network of cables. The servers, J, K, L, M, N, O and P, are shown as vertices on the graph.

The edges on the graph represent the cables that could connect adjacent computer servers. The numbers on the edges show the cost, in dollars, of installing each cable.

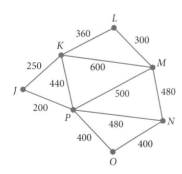

a What is the cost, in dollars, of installing the cable between server L and server M? (1 mark)

b What is the cheapest cost, in dollars, of installing cables between server K and server N? (1 mark)

c An inspector checks the cables by walking along the length of each cable in one continuous path. To avoid walking along any of the cables more than once, at which vertex should the inspector start and where would the inspector finish? (1 mark)

d The computer servers will be able to communicate with all the other servers as long as each server is connected by cable to at least one other server.

   i The cheapest installation that will join the seven computer servers by cable in a connected network follows a minimum spanning tree. Copy the plan shown and draw the minimum spanning tree on it.   **(1 mark)**

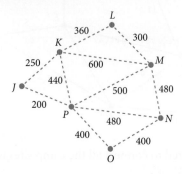

   ii The factory's manager has decided that only six connected computer servers will be needed, rather than seven. How much would be saved in installation costs if the factory removed computer server *P* from its minimum spanning tree network?   **(1 mark)**

14 **(2 marks)** While on holiday, four friends visit a theme park where there are nine rides. On the graph shown, the positions of the rides are indicated by the vertices. The numbers on the edges represent the distances, in metres, between rides.

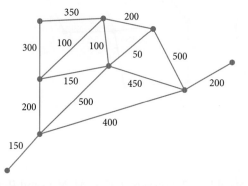

Electrical cables are required to power the rides. These cables will form a connected graph. The shortest total length of cable will be used.

a Give a mathematical term to describe a graph that represents these cables.   **(1 mark)**

b Copy the following diagram and use it to draw the graph that represents these cables.   **(1 mark)**

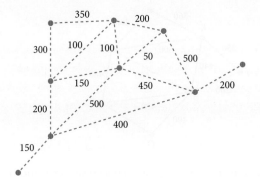

8.2 Flow networks

The applications of flow networks range from transporting water through a network of water pipes to moving people or products by road or a rail network.

Flow capacity and maximum flow

The first vertex of the network is called the **source** and the final vertex of the network is called the **sink**.

> ### Flow capacity and maximum flow
>
> The **inflow capacity** is the total flow capacity entering a vertex and the **outflow capacity** is the total flow capacity leaving the vertex.
>
> The **maximum flow through a vertex** is the smaller of the total inflow capacity of the vertex and the total outflow capacity of the vertex.

Video playlist
Flow networks

Worksheet
Network flow capacity

WORKED EXAMPLE 5	Finding the inflow, outflow and maximum capacity of a vertex

The directed graph shows the flow capacities in litres per minute. Determine the

a inflow capacity of vertex A

b outflow capacity of vertex A

c maximum flow out of vertex A.

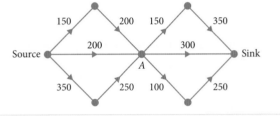

Steps	**Working**
a The total inflow capacity of vertex A can be found by adding the values on the edges entering vertex A.	Total inflow capacity of vertex A $= 200 + 200 + 250$ $= 650$ litres per minute
b The total outflow capacity of vertex A can be found by adding the values on the edges leaving vertex A.	Total outflow capacity of vertex A $= 150 + 300 + 100$ $= 550$ litres per minute
c The maximum flow out of vertex A is the smaller of inflow capacity and outflow capacity.	The flow out of vertex A is 550 litres per minute.

> ### The capacity of a cut
>
> A **cut** through a network must stop all flow from the start (*source*) to the end (*sink*).

One method of finding the maximum flow through a network is to use the minimum capacity of the cuts. In the network shown, Cut 1 and Cut 2 are both valid. The cuts sever the pipes in the network in such a way that there is no flow path from the source (A) to the sink (E). Cut 3 is **not** a valid cut because it does not stop all the flow between the source and the sink. It is still possible for flow to occur from A to D to E.

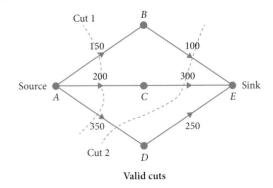

Valid cuts

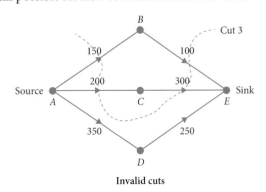

Invalid cuts

The capacity of a cut

The **capacity of a cut** can be determined by adding all the flow capacities that pass across the cut in the direction from source to sink.

WORKED EXAMPLE 6	Finding the capacity of a cut

In the network, the numbers on the edges show the maximum possible flow between the vertices. The direction of the arrow indicates the direction of the flow. A cut separating the sink from the source is also shown. Determine the capacity of the cut.

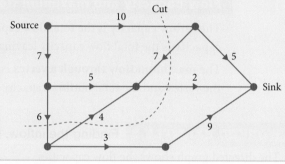

Steps	Working
Add the flows where the arrow crosses the cut in the direction from the source side of the cut to the sink side of the cut.	flow of cut = 6 + 2 + 10 = 18

In the diagram below, the source side of the cut has been shaded. The flows with capacity 7 and 4 are not included in the cut capacity as they flow towards the shaded source side instead of towards the unshaded sink side of the cut.

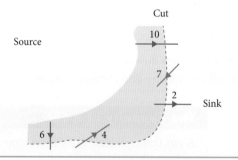

Maximum flow – minimum cut

The **maximum flow** from source to sink can be determined by finding the cut that produces the minimum value.

Determining the maximum flow for a network

To determine the maximum flow for a network:

1 identify cuts through the network

2 find the capacity of each cut

3 maximum flow = capacity of the minimum cut.

a Find the maximum flow for the network.

b Show the flow that would be directed along each edge to achieve the maximum flow.

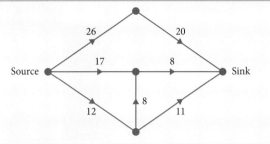

Steps	Working
a **1** Identify cuts that stop the flow from source to sink.	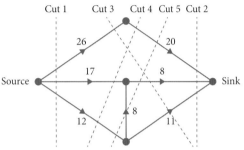
2 Find the capacity of each cut.	Cut 1 capacity = 26 + 17 + 12 = 55
Note: The vertical edge with a flow capacity of 8 is not included in the cut 5 capacity because its flow direction is from sink to source.	Cut 2 capacity = 20 + 8 + 11 = 39
	Cut 3 capacity = 26 + 8 + 11 = 45
	Cut 4 capacity = 12 + 17 + 20 = 49
	Cut 5 capacity = 12 + 8 + 20 = 40
3 Maximum flow = minimum cut	The maximum flow from source to sink is 39.

b The flow leaving the source and entering the sink must be 39.

The **maximum flow through a vertex** is the smaller of the total inflow capacity of the vertex and the total outflow capacity of the vertex.

Calculate the maximum flow through each vertex.

Adjust the inflow or outflow at each vertex to achieve the maximum flow.

Vertex	Inflow capacity	Outflow capacity	Actual flow
A	26	20	20
B	12	8 + 11 = 19	12
C	17 + 8 = 25	8	8

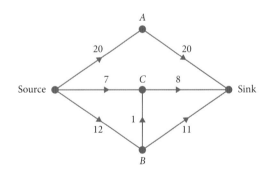

EXERCISE 8.2 Flow networks ANSWERS p. 451

Recap

1 Consider the weighted graph shown.

  a Find the number of edges in the spanning tree.

  b Use Prim's algorithm to find the weight of the minimum spanning tree.

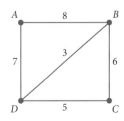

2 Use the table method of Prim's algorithm to find the weight of the minimum spanning tree.

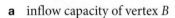

	A	B	C	D	E
A		25	35	15	
B	25			10	20
C	35			25	
D	15	10	25		30
E		20		30	

Mastery

3 WORKED EXAMPLE 5 The traffic capacity, in vehicles per hour, of roads in and out of an intersection, labelled vertex B, is shown. Determine the

  a inflow capacity of vertex B

  b outflow capacity of vertex B

  c maximum flow out of vertex B.

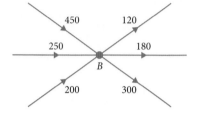

4 WORKED EXAMPLE 6 Determine the capacity of the cut.

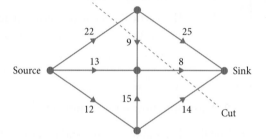

5 WORKED EXAMPLE 7

  a Find the maximum flow for the network.

  b Show the flow that would be directed along each edge to achieve the maximum flow.

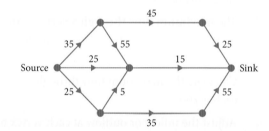

Calculator-free

6 (2 marks) The flow of water through a series of pipes is shown in the network. The numbers on the edges show the maximum flow through each pipe in litres per minute.

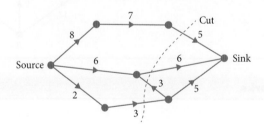

Determine the

  a capacity of cut Q, in litres per minute   (1 mark)

  b maximum flow from source to sink.   (1 mark)

7 (1 mark) In the directed graph below, the weight of each edge is non-zero. Find the capacity of the cut shown.

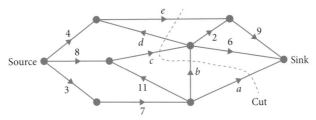

8 (2 marks) The directed graph shows the flow of water, in litres per minute, in a system of pipes connecting the source to the sink.

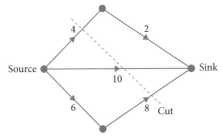

Determine the

a capacity of the cut (1 mark)

b maximum flow, in litres per minute, from the source to the sink. (1 mark)

9 (6 marks) The flow of oil through a series of pipelines, in litres per minute, is shown in the network.

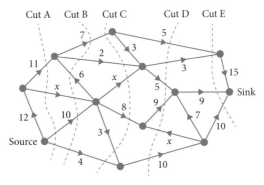

The weightings of three of the edges are labelled x. Five cuts labelled A–E are shown on the network.

Find the

a flow capacity of each cut (5 marks)

b maximum flow of oil from the source to the sink, in litres per minute, if $x = 2$. (1 mark)

10 (3 marks) On the directed graph, the values on the edges give the maximum flow between nodes in the direction of the arrows. Five cuts have been made on the diagram.

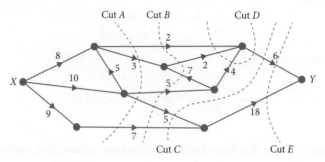

a Find the flow capacity of cut *B*. (1 mark)

b Which cut allows you to find the maximum flow from point *X* to point *Y*? (2 marks)

11 (3 marks) In the network below, the values on the edges give the maximum flow possible between each pair of vertices. The arrows show the direction of flow. A cut that separates the source from the sink in the network is also shown.

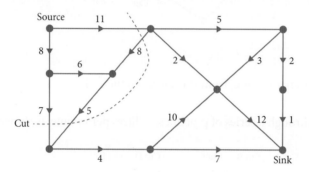

Determine the

a capacity of the cut (1 mark)

b maximum flow between source and sink through the network. (2 marks)

12 (2 marks) Vehicles from a town can drive onto a freeway along a network of one-way and two-way roads, as shown in the network diagram.

The numbers indicate the maximum number of vehicles per hour that can travel along each road in this network. The arrows represent the permitted direction of travel.

One of the four dotted lines shown on the diagram is the minimum cut for this network.

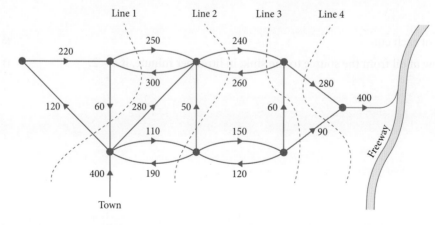

Find the maximum number of vehicles per hour that can travel through this network from the town onto the freeway.

9780170476959

13 © SCSA | MA2017 Q13 | (8 marks) The traffic flow (in hundreds of cars per hour) through a road network (*F* to *H*) is shown below.

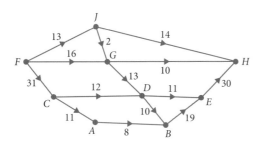

8.2

a By listing the different paths and their flow rate, determine the maximum flow through the network. (4 marks)

b Copy the diagram and verify the maximum flow obtained in part **a** by showing the minimum cut on the given network. (1 mark)

c i If **one** road is to be widened to allow for more traffic, which road should be chosen to increase the maximum flow the most? (1 mark)

ii How much more traffic should this road allow to flow and what would be the new maximum flow for the network? (2 marks)

14 © SCSA | MA2020 Q15 | (8 marks) The graph below shows a network of sewage pipes. The numbers on the edges indicate the number of litres per minute that can flow along each pipe.

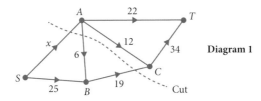

Diagram 1

a Show that the value of *x* is 38, given that the value of the cut is 57. (2 marks)

b Calculate the value of the maximum flow through the network. (2 marks)

c Copy **Diagram 2** below and indicate a possible flow along each pipe corresponding to the maximum flow calculated in part **b**. (2 marks)

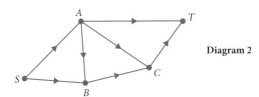

Diagram 2

Extra pipes, *BD*, *DC* and *DT*, are added to form a new system shown below. The capacities of the new pipes are indicated on the diagram. The original pipes have the same capacity as shown in Diagram 1.

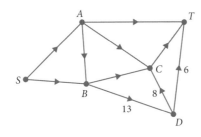

d How will the addition of these pipes affect the new maximum flow through the system? (2 marks)

15 ©SCSA MA2016 Q9 (8 marks) The network below shows the maximum rate of water flow (in litres per minute) through a system of water pipes from a source at A.

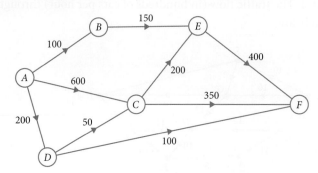

a What is the maximum amount of water that could be delivered to *F*, in litres per minute? (List each path used and the corresponding flow.)   (3 marks)

b Copy the above diagram and verify the maximum flow obtained in part **a** by showing a minimum cut on the given network.   (1 mark)

c Copy the following network and relabel it to show the flow you would direct along each pipe in order to achieve the maximum flow found in part **a** to point *F*.   (1 mark)

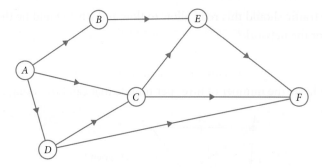

d When the maximum flow occurs from *A* to *F*, how much of the water, in litres per minute, passes through *C*?   (1 mark)

e The water flow through *C*, as calculated in part **d**, is reduced to a maximum of 480 litres per minute. In order to maintain the same maximum flow as that obtained in part **a**, the capacity of a single pipe (arc) is to be increased by the least amount. Which pipe should be chosen, and by how much should its capacity be increased?   (2 marks)

16 ©SCSA MA2018 Q17 (11 marks) Diagram 1 shows a network of pipes. The number on each edge gives the capacity of that pipe in L/min.

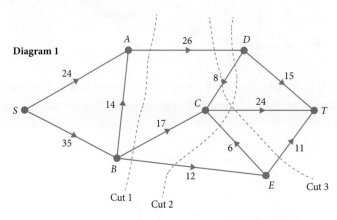

a State the capacities of the three cuts in Diagram 1.   (3 marks)

Diagram 2 shows a possible flow for the network of pipes.

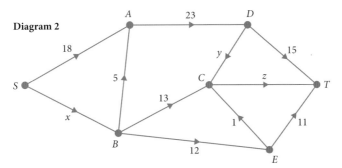

Diagram 2

b **i** Explain why the value of x is 30. (1 mark)

   ii Calculate the values of y and z. (2 marks)

c State which of the pipes are at full capacity in Diagram 2. (2 marks)

d State the value of the flow for the network in Diagram 2. (1 mark)

e **i** The value of the flow for Diagram 2 can be increased by 2 L/min. List the series of pipes that could be used to achieve this. (1 mark)

   ii Show that the increased flow in part **e i** is a maximum for this network of pipes. (1 mark)

17 (4 marks) The rangers at a wildlife park restrict access to the walking tracks through areas where the animals breed.

The edges on the directed network diagram represent one-way tracks through the breeding areas. The direction of travel on each track is shown by an arrow. The numbers on the edges indicate the maximum number of people who are permitted to walk along each track each day.

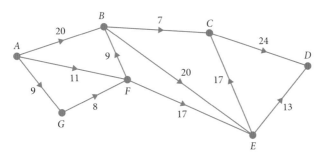

a Starting at A, how many people, in total, are permitted to walk to D each day? (1 mark)

On one day all the available walking tracks will be used by students on a school excursion. The students will start at A and walk in four separate groups to D.

Students must remain in the same groups throughout the walk.

b **i** Group 1 will have 17 students. This is the maximum group size that can walk together from A to D. Write down the path that group 1 will take. (1 mark)

   ii Groups 2, 3 and 4 will each take different paths from A to D.

   Copy the table below and complete the six missing entries. (2 marks)

Group	Maximum group size	Path taken from A to D
1	17	answered in part **b i**
2		
3		
4		

18 (3 marks) As an attraction for young children, a miniature railway runs throughout a new housing estate. The trains travel through stations that are represented by nodes on the directed network diagram.

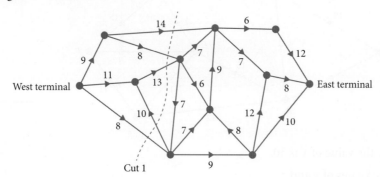

The number of seats available for children, between each pair of stations, is indicated beside the corresponding edge.

Cut 1, through the network, is shown in the diagram above.

a Determine the capacity of Cut 1. (1 mark)

b Determine the maximum number of seats available for children for a journey that begins at the West terminal and ends at the East terminal. (1 mark)

On one particular train, 10 children set out from the West terminal. No new passengers board the train on the journey to the East terminal.

c Determine the maximum number of children who can arrive at the East terminal on this train. (1 mark)

19 (3 marks) Simon built his holiday home on an estate. The estate has one-way streets between the entrance and the exit. There are restrictions on the number of trucks that are allowed to travel along each street per day. On the directed graph, the vertices represent the intersections of the one-way streets. The number on each edge is the maximum number of trucks that are allowed to travel along that street per day.

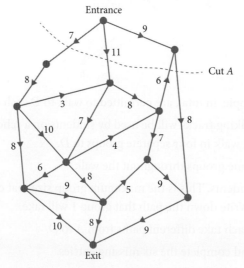

When considering the possible flow of trucks through this network, many different cuts can be made.

a Determine the capacity of Cut A, shown above. (1 mark)

b Find the maximum number of trucks that could travel from the entrance to the exit per day. (1 mark)

c A company would like to send one group of trucks from the entrance to the exit. All trucks in this group must follow each other and travel along the same route. The trucks in this group will be the only trucks to use these streets on that day. What is the maximum number of trucks that could be in this group? (1 mark)

 9780170476959

8.3 The assignment problem and bipartite graphs

The **assignment problem** (or **allocation problem**) involves finding the best way of matching the elements in two sets. For example, a group of workers could be allocated to a set of tasks in order to optimise a stated objective such as minimising cost, distance or time.

Bipartite graphs

A **bipartite graph** can be used to display the two sets of elements in an assignment problem. A bipartite graph has its vertices in two distinct sets and the edges join elements in the first set to elements in the second set.

Bipartite graphs can be undirected or directed.

In the bipartite graph on the right, there are three tasks {1, 2, 3} that can be performed by three people {A, B, C}. The edges show which people are qualified to perform which tasks.

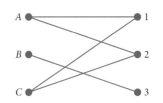

The bipartite graph shows:

- A can perform tasks 1 and 2
- B can perform task 3
- C can perform tasks 1 and 2.

Video playlist
The assignment problem and bipartite graphs

WORKED EXAMPLE 8	Finding the allocation from a bipartite graph

The bipartite graph shows the tasks that each of the four people is able to undertake. If each person must complete one task, find a valid allocation.

Steps	**Working**
1 Identify tasks that have the smallest number of links.	Cam is the only person who can do task 4.
2 Cam is the only person who can do task 4. Allocate Cam to task 4 then eliminate links from Cam to tasks 1, 2 and 3.	
3 Task 1 can only be done by Allan so allocate Allan to task 1 then eliminate links from Allan to task 3.	

---

4 Allocate Ben to task 3, which leaves Dan allocated task 2.

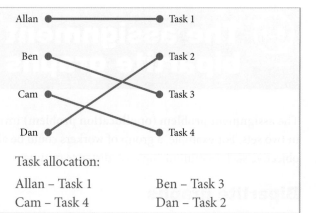

Task allocation:

Allan – Task 1	Ben – Task 3
Cam – Task 4	Dan – Task 2

Adjacency matrices and bipartite graphs

Adjacency matrix representation for a bipartite graph

An **adjacency matrix** can be used to represent a bipartite graph.

In the matrix, 1 represents a connection and 0 indicates that there is no connection.

This bipartite graph

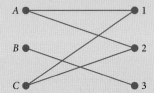

is represented by this matrix:

$$\begin{array}{c} \\ A \\ B \\ C \end{array} \begin{array}{ccc} 1 & 2 & 3 \\ \begin{bmatrix} 1 & 1 & 0 \\ 0 & 0 & 1 \\ 1 & 1 & 0 \end{bmatrix} \end{array}$$

WORKED EXAMPLE 9 | Drawing a bipartite graph from an adjacency matrix

Four workers A, B, C and D need to be allocated one task from tasks 1, 2, 3 and 4. The workers are only qualified to perform certain tasks and this is indicated in the matrix on the right. Draw a bipartite graph from the adjacency matrix and determine a valid allocation.

$$\begin{array}{c} \\ A \\ B \\ C \\ D \end{array} \begin{array}{cccc} 1 & 2 & 3 & 4 \\ \begin{bmatrix} 0 & 0 & 1 & 0 \\ 1 & 1 & 0 & 0 \\ 0 & 0 & 0 & 1 \\ 1 & 0 & 1 & 1 \end{bmatrix} \end{array}$$

Steps	**Working**
1 Draw a connection for every 1 in the matrix. $A3$, $B1$, $B2$, $C4$, $D1$, $D3$ and $D4$	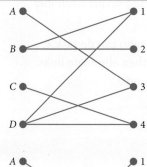
2 Start with the tasks with the smallest number of links. Allocate B to task 2 and remove the other link from B to 1.	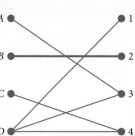

9780170476959

3 Task 1 can only be done by *D* so allocate
D to task 1 and remove any other links
from *D* to tasks 3 and 4.

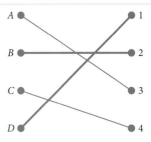

The allocation is *A*3, *B*2, *C*4 and *D*1.

The allocation problem – finding an optimum allocation

The aim of the allocation problem is to find a way of assigning the elements in the first set to the elements in the second set that meets an objective such as minimising cost or time. The method used to find the optimum allocation is called the **Hungarian algorithm** and is completed in two stages. Stage 1 is a row and column reduction. In most cases, this is sufficient to find a solution.

The Hungarian algorithm (Stage 1) – row and column reduction

1 Choose the smallest number in each row and subtract it from every element in the same row.

2 For every column that does not have a zero value, choose the smallest number in the column and subtract it from every element in the same column.

3 Cover all the zeros with the smallest number of lines (horizontal or vertical). The process is complete if the number of lines is equal to the number of rows.

WORKED EXAMPLE 10 | Applying Stage 1 of the Hungarian algorithm

Three workers *A*, *B* and *C* need to be allocated one task from tasks 1, 2 and 3. The time in hours that each worker takes to complete the tasks is shown in the table.

	1	2	3
A	8	10	9
B	9	11	7
C	7	9	8

If the tasks must be completed in the minimum time, find how the tasks are assigned.

Steps	Working

1 Identify the smallest number in each row and subtract it from the other numbers in the same row.

    Row *A* subtract 8

    Row *B* subtract 7

    Row *C* subtract 7

	1	2	3
A	0	2	1
B	2	4	0
C	0	2	1

2 For every column that does not have a zero value, choose the smallest number in the column and subtract it from every element in the same column.

    Column 1 has a zero

    Column 2 subtract 2

    Column 3 has a zero

	1	2	3
A	0	0	1
B	2	2	0
C	0	0	1

3 Cover all the zeros with the smallest number of lines (horizontal or vertical).

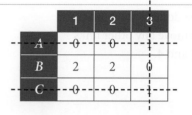

There are three lines and three rows, so a solution has been found.

4 The zeros indicate the allocations. Draw a bipartite graph connecting $A1$, $A2$, $B3$, $C1$ and $C2$.

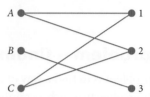

B needs to complete task 3, but A and C can complete either task 1 or task 2. So there are two possible allocations:

$A2$, $B3$, $C1$ and $A1$, $B3$, $C2$

The Hungarian algorithm – Stage 2

Stage 2 of the Hungarian algorithm is an extension to row and column reduction when this does **not** produce an optimum solution.

The Hungarian algorithm – Stage 2

1 Perform a row and column reduction and cover all the zeros with the smallest number of horizontal or vertical lines possible. If the number of lines is equal to the number of rows, a solution has been found.

2 If the number of lines does not equal the number of rows, find the smallest uncovered number and add it to every covered number. Numbers that are covered twice have this number added twice.

3 Subtract the smallest number in the matrix or table from all of the numbers in the table.

4 Cover all the zeros with the smallest number of horizontal or vertical lines and if the number of lines is equal to the number of rows, the process is complete. If this is not the case, repeat steps 2, 3 and 4 until a solution is found.

5 Draw a bipartite graph and use it to determine the optimum allocation.

WORKED EXAMPLE 11 | Applying stages 1 and 2 of the Hungarian algorithm

Four workers A, B, C and D all provide quotes for jobs 1, 2, 3 and 4. Their quotes are shown in the matrix, where the rows represent the workers A, B, C and D and the columns represent the jobs 1, 2, 3 and 4.

$$\begin{array}{c} \\ A \\ B \\ C \\ D \end{array} \begin{array}{cccc} 1 & 2 & 3 & 4 \\ \left[\begin{array}{cccc} 15 & 24 & 40 & 21 \\ 24 & 18 & 19 & 20 \\ 28 & 13 & 30 & 22 \\ 10 & 18 & 17 & 13 \end{array} \right] \end{array}$$

a Determine the best allocation of workers that minimises the total quote.

b Find the cost of the four jobs.

Steps	Working

a 1 Perform a row reduction.

Row A subtract 15

Row B subtract 18

Row C subtract 13

Row D subtract 10

	1	2	3	4
A	0	9	25	6
B	6	0	1	2
C	15	0	17	9
D	0	8	7	3

2 Perform a column reduction.

Column 3   subtract 1

Column 4   subtract 2

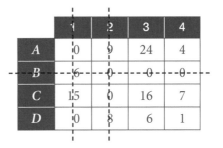

	1	2	3	4
A	0	9	24	4
B	6	0	0	0
C	15	0	16	7
D	0	8	6	1

3 Cover the zeros.

There are three lines and four rows.

As the number of lines does not equal the number of rows, Stage 2 of the Hungarian algorithm must be used.

	1	2	3	4
A	0	9	24	4
B	6	0	0	0
C	15	0	16	7
D	0	8	6	1

4 The smallest uncovered number is 1. Add this to every covered number but add 2 to the numbers covered twice.

	1	2	3	4
A	1	10	24	4
B	8	2	1	1
C	16	1	16	7
D	1	9	6	1

5 The smallest value is 1 so subtract this from every element in the matrix.

Now cover the zeros. There are now four lines and four rows, so the allocation is complete.

	1	2	3	4
A	0	9	23	3
B	7	1	0	0
C	15	0	15	6
D	0	8	5	0

6 Draw the bipartite graph. The zeros in the matrix represent the allocations.

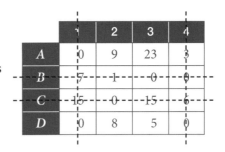

7 Determine the allocation.

Two jobs have a single worker connected.

Worker *C* for job 2

Worker *B* for job 3

This leaves

Worker *D* for job 4

Worker *A* for job 1

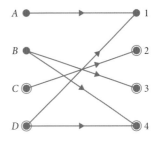

Allocation

*A*1, *B*3, *C*2, *D*4

b Look at the original matrix to determine the cost:

*A*1 = 15, *B*3 = 19, *C*2 = 13, *D*4 = 13

Cost = \$15 + \$19 + \$13 + \$13

    = \$60

The Hungarian algorithm – maximisation

The Hungarian algorithm can also be used to find the maximum weight matching. The assignment problem is transformed into a minimisation problem by subtracting every element in the matrix from the highest matrix element.

> ### The Hungarian algorithm – maximum allocation
>
> 1 Subtract every element in the matrix from the highest matrix element.
>
> 2 Use the Hungarian algorithm on the resulting matrix.

WORKED EXAMPLE 12 | **Maximum allocation problems**

Four workers A, B, C and D work at stations 1, 2, 3 and 4 for a four-hour shift. The number of items produced are shown in the matrix, where the rows represent the workers A, B, C and D and the columns represent the stations 1, 2, 3 and 4.

$$\begin{array}{c} \\ A \\ B \\ C \\ D \end{array} \begin{array}{cccc} 1 & 2 & 3 & 4 \\ \left[\begin{array}{cccc} 32 & 38 & 40 & 28 \\ 40 & 24 & 28 & 21 \\ 41 & 27 & 33 & 36 \\ 22 & 38 & 41 & 30 \end{array}\right] \end{array}$$

Determine the best allocation of workers that maximise the total number of items produced and the maximum total number of produced items.

Steps	Working
1 The highest element in the matrix is 41. Subtract every element in the matrix from 41. Row A: $41 - 32 = 9$, $41 - 38 = 3$, $41 - 40 = 1$ and $41 - 28 = 13$ Repeat this process for rows B, C and D.	$$\begin{array}{c} \\ A \\ B \\ C \\ D \end{array} \begin{array}{cccc} 1 & 2 & 3 & 4 \\ \left[\begin{array}{cccc} 9 & 3 & 1 & 13 \\ 1 & 17 & 13 & 20 \\ 0 & 14 & 8 & 5 \\ 19 & 3 & 0 & 11 \end{array}\right] \end{array}$$
2 Perform a row and column reduction.	Row reduction $$\begin{array}{c} \\ A \\ B \\ C \\ D \end{array} \begin{array}{cccc} 1 & 2 & 3 & 4 \\ \left[\begin{array}{cccc} 8 & 2 & 0 & 12 \\ 0 & 16 & 12 & 19 \\ 0 & 14 & 8 & 5 \\ 19 & 3 & 0 & 11 \end{array}\right] \end{array}$$ Column reduction $$\begin{array}{c} \\ A \\ B \\ C \\ D \end{array} \begin{array}{cccc} 1 & 2 & 3 & 4 \\ \left[\begin{array}{cccc} 8 & 0 & 0 & 7 \\ 0 & 14 & 12 & 14 \\ 0 & 12 & 8 & 0 \\ 19 & 1 & 0 & 6 \end{array}\right] \end{array}$$
3 Cover the zeros with the smallest number of lines. As the number of lines is equal to the number of rows, the process is complete.	$$\begin{array}{c} \\ A \\ B \\ C \\ D \end{array} \begin{array}{cccc} 1 & 2 & 3 & 4 \\ \left[\begin{array}{cccc} 8 & 0 & 0 & 7 \\ 0 & 14 & 12 & 14 \\ 0 & 12 & 8 & 0 \\ 19 & 1 & 0 & 6 \end{array}\right] \end{array}$$
4 Draw a bipartite graph and determine the maximum matching.	 The maximum allocation is $A2$, $B1$, $C4$, $D3$.
5 Use the original matrix to find the total production.	The maximum total production is $38 + 40 + 36 + 41 = 155$.

9780170476959

The Hungarian algorithm – unbalanced allocations

When the cost matrix is not a square matrix, where the number of rows equals the number of columns, the assignment problem is called an unbalanced assignment problem. In problems of this type, dummy rows or columns are added to the matrix to transform it into a square matrix. The elements in the dummy row or column are all zeros.

> **The Hungarian algorithm – unbalanced allocations**
>
> 1 Add rows or columns of zeros to transform the matrix into a square matrix.
>
> 2 Use the Hungarian algorithm on the resulting matrix.

WORKED EXAMPLE 13	Unbalanced allocation problems

The time taken for machines A, B, C and D to complete three tasks 1, 2 and 3 are shown in the matrix. If each machine can complete only one task, find the allocation that minimises the total time taken.

$$\begin{array}{c} & \begin{array}{ccc} 1 & 2 & 3 \end{array} \\ \begin{array}{c} A \\ B \\ C \\ D \end{array} & \left[\begin{array}{ccc} 5 & 7 & 11 \\ 8 & 5 & 5 \\ 6 & 7 & 10 \\ 10 & 4 & 4 \end{array}\right] \end{array}$$

Steps	**Working**

1 Add a dummy column of zeros to transform the matrix into a square 4×4 matrix.

$$\begin{array}{c} & \begin{array}{cccc} 1 & 2 & 3 & 4 \end{array} \\ \begin{array}{c} A \\ B \\ C \\ D \end{array} & \left[\begin{array}{cccc} 5 & 7 & 11 & 0 \\ 8 & 5 & 5 & 0 \\ 6 & 7 & 10 & 0 \\ 10 & 4 & 4 & 0 \end{array}\right] \end{array}$$

2 Perform a row and column reduction.

The row reduction will not alter any rows because each row contains a zero.

Column reduction

$$\begin{array}{c} & \begin{array}{cccc} 1 & 2 & 3 & 4 \end{array} \\ \begin{array}{c} A \\ B \\ C \\ D \end{array} & \left[\begin{array}{cccc} 0 & 3 & 7 & 0 \\ 3 & 1 & 1 & 0 \\ 1 & 3 & 6 & 0 \\ 5 & 0 & 0 & 0 \end{array}\right] \end{array}$$

3 Cover the zeros with the smallest number of lines. There are three lines but four rows so we must complete Stage 2 of the Hungarian algorithm.

$$\begin{array}{c} & \begin{array}{cccc} 1 & 2 & 3 & 4 \end{array} \\ \begin{array}{c} A \\ B \\ C \\ D \end{array} & \left[\begin{array}{cccc} 0 & 3 & 7 & 0 \\ 3 & 1 & 1 & 0 \\ 1 & 3 & 6 & 0 \\ 5 & 0 & 0 & 0 \end{array}\right] \end{array}$$

4 The smallest uncovered number is 1. Add 1 to every covered number but 2 to numbers covered twice.

Continue until Stage 2 is complete.

$$\begin{array}{c} & \begin{array}{cccc} 1 & 2 & 3 & 4 \end{array} \\ \begin{array}{c} A \\ B \\ C \\ D \end{array} & \left[\begin{array}{cccc} 1 & 4 & 8 & 2 \\ 3 & 1 & 1 & 1 \\ 1 & 3 & 6 & 1 \\ 6 & 1 & 1 & 2 \end{array}\right] \end{array}$$

$$\begin{array}{c} & \begin{array}{cccc} 1 & 2 & 3 & 4 \end{array} \\ \begin{array}{c} A \\ B \\ C \\ D \end{array} & \left[\begin{array}{cccc} 0 & 3 & 7 & 1 \\ 2 & 0 & 0 & 0 \\ 0 & 2 & 5 & 0 \\ 5 & 0 & 0 & 1 \end{array}\right] \end{array}$$

5 Use a bipartite graph to show the matching of machines to tasks. The possible solutions are A1, B2, C4, D3 or A1, B3, C4, D2.

Task 4 is a dummy task so it is not listed as part of the allocation.

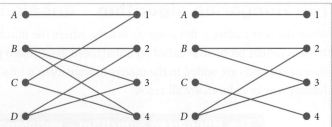

The best allocation of machines to tasks is A1, B2, D3 or A1, B3, D2.

Machine C will not be allocated a task because task 4 matches the dummy column.

Vdeo
WACE question analysis: Trees, flow and assignment

WACE QUESTION ANALYSIS

© SCSA | MA2021 Q15 | Calculator-assumed (7 marks)

A number of streets connecting locations *A* to *G* inclusive have been identified as needing lighting upgrades with energy-efficient LED lamps. The council has decided that each location must have at least one connecting street that receives the lighting upgrade. The table below shows the cost, in dollars, of completing the upgrade in each street.

	A	B	C	D	E	F	G
A	–	7 900	10 300	–	7 100	–	–
B	7 900	–	7 600	6 200	6 500	–	15 400
C	10 300	7 600	–	8 500	–	12 200	9 200
D	–	6 200	8 500	–	8 000	4 700	9 800
E	7 100	6 500	–	8 000	–	4 800	–
F	–	–	12 200	4 700	4 800	–	10 100
G	–	15 400	9 200	9 800	–	10 100	–

The council has a limited budget, so it needs to complete the upgrades at minimum cost.

a Copy the following diagram. Demonstrate the use of Prim's algorithm on the table above to determine the minimum spanning tree and draw it on your diagram. (3 marks)

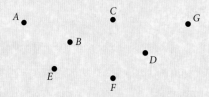

b The council has set aside $42 000 to complete the lighting upgrades. Does it have enough in its budget to make the necessary upgrades? Justify your answer. (2 marks)

c Due to the location of the police station, the upgrade from *D* to *C* must be included. What effect will this have on the minimum cost and the spanning tree? (2 marks)

Reading the question

- Take note of the information in the table that represents a network: the numbers in the table representing the cost of lighting in particular streets and each number is the weight of an edge in the graph.
- Highlight the type of answer required for each question. This requires completing an algorithm in a table and interpreting the result by drawing a spanning tree.
- The question does not specify the graphing or table methods of Prim's algorithm, so choose the method that best suits the problem.

Thinking about the question

- The question requires an understanding of minimum spanning trees and Prim's algorithm.
- You will need to know how to use Prim's algorithm in a table.
- You will need to know how to determine the weight of a minimum spanning tree.

Worked solution ($\checkmark$ = 1 mark)

a Prim's algorithm

	A	B	C	D	E	F	G
A	~	7900	10300	~	7100	~	~
B	7900	~	7600	6 200	6500	~	15400
C	10300	7 600	~	8500	~	12200	9200
D	~	6200	8500	~	8000	4 700	9800
E	7 100	6500	~	8000	~	4800	~
F	~	~	12200	4700	4 800	~	10100
G	~	15400	9 200	9800	~	10100	~

correctly demonstrates use of Prim's algorithm on the table $\checkmark$

obtains correct solution using Prim's algorithm $\checkmark$

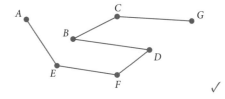

$\checkmark$

b Cost = 7100 + 4800 + 4700 + 6200 + 7600 + 9200

   = **\$39 600** $\checkmark\checkmark$

c 39 600 − 7600 (BC) = 32 000

   32 000 + 8500 (DC) = 40 500

   This means the cost now **increases by \$900 to \$40 500**. $\checkmark$

   The spanning tree will **not include BC**. $\checkmark$

Recap

Use the following information to answer Questions 1 and 2.

In this directed graph, the values on the edges give the maximum flow between vertices in the direction of the arrows.

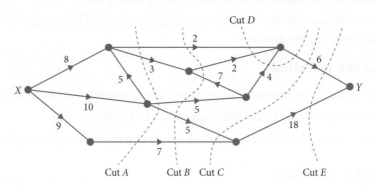

1 The capacity of cut *B* is

   A 18   **B** 21   **C** 23   **D** 24   **E** 28

2 The invalid cut is

   A Cut *A*   **B** Cut *B*   **C** Cut *C*   **D** Cut *D*   **E** Cut *E*

Mastery

3 WORKED EXAMPLE 8   Determine the allocation of tasks from the following bipartite graph.

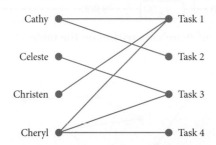

4 WORKED EXAMPLE 9   Workers *A*, *B*, *C* and *D* are each to be assigned one task from tasks 1, 2, 3 and 4. The matrix representing the workers that are qualified to perform the tasks is shown.

Draw a bipartite graph and, hence, determine a valid allocation.

$$\begin{array}{c} \\ A \\ B \\ C \\ D \end{array} \begin{array}{c} 1 \quad 2 \quad 3 \quad 4 \\ \begin{bmatrix} 0 & 0 & 1 & 0 \\ 1 & 0 & 1 & 1 \\ 0 & 0 & 1 & 1 \\ 0 & 1 & 0 & 1 \end{bmatrix} \end{array}$$

5 WORKED EXAMPLE 10   Perform row and column reductions on the matrix that represents the time, in hours, for workers *A*, *B*, *C* and *D* to complete tasks 1, 2, 3 and 4 and, hence, determine the optimum allocation and the minimum time to complete all four tasks.

$$\begin{array}{c} \\ A \\ B \\ C \\ D \end{array} \begin{array}{c} 1 \quad\; 2 \quad\; 3 \quad\; 4 \\ \begin{bmatrix} 25 & 14 & 7 & 15 \\ 6 & 10 & 8 & 9 \\ 13 & 5 & 9 & 5 \\ 8 & 5 & 10 & 11 \end{bmatrix} \end{array}$$

6 🔗 WORKED EXAMPLE 11   A taxi company has three taxis A, B and C available and there are three customers 1, 2 and 3 who require a taxi. The distance (km) that each taxi must travel to get to the customers is shown in the table.

Use the Hungarian algorithm to find the optimum allocation that minimises the distance travelled.

	1	2	3
A	11	19	17
B	21	15	13
C	15	18	21

7 🔗 WORKED EXAMPLE 12   Four employees A, B, C and D need to be allocated to four sales districts 1, 2, 3 and 4. Their estimated sales in each district, in a month, are shown in the matrix. Determine the best allocation of employees to districts that maximises the total number of items sold and the maximum total sales.

$$\begin{array}{c} \\ A \\ B \\ C \\ D \end{array} \begin{array}{cccc} 1 & 2 & 3 & 4 \\ \left[\begin{array}{cccc} 33 & 40 & 35 & 39 \\ 27 & 33 & 30 & 37 \\ 38 & 40 & 28 & 40 \\ 24 & 28 & 21 & 36 \end{array}\right] \end{array}$$

8 🔗 WORKED EXAMPLE 13   The times taken by three workers A, B and C to complete four tasks 1, 2, 3 and 4 are shown in the matrix. If each worker can complete only one task, find the allocation that minimises the total time taken.

$$\begin{array}{c} \\ A \\ B \\ C \end{array} \begin{array}{cccc} 1 & 2 & 3 & 4 \\ \left[\begin{array}{cccc} 21 & 26 & 23 & 29 \\ 16 & 19 & 24 & 18 \\ 20 & 18 & 22 & 25 \end{array}\right] \end{array}$$

Calculator-free

9 © SCSA   MA2017 Q2   (7 marks) A supermarket provides a delivery service to its customers. This morning, there are four deliveries (1, 2, 3 and 4) to be made. Each of four drivers, John, Kerry, Liam and Max, is available to do one of the deliveries.

Table 1 shows the time, in minutes, that each driver would take to complete each of the four deliveries.

Table 1

		Delivery driver			
		John	Kerry	Liam	Max
Deliveries	1	35	31	41	36
	2	25	26	33	36
	3	32	28	25	24
	4	27	30	31	28

The store manager will allocate the deliveries so that the total delivery time is at a minimum. He decides to use the Hungarian algorithm to determine the allocation of deliveries to the drivers.

His first step is to subtract the minimum entry in each row from each element, ensuring that each row contains at least one zero.

Table 2

		Delivery driver			
		John	Kerry	Liam	Max
Deliveries	1	4	0	10	5
	2	0	1		11
	3	8	4	1	0
	4	0	3	4	1

a What is the number missing from the shaded cell? (1 mark)

The second step is to ensure that all columns contain at least one zero. The numbers that result from this step are shown in Table 3

Table 3

		Delivery driver			
		John	Kerry	Liam	Max
Deliveries	1	4	0	9	5
	2	0	1	7	11
	3	8	4	0	0
	4	0	3	3	1

b The smallest number of horizontal and vertical lines that can be drawn to cover all the zeros is three.

   i Copy Table 3 and draw in these lines. (1 mark)

   ii State why an allocation of delivery drivers cannot be made yet. (1 mark)

c Continue the steps of the Hungarian algorithm to determine the optimum allocation of deliveries to the drivers. Copy and complete the table below and state the minimum total delivery time. (4 marks)

Delivery driver	John	Kerry	Liam	Max
Delivery				

10 ⓒ SCSA   MA2016 Q3   (8 marks) A foreman in a factory has four workers, Adam, Ben, Cate and Demi, and four jobs to complete. The time, in hours, each worker can complete a particular job is given in the weighted bipartite graph below.

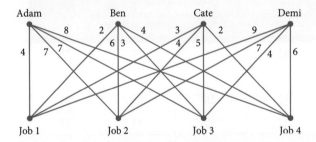

a Copy and complete the matrix associated with the bipartite graph above. (2 marks)

$$\begin{array}{c} \\ A \\ B \\ C \\ D \end{array} \begin{array}{cccc} J1 & J2 & J3 & J4 \end{array} \\ \left[\begin{array}{cccc} 4 & & 7 & \\ & 6 & 3 & \\ 3 & & & 2 \\ & 7 & & 6 \end{array} \right]$$

b Using the Hungarian algorithm, determine which job the foreman should assign to each of his workers so that the total time is minimised. (6 marks)

11 ©SCSA MA2020 Q3 (8 marks) The bipartite graph below shows the average number of sales per day of four specialist pieces of machinery by four salespersons. The salespersons went through a rotation process of spending a number of full days selling only one type of machine before moving to the next. The company that makes these machines wants to maximise the sales by allocating the sales of each machine to only one person.

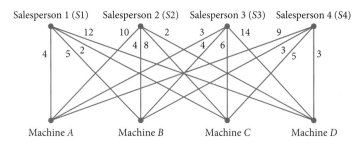

a Copy and complete the matrix below showing the information from the bipartite graph. (1 mark)

$$\begin{array}{c} \\ S1 \\ S2 \\ S3 \\ S4 \end{array} \begin{array}{cccc} A & B & C & D \\ \left[\begin{array}{cccc} 4 & 5 & 2 & 12 \\ 10 & 4 & 8 & 2 \\ 3 & 4 & & \\ & & & \end{array}\right] \end{array}$$

b Show use of the Hungarian algorithm to allocate each salesperson to a specialist machine in order to maximise sales by copying and completing the table below. (5 marks)

Salesperson	1	2	3	4
Machine				

c How many more sales are made by using the allocation in part **b** compared to the allocation: S1 – A, S2 – B, S3 – C and S4 – D? (2 marks)

12 ©SCSA MA2018 Q4 (8 marks) A company produces rolls of shade cloth. Today, there are three different machines that can be used (1, 2 and 3) and four workers who can operate these machines (Peter, Quentin, Rebecca and Sasha). Each machine will have one worker assigned to it for the whole day.

The table below shows the number of metres of shade cloth that can be produced in a day by each machine operator.

		Machines		
		1	2	3
Workers	Peter	300	250	270
	Quentin	290	410	320
	Rebecca	190	240	120
	Sasha	310	410	280

a Draw the weighted bipartite graph by copying the below and showing the possible allocations for each of the workers. (2 marks)

Peter   Quentin   Rebecca   Sasha

•   •   •   •

•   •   •

Machine 1   Machine 2   Machine 3

The company manager wants to allocate the workers to the machines so that the production for the day is at a maximum. She decides to use the Hungarian algorithm to determine the allocation. Her first step is to rewrite the information in matrix form, adding in a column containing all zeros.

$$\begin{bmatrix} 300 & 250 & 270 & 0 \\ 290 & 410 & 320 & 0 \\ 190 & 240 & 120 & 0 \\ 310 & 410 & 280 & 0 \end{bmatrix}$$

b Why has she added the column of zeros? (1 mark)

c Continue the steps of the Hungarian algorithm, showing the optimum allocation of workers to machines in a table like the table below. State the maximum total length of shade cloth that can be produced in the day. (5 marks)

Worker	Peter	Quentin	Rebecca	Sasha
Machine				

13 (6 marks) Kate, Lexie, Mei and Nasim enter a competition as a team. In this competition, the team must complete four tasks, W, X, Y and Z, as quickly as possible.

The table shows the time, in minutes, that each person would take to complete each of the four tasks.

	Kate	Lexie	Mei	Nasim
W	6	3	4	6
X	4	3	5	5
Y	5	7	9	6
Z	3	2	3	2

If each team member is allocated one task only, determine the minimum time in which this team would complete the four tasks.

Calculator-assumed

14 (3 marks) George, Harriet, Ian, Josie and Keith are a group of five musicians. They are forming a band where each musician will fill one position only. The bipartite graph illustrates the positions that each is able to fill.

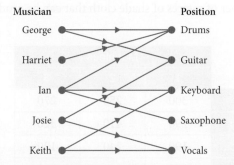

a Which musician **must** play the guitar? (1 mark)

b Copy and complete the table showing the positions that the following musicians **must** fill in the band. (2 marks)

Person	Position
Harriet	
Ian	
Keith	

9780170476959

15 (5 marks) Four tasks, *W*, *X*, *Y* and *Z*, must be completed. Four workers, Julia, Ken, Lana and Max, will each do one task.

Table 1 shows the time, in minutes, that each person would take to complete each of the four tasks.

Table 1

		Worker			
		Julia	Ken	Lana	Max
Task	W	26	21	22	25
	X	31	26	21	38
	Y	29	26	20	27
	Z	38	26	26	35

The tasks will be allocated so that the total time of completing the four tasks is a minimum. The Hungarian algorithm will be used to find the optimal allocation of tasks.

Step 1 of the Hungarian algorithm is to subtract the minimum entry in each row from each element in the row.

a Complete step 1 for task *X* by determining the number missing from the shaded cell in Table 2. (1 mark)

Table 2

		Worker			
		Julia	Ken	Lana	Max
Task	W	5	0	1	4
	X	10	5	0	
	Y	9	6	0	7
	Z	12	0	0	9

The second step of the Hungarian algorithm ensures that all columns have at least one zero. The numbers that result from this step are shown in Table 3.

Table 3

		Worker			
		Julia	Ken	Lana	Max
Task	W	0	0	1	0
	X	5	5	0	13
	Y	4	6	0	3
	Z	7	0	0	5

b Following the Hungarian algorithm, the smallest number of lines that can be drawn to cover the zeros is shown dashed in Table 3. These dashed lines indicate that an optimal allocation cannot be made yet. Give a reason why. (1 mark)

c Complete the steps of the Hungarian algorithm to produce a table from which the optimal allocation of tasks can be made. (1 mark)

d State the name of the task that each person should do for the optimal allocation of tasks. (2 marks)

16 (3 marks) Fencedale High School offers students a choice of four sports, football, tennis, athletics and basketball.

The bipartite graph illustrates the sports that each student can play.

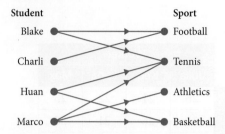

Each student will be allocated to only one sport.

a Copy and complete the table below by allocating the appropriate sport to each student. (1 mark)

Student	Sport
Blake	
Charli	
Huan	
Marco	

b The school medley relay team consists of four students, Anita, Imani, Jordan and Lola. The medley relay race is a combination of four different sprinting distances: 100 m, 200 m, 300 m and 400 m, run in that order.

The following table shows the best time, in seconds, for each student for each sprinting distance.

Student	Best time for each sprinting distance (seconds)			
	100 m	200 m	300 m	400 m
Anita	13.3	29.6	61.8	87.1
Imani	14.5	29.6	63.5	88.9
Jordan	13.3	29.3	63.6	89.1
Lola	15.2	29.2	61.6	87.9

The school will allocate each student to one sprinting distance in order to minimise the total time taken to complete the race.

To which distance should each student be allocated? (2 marks)

17 (4 marks) Each child is to be driven by his or her parents to one of four different concerts. The following table shows the distance that each car would have to travel, in kilometres, to each of the four concerts.

	Concert 1	Concert 2	Concert 3	Concert 4
James	10	16	18	20
Dante	9	14	19	15
Tahlia	15	13	20	18
Chanel	10	15	21	16

The concerts will be allocated so as to minimise the total distance that must be travelled to take the children to the concerts. The Hungarian algorithm is to be used to find this minimum value.

9780170476959

a Step 1 of the Hungarian algorithm is to subtract the minimum entry in each row from each element in the row. Copy the table below and **complete** step 1 for Tahlia by writing the missing values. (1 mark)

	Concert 1	Concert 2	Concert 3	Concert 4
James	0	6	8	10
Dante	0	5	10	6
Tahlia				
Chanel	0	5	11	6

After further steps of the Hungarian algorithm have been applied, the table is as follows.

	Concert 1	Concert 2	Concert 3	Concert 4
James	0	5	0	4
Dante	0	4	2	0
Tahlia	3	0	0	0
Chanel	0	4	3	0

It is now possible to allocate each child to a concert.

b Explain why this table shows that Tahlia should attend Concert 2. (1 mark)

c Determine the concerts that could be attended by James, Dante and Chanel to minimise the total distance travelled. Copy the table and write in your answers. (1 mark)

	Concert
James	
Dante	
Tahlia	2
Chanel	

d Determine the minimum total distance, in kilometres, travelled by the four cars. (1 mark)

18 (2 marks) Bai joins his friends Agatha, Colin and Diane when he arrives for the holiday in Seatown. Each person will plan one tour that the group will take.

Table 1 shows the time, in minutes, it would take each person to plan each of the four tours.

Table 1

	Agatha	Bai	Colin	Diane
Tour 1	13	7	13	12
Tour 2	14	9	8	7
Tour 3	19	25	21	18
Tour 4	10	7	11	10

The aim is to minimise the total time it takes to plan the four tours. Agatha applies the Hungarian algorithm to Table 1 to produce Table 2. Table 2 shows the final result of all her steps of the Hungarian algorithm.

Table 2

	Agatha	Bai	Colin	Diane
Tour 1	3	0	3	3
Tour 2	6	4	0	0
Tour 3	0	9	2	0
Tour 4	0	0	1	1

a In Table 2 there is a zero in the column for Colin.

When all values in the table are considered, what conclusion about minimum total planning time can be made from this zero? (1 mark)

b Determine the minimum total planning time, in minutes, for all four tours. (1 mark)

Trees and spanning trees

- A **tree** is a connected graph with no loops, multiple edges or cycles.
- The number of edges in a tree is always one less than the number of vertices.
- A **spanning tree** is a tree subgraph that connects all the vertices of the original graph.
- Every connected graph has at least one spanning tree.
- A **minimum spanning tree** is the spanning tree with the smallest total weight.

Finding a minimum spanning tree

- The minimum spanning tree can be found by inspection if the graph isn't complex.

Use **Prim's algorithm** for complex graphs:

1   Start at any vertex and choose the edge with the lowest weight connected to this vertex.

2   Look at *all* the edges connecting to the vertices you've chosen so far (*not just the last vertex connected*) and choose the edge with the lowest weight that doesn't connect to a vertex already in the tree. If there are edges with equal lowest weights, choose one of them.

Prim's algorithm from a table

Prim's algorithm for finding a minimum spanning tree using a table:

1   Pick any column vertex to start the spanning tree and highlight the column header and strike out the row with the same letter.

2   Select and circle the lowest weight in the column.

3   If the lowest weight is in column *A* row *B*, highlight column header *B* and strike out row *B*.

4   Locate and circle the next lowest weight edge, selecting only from highlighted columns. You cannot select from the rows which have been eliminated by a strike-out line.

5   If there is more than one edge with the lowest weight, pick any of them.

6   Identify the row letter of the circled weight and highlight the column header, then strike out the row with the same letter.

7   Continue until all the columns are highlighted and all the rows have been crossed out. The selected weights identify the edges in the minimum spanning tree.

Flow capacity and maximum flow

- The first vertex of the network is called the **source** and the final vertex of the network is called the **sink**.
- The **inflow capacity** is the total flow capacity entering a vertex and the **outflow capacity** is the total flow capacity leaving the vertex.
- The **maximum flow through a vertex** is the smaller of the total inflow capacity of the vertex and the total outflow capacity of the vertex.

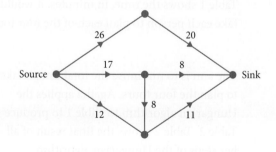

Maximum flow – minimum cut

- The maximum flow from source to sink can be determined by finding the cut that produces the minimum value.
- The maximum flow from source to sink = the capacity of the minimum cut.
- To determine the maximum flow for a network
  1. Identify cuts through the network.
  2. Find the capacity of each cut.
  3. Maximum flow = capacity of the minimum cut.

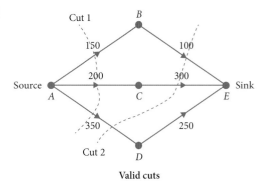

Valid cuts

Bipartite graphs

- A **bipartite graph** has its vertices in two distinct sets and the edges join elements in the first set to elements in the second set.

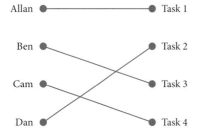

The Hungarian algorithm (stage 1) – row and column reduction

1. Choose the smallest number in each row and subtract it from every element in the same row.
2. For every column that does not have a zero value, choose the smallest number in the column and subtract it from every element in the same column.
3. Cover all the zeros with the smallest number of lines (horizontal or vertical). The process is complete if the number of lines is equal to the number of rows.

The Hungarian algorithm – Stage 2

1. Perform a row and column reduction and cover all the zeros with the smallest number of horizontal or vertical lines possible. If the number of lines is equal to the number of rows, a solution has been found.
2. If the number of lines does not equal the number of rows, find the smallest uncovered number and add it to every covered number. Numbers that are covered twice have this number added twice.
3. Subtract the smallest number in the matrix or table from all the numbers in the table.
4. Cover all the zeros with the smallest number of horizontal or vertical lines and if the number of lines is equal to the number of rows, the process is complete. If this is not the case, repeat steps 2, 3 and 4 until a solution is found.
5. Draw a bipartite graph and use it to determine the optimum allocation.

The Hungarian algorithm – maximisation

1. Subtract every element in the matrix from the highest matrix element.
2. Use the Hungarian algorithm on the resulting matrix.

The Hungarian algorithm – unbalanced allocations

1. Add rows or columns of zeros to transform the matrix into a square matrix.
2. Use the Hungarian algorithm on the resulting matrix.

Cumulative examination: Calculator-free

Total number of marks: 28   Reading time: 3 minutes   Working time: 28 minutes

1 (5 marks) The height of a woodland elf was 40 cm at birth. Her increase in height in each year, in cm, is modelled by the arithmetic sequence 18 cm, 15 cm, 12 cm …

  a Write a recursive rule representing this sequence. (2 marks)

  b In what year did the elf not grow any taller? (2 marks)

  c Find the maximum height of the elf. (1 mark)

2 (3 marks) A graph containing five vertices, *A*, *B*, *C*, *D* and *E*, is shown.

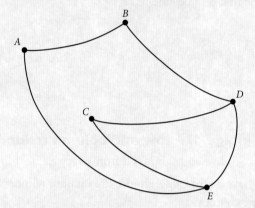

State the appropriate mathematical term that best describes each of the following walks.

  a *A–B–D–C–E* (1 mark)

  b *C–E–A–B–D–C* (1 mark)

  c *D–B–A–E–D–C–E* (1 mark)

3 (6 marks) Laura was keen to compare interest rates offered by different banks, so she decided to construct a table showing the effective annual rates of interest (%). Part of her table is shown below.

Compounding period	Rate of interest (p.a.)				
	3.50%	4%	4.50%	5%	5.50%
Quarterly	3.553	4.06	4.577	5.095	5.614
Monthly	3.565	4.074	4.594	5.116	5.641
Daily	3.57	4.081	4.602	5.127	5.654

  a Laura wants to borrow $7000 to purchase a second-hand car. A bank offers to lend her the money at the rate of 5% p.a. for one year. She plans to pay off the entire loan (including the interest) at the end of the year. Which compounding period should she sign up for? Justify your decision. (2 marks)

  b Laura is curious to know how much interest she would earn by investing $200 for a year, earning 3.5% p.a. with interest compounded quarterly. Determine the interest she would earn. (2 marks)

  c Laura's brother has $10 000 to invest for a year. He has been offered a rate of 5.5% p.a., with interest compounded monthly. Determine the value of the investment at the end of the year. (2 marks)

4 © SCSA MA2018 Q1 (4 marks) The weighted network below represents an orienteering map where the vertices represent the various stations and the edges represent bush tracks joining the stations. The distances on the edges are in metres. The organisers wish to install freshwater fountains at each station, using the minimum length of piping necessary to connect the stations along the bush tracks.

a Copy the diagram below and highlight the bush tracks where the pipes should be installed. (2 marks)

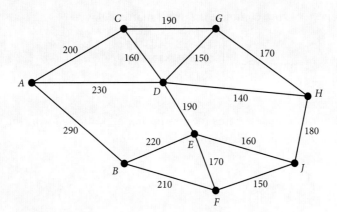

b Calculate the minimum length of piping required. (2 marks)

5 © SCSA MA2019 Q3 (10 marks) A company has four small workshops that each produce four different types of outdoor furniture. The annual cost of production of the furniture at each workshop is shown in the table below, with all values in thousands of dollars.

	Type 1 $'000	Type 2 $'000	Type 3 $'000	Type 4 $'000
Workshop A	25	43	50	39
Workshop B	33	31	56	39
Workshop C	28	47	59	38
Workshop D	36	32	56	41

The cost matrix is given by

$$\begin{bmatrix} 25 & 43 & 50 & 39 \\ 33 & 31 & 56 & 39 \\ 28 & 47 & 59 & 38 \\ 36 & 32 & 56 & 41 \end{bmatrix}$$

The company is interested in knowing what the minimum annual cost would be if each furniture type was allocated to its own individual workshop. The Hungarian algorithm is to be used to determine the allocation and the minimum annual cost. The first step of the Hungarian algorithm, where the smallest number in each row is subtracted from all other numbers in that row, is shown below.

$$\begin{bmatrix} 0 & 18 & 25 & 14 \\ 2 & 0 & 25 & 8 \\ 0 & 19 & 31 & 10 \\ 4 & 0 & 24 & 9 \end{bmatrix}$$

a Copy the following table and continue the steps of the Hungarian algorithm to determine the appropriate allocation of workshops to furniture type, and state the **minimum** annual cost. (5 marks)

Type	Type 1	Type 2	Type 3	Type 4
Workshop				

The revenue matrix, in thousands of dollars, for the sale of the furniture produced annually at each workshop is given by

$$\begin{bmatrix} 37 & 61 & 60 & 53 \\ 45 & 52 & 73 & 50 \\ 38 & 65 & 75 & 55 \\ 44 & 54 & 76 & 45 \end{bmatrix}$$

b Given that *profit = revenue − cost*, complete the profit matrix below. (1 mark)

$$\text{Profit matrix} = \begin{bmatrix} 12 & 18 & 10 & 14 \\ 12 & 21 & 17 & 11 \\ 10 & 18 & & \\ & & & \end{bmatrix}$$

c Use the Hungarian algorithm to determine the appropriate allocation of workshops to furniture type that will produce the **maximum** annual profit. (4 marks)

Cumulative examination: Calculator-assumed

Total number of marks: 35   Reading time: 4 minutes   Working time: 35 minutes

1 (2 marks) The table shows the *mean age*, in years, and the *mean height*, in centimetres, of 648 women from seven different age groups.

	Age group						
	Twenties	Thirties	Forties	Fifties	Sixties	Seventies	Eighties
Mean age (years)	26.3	35.2	45.2	55.3	65.1	74.8	83.1
Mean height (cm)	167.1	164.9	164.8	163.4	161.2	158.4	156.7

Data: J Sorkin et al., 'Longitudinal change in height of men and women: Implications for interpretation
of the body mass index', *American Journal of Epidemiology*, vol. 150, no. 9, 1999, p. 971

A scatterplot displaying this data shows an association between the *mean height* and the *mean age* of these women. In an initial analysis of the data, a line is fitted to the data by eye, as shown.

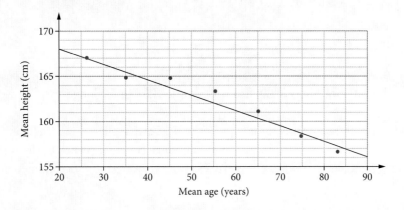

a Describe this association in terms of strength and direction.   (1 mark)

b In a further analysis of the data, a least-squares line was fitted. The associated residual plot that was generated is shown below.

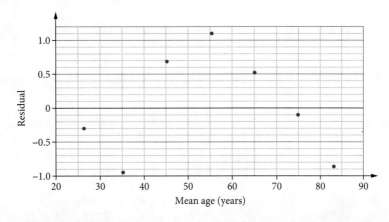

Determine if a linear model is appropriate for this graph by referring to the residual plot.   (1 mark)

2 (4 marks) The population of fish in a lake decreases, without intervention, at a rate of 6% per month. At the end of each month, an extra 15 fish are added to the lake. At the beginning of January 2025, the lake holds 720 fish.

Let P_n be the population of fish in the lake at the beginning of month n, where $n = 1$ corresponds to January 2025.

  a Copy the axes below and graph the population of fish at the beginning of January each year from January 2025 up to and including January 2030.

(2 marks)

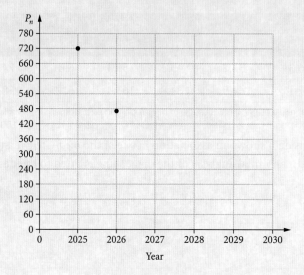

  b To the nearest whole number, how many fish will be in the lake in the long run?

(2 marks)

3 (3 marks) An area of a property contains eight large bushes that are labelled A to H, as shown on the graph. The farmer's dog enjoys running around this area, stopping at each bush on the way. The numbers on the edges joining the vertices give the shortest distance, in metres, between bushes.

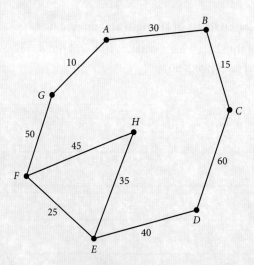

  a Explain why the dog could not follow a closed Eulerian trail through this network.

(1 mark)

  b If the dog follows the shortest Hamiltonian path, name a bush at which the dog could start and a bush at which the dog could finish.

(1 mark)

  c The sum of all distances shown on the graph is 310 m. The dog starts and finishes at bush F and runs along every edge in the network. What is the shortest distance, in metres, that the dog could have run?

(1 mark)

4 (3 marks) Samantha has borrowed $12 000 from a bank at a reducible interest rate of 15% per annum with interest accrued and repayments made monthly. Standard repayments are set at $400 per month.

The table below shows the progress of the loan for the first 3 months. All values have been rounded to the nearest cent.

Month	Amount ($) owing at beginning of month	Interest for the month	Repayment	Amount ($) owing at end of month
1	12 000.00	150	400	11 750.00
2	11 750.00	146.88	400	11 496.88
3	11 496.88	C	400	D

a What is the monthly interest rate? (1 mark)

b Determine the values of C and D. (2 marks)

5 ©SCSA MA2019 Q15 (11 marks) The directed network below shows the maximum available capacity for transferring power between different sub-stations on a small island. The number on each edge gives the capacity in kilovolts (kV).

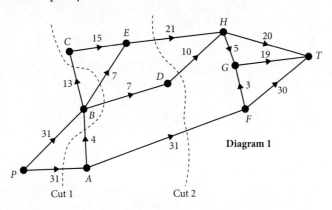

Diagram 1

a State the capacity of each cut in Diagram 1. (2 marks)

Diagram 2 shows a possible flow through the same network.

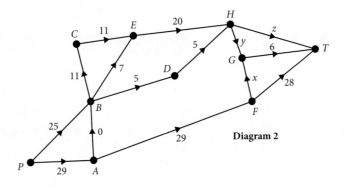

Diagram 2

b Determine the initial flow in Diagram 2. (1 mark)

c Calculate the value of x, y and z in Diagram 2. (3 marks)

d Determine the maximum flow for the original network (Diagram 1). (2 marks)

Engineers wish to increase the maximum capacity to sub-station T. They propose to add a new transmission line from E to T of capacity 3 kV **or** a new transmission line from D to G of capacity 3 kV.

e Determine which of these proposals will increase the maximum capacity to sub-station T. Justify your answer. (3 marks)

6 © SCSA MA2021 Q16 (10 marks) The graph below shows a network of water pipes. The water source and main pumping station are located at *S*. The distribution centre is at *T* and the other vertices are intermediate pumping stations. The weights on the edges show the capacities in kilolitres per hour that can flow through each pipe.

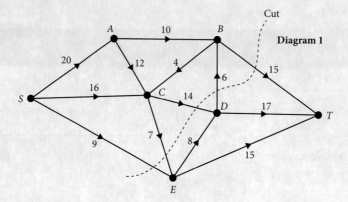

Diagram 1

a **i** Determine the value of the cut shown in Diagram 1. (1 mark)

ii Using your answer to part **a i**, what can be said about the maximum flow of water through the network? (1 mark)

b State the maximum possible flows along the paths *SABT* and *SCDT*. (2 marks)

c Determine the maximum flow from *S* to *T*, listing each path and the corresponding flow. (3 marks)

d Copy Diagram 2 below, and indicate a possible flow along each pipe corresponding to the maximum flow calculated in part **c**. (2 marks)

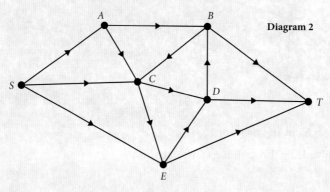

Diagram 2

e Determine the minimum cut that corresponds to the maximum flow. Copy Diagram 1 (excluding the shown cut) and illustrate this cut on the diagram. (1 mark)

7 (2 marks) The total length of pipe that supplies water from a pump to eight locations on a farm (including the house) is a minimum. This minimum length of pipe is laid along some of the edges in the network shown.

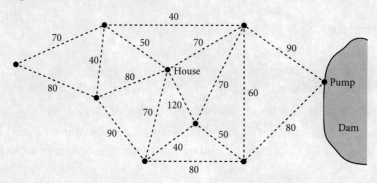

a Copy the diagram and draw in the minimum length of pipe that is needed to supply water to all locations on the farm. (1 mark)

b What is the mathematical term that is used to describe this minimum length of pipe in part **a**? (1 mark)

CHAPTER

CRITICAL PATH ANALYSIS

9

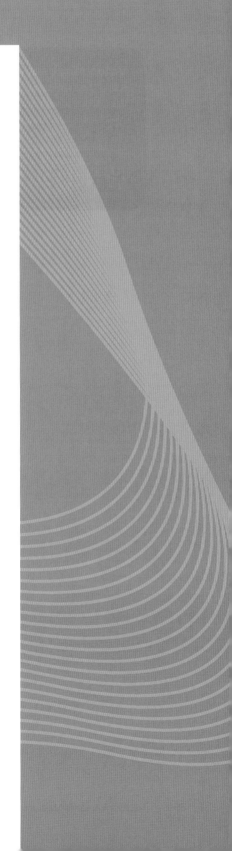

9.1 The scheduling problem

Directed graphs

Directed graphs can be used to schedule tasks made up of many activities. Each vertex is shown by a circle representing the start and finish of a particular event. An **immediate predecessor** is any activity that must be completed before the current activity can commence.

Different types of connections in a directed graph

One preceding event

Event B is preceded by A.

More than one activity with the same preceding event

Events B and C are preceded by A.

An activity with more than one preceding event

Event I is preceded by F, G and H.

Video playlist
The scheduling problem

Worksheets
Project networks

Drawing directed graphs

Digraphs with direction and weight

In Chapter 4, we learned about weighted graphs that show extra information on the arcs. A weighted directed graph can be used to represent a task or project. When a large number of activities are involved in a task, it is often useful to draw a weighted directed graph to show the sequence of activities and the time required for each activity.

Each **activity** is represented by an arc and is labelled with a letter and a weighting, which represents the time taken to complete the activity. The unit of time will be given in the question in minutes, hours, days etc. In the weighted digraph shown on the right, Activity A takes 5 hours to complete.

Start of activity A •———$A\ 5$———→• End of activity A
Activity

Activity tables

An **activity table** shows the order and estimated time for each activity. The table below shows eight activities (A to H) that must be completed in a task. The directed graph that corresponds to the activity table is shown to the right of the table.

Activity	Activity time (hours)	Immediate predecessor
A	7	–
B	3	–
C	4	B
D	4	B
E	6	A, C
F	4	A, C
G	4	E
H	6	D, F

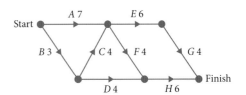

Guidelines to follow when drawing a network diagram

- Use a vertex to represent the start of the network and label it as 'start'.
- Look for activities that do not have any predecessors. These will be your starting activities.
- Multiple predecessors to an activity will all end at the same vertex.
- An activity should not be represented by more than one arc in the network.
- Two vertices can be connected by one arc only.
- A vertex indicating the completion of the project needs to be included in the network and labelled as 'finish'.

WORKED EXAMPLE 1 | Drawing a directed graph from an activity table

Draw the directed graph for the activity table.

Activity	Activity time (hours)	Immediate predecessor
A	3	–
B	4	A
C	3	A
D	8	A
E	3	B
F	5	C
G	7	E, D, F

Steps	Working
1 Activity A has no predecessors. Label the first vertex 'Start' and from this draw an arc labelled A 3.	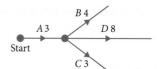
2 Activities B, C and D are all preceded by A. From the vertex at the end of arc A, draw three arcs labelled B 4, C 3 and D 8.	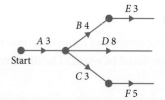
3 Activity E is preceded by B. From the vertex at the end of arc B, draw an arc labelled E 3. Activity F is preceded by C. From the vertex at the end of C, draw an arc labelled F 5.	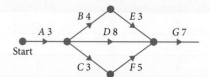
4 Activity G is preceded by activities E, F and D. Draw a vertex that is connected to the arcs E, F and D. From this vertex, draw an arc labelled G 7.	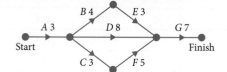
5 The final vertex is drawn at the end of arc G and labelled 'Finish'.	

Dummy activities

A **dummy activity** is an imaginary or redundant activity that is added to a directed graph to ensure that no two vertices are connected by multiple arcs or to maintain precedence structure.

A dummy activity has zero time and is shown as an arc with a broken line.

1 Multiple arcs

A directed graph for a project cannot be drawn with two activities that have the same beginning and the same end. This causes a loss of identity to the activities and results in errors during network computations. The problem can be solved using a dummy activity.

In the example below, the directed graph cannot be drawn with two vertices both connected by activities D and E. A dummy activity needs to be included. It can occur before or after activities D or E to overcome this problem.

Activity	Immediate predecessor
D	C
E	C
F	D, E

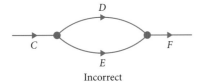
Incorrect

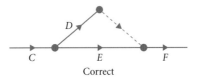
Correct

2 Logic difficulties in the precedence structure

The second use of dummy activities is to ensure the logic of the network is maintained. The first network has activity G preceded by activities C, D and E, which does not follow the logic of the activity table. A second vertex and a dummy activity needs to be included so activities F and G do not both start from the same vertex, and activities C, D and E do not finish at the same vertex.

Activity	Immediate predecessor
F	C, D, E
G	E

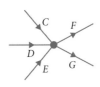
Incorrect

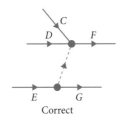
Correct

Draw the directed graph for the project shown below, including dummy activities where required.

Activity	Activity time (hours)	Immediate predecessor
A	8	–
B	6	–
C	4	A, B
D	5	A, B
E	7	C, D

Steps | **Working**

1 Activities A and B have no predecessors.

Label the first vertex 'Start' and from this draw arcs labelled A 8 and B 6.

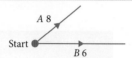

2 Activities C and D are both preceded by A and B; however, two vertices cannot be connected by multiple arcs. A dummy activity must be included after activity A or activity B.

If the dummy activity is placed at the end of activity A, connect the end of activity B and the dummy to the same vertex. Draw two arcs labelled C 4 and D 5 from this vertex.

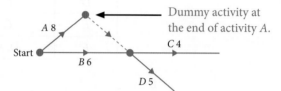

Dummy activity at the end of activity A.

The dummy activity could also be placed at the end of activity B, which would produce

Dummy activity at the end of activity B.

3 Activity E is preceded by activities C and D. A dummy activity is required so that two vertices are not connected by multiple edges.

Using the directed graph with dummy activity after A, draw an arc labelled E 7 from the vertex at the end of activity C.

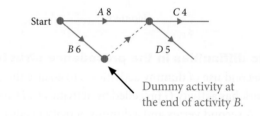

4 Draw a vertex labelled 'Finish' at the end of activity E.

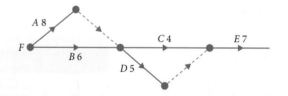

🔓 Exam hack

Always check your directed graph to ensure that it:

- contains all of the connections and times specified in the activity table
- doesn't have two vertices connected with multiple arcs
- has a start vertex and a finish vertex.

Mastery

1   🔗 WORKED EXAMPLE 1   For a particular project, nine activities must be completed.

These activities, the activity times and their immediate predecessors are given in the following table.

Activity	Activity time (days)	Immediate predecessor
A	1	–
B	4	–
C	6	A
D	4	A
E	7	B, C
F	5	B, C
G	7	F
H	6	D
I	2	G

Draw the directed graph for this project.

2   🔗 WORKED EXAMPLE 2   Draw the directed graph for the project shown, including dummy activities where required.

Activity	Activity time (hours)	Immediate predecessor
A	8	–
B	4	–
C	7	–
D	5	A, B, C
E	7	A, B, C

Calculator-free

3 (2 marks) For a particular project there are ten activities that must be completed.

These activities and their immediate predecessors are given in the following table. Draw the directed graph for this network.

Activity	Immediate predecessors
A	–
B	–
C	–
D	A
E	B
F	D, E
G	C
H	C
I	F, G
J	H, I

4 (2 marks) The directed graph below shows the sequence of activities required to complete a project. All times are in hours.

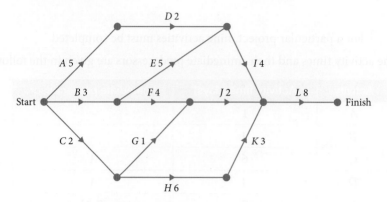

a List the immediate predecessors of activity *L*. (1 mark)

b Determine the number of activities that have exactly two immediate predecessors. (1 mark)

5 (2 marks) The network shows the activities that are needed to complete a particular project.

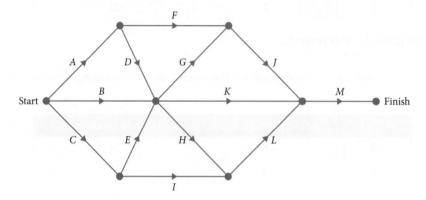

a List the immediate predecessors of activity *G*. (1 mark)

b Determine the total number of activities that need to be completed before activity *L* may begin. (1 mark)

6 (2 marks) A project involves nine activities, *A* to *I*. The immediate predecessor(s) of each activity is shown in the table.

Activity	Immediate predecessor(s)
A	–
B	*A*
C	*A*
D	*B, C*
E	*B, C*
F	*D*
G	*D*
H	*E, F*
I	*G, H*

A directed network for this project will require a dummy activity.

Copy and complete the network diagram including the dummy activity where required.

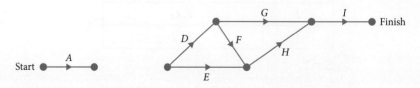

7 © SCSA MA2018 Q6a (3 marks) Yana has booked three gardeners to landscape her garden. The table below shows the required activities, together with the times taken (in hours) and the immediate predecessors for each activity.

Activity	P	Q	R	S	T	U	V	W	X
Time (hours)	2.5	1.5	3	1.5	1.5	2	2.5	2	2.5
Immediate predecessors	–	–	–	P	S, Q, U	R	T, X	P	R

Copy and complete the network diagram below, showing all tasks and durations.

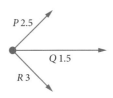

Calculator-assumed

8 © SCSA MA2017 Q11a (3 marks) The following table, consisting of 11 activities, contains information for a project in a small manufacturing company.

Activity	Immediate predecessors	Time (hours)
A	–	4
B	–	5
C	A	14
D	A	7
E	–	7
F	–	5
G	B, C	7
H	D	6
J	E, F	9
K	H, J	10
L	G, K	6

Copy and complete the project network below.

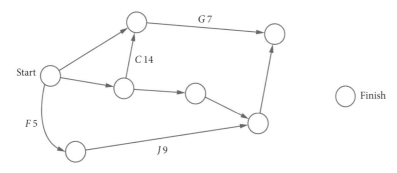

9 © SCSA | MA2020 Q16a | (3 marks) Valeska and her sister Katrin are planning a small building project. The table below shows the required activities, together with the times taken (in days) and the immediate predecessors for each activity.

Activity	A	B	C	D	E	F	G	H
Time (days)	2	4	5	3	6	3	8	2
Immediate predecessors	–	–	A, B	B	C	C	D	D, F

Copy and complete the project network below, showing all activities and durations.

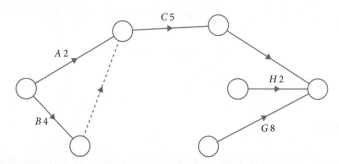

10 (2 marks) A project will be undertaken in a wildlife park. This project involves the 13 activities shown in the table below. The duration, in hours, and predecessor(s) of each activity are also included in the table.

Activity	Duration (hours)	Predecessor(s)
A	5	–
B	7	–
C	4	–
D	2	C
E	3	C
F	15	A
G	4	B, D, H
H	8	E
I	9	F, G
J	9	B, D, H
K	3	J
L	11	J
M	8	I, K

Activity G is missing from the network diagram for this project, which is shown below.

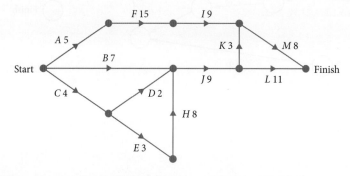

a Copy and complete the network diagram above by inserting activity G. (1 mark)

b List all the activities that must be completed before activity J can commence. (1 mark)

9.2 Critical path analysis

Critical path analysis is a step-by-step project management technique that is used to examine every activity in a project and how each activity affects the project completion time.

Forward scanning to determine EST

Forward scanning through a network enables us to determine the **earliest starting time (EST)** for every activity in the network. The EST is the earliest time it is possible to start an activity.

Method for finding the EST

To determine the ESTs:

1 Draw an open circle with horizontal diameter at each vertex. At the start vertex, label the top semicircle EST and the bottom LST.

2 Begin with the first vertex that has an EST of zero. We will use the convention of writing the earliest starting times in the top box next to each vertex.

3 ESTs are calculated from left to right.

4 Add the activity time to the EST of the previous vertex. If more than one activity leads to the vertex, the **highest** figure obtained becomes the new EST.

5 Continue until the finish is reached.

Video playlist
Critical path analysis

WORKED EXAMPLE 3 | Finding earliest starting times

Determine the earliest starting times (ESTs) for each activity in the network shown. Activity times shown are in hours.

Steps	Working
1 Draw an open circle at each vertex, including a diameter as shown. At the start vertex, label the top half EST and the bottom LST (LSTs are covered in the next section). The value of the EST at the start vertex is zero.	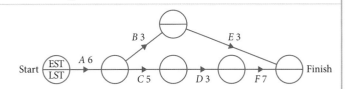
2 To find the EST for the second vertex, add the activity time of 6 to the previous EST. This gives an earliest starting time for activities *B* and *C* as 6 hours.	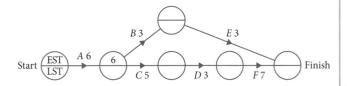

3 The EST for activity E is $6 + 3 = 9$.
This is written in the top semicircle at the start of activity E.

This process is repeated:

EST for activity $D = 6 + 5 = 11$

EST for activity $F = 11 + 3 = 14$

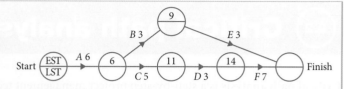

4 To calculate the EST for the finish, find the total for the paths containing activity E and the path containing activity F. The highest figure obtained is the final EST.

Activity E: $9 + 3 = 12$

Activity F: $14 + 7 = 21$

Therefore, 21 hours is the final EST.

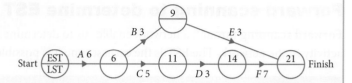

Backward scanning to determine LST

Backward scanning is the process used to find the **latest starting times (LSTs)** for an activity. This is the latest time an activity can start without affecting the project completion time.

Latest starting times

LST for an activity = LST at right vertex − activity time

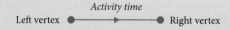

Method for finding the LST

To determine the latest starting times (LSTs):

1 Commence LST calculations at the 'finish' vertex. At the 'finish' vertex, LST = EST.

2 Work backwards from right to left. LSTs are written in the bottom semicircle of each vertex.

3 To find the LST at the left vertex, work backwards subtracting the activity time from the LST at the right vertex. Where there is more than one path to the previous vertex, take the smallest result as the LST for the node.

4 The LST value at the first vertex must be zero.

5 The LST values for each activity can then be calculated using the formula:

LST for an activity = LST at right vertex − activity time

 Exam hack

The LST value for an activity is the same as the LST for the node or vertex only when exactly one activity starts at the vertex.

WORKED EXAMPLE 4 | Finding latest starting times

Determine the latest starting times (LSTs) for each activity in the network. Activity times shown are in hours.

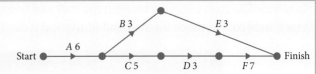

Steps	Working
1 First calculate ESTs for each activity. These were completed in the previous example.	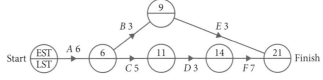
2 Work backwards from right to left. The LST for the 'finish' is equal to the EST.	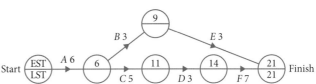
3 The LST for activity E is $21 - 3 = 18$. This is written in the bottom semicircle at the start of activity E. The LST for: activity $F = 21 - 7 = 14$ activity $D = 14 - 3 = 11$	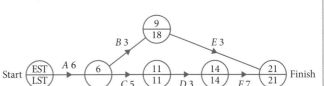
4 Calculate the LST for activities B and C. activity B: $18 - 3 = 15$ activity C: $11 - 5 = 6$ The LST for the vertex at the start of activities B and C is the smallest of these values. Therefore, 6 hours is the LST for the second vertex. The start LST $= 6 - 6 = 0$	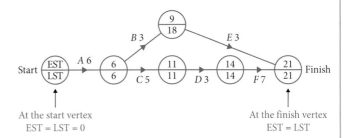

5 Complete a table showing the EST and LST values for each activity. LST values are found using the formula:

LST for activity =
LST at right vertex – activity time

Activity	EST	LST
A	0	0
B	6	$18 - 3 = 15$
C	6	$11 - 5 = 6$
D	11	$14 - 3 = 11$
E	9	$21 - 3 = 18$
F	14	$21 - 7 = 14$

Worksheet
Calculating EST and LST

Identifying critical paths

There are often multiple paths between the start and finish *vertices* of a network. The **critical path** is the longest time path between the start and finish, and it determines the project completion time.

The critical path

On the critical path, activities have equal EST and LST values.

Activities on the critical path are called critical activities.

Activities not on the critical path are called non-critical activities.

🔓 Exam hack

If two or more activities start at the same vertex, the latest starting time (LST) for each activity needs to be calculated using the formula:

LST for activity
= LST at right vertex – activity time

Method for finding the critical path

To determine the critical path:

1 use forward scanning to determine the ESTs for each activity

2 identify the path or paths that produce the final EST

3 use backward scanning to determine the LSTs for each activity

4 activities for which EST = LST are on the critical path.

WORKED EXAMPLE 5 | Finding the critical path using EST and LST

Determine the earliest starting times (EST) and latest starting times (LST) for each activity in the network and, hence, determine the critical path. Activity times shown are in hours.

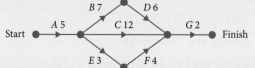

Steps	Working

1 First calculate ESTs for each activity.

2 Calculate LSTs for each activity.

3 Complete a table showing the EST and LST values for each activity.

Activity	EST	LST
A	0	0
B	5	12 − 7 = 5
C	5	18 − 12 = 6
D	12	18 − 6 = 12
E	5	14 − 3 = 11
F	8	18 − 4 = 14
G	18	20 − 2 = 18

4 Activities where EST = LST are critical activities and are on the critical path. These are highlighted on the table.

The critical path is *A*–*B*–*D*–*G*.

Activity float time

The **float time** for an activity is the maximum time an activity can be extended or postponed without affecting the project completion time. Activities on the critical path all have float times of zero.

In the diagram shown:

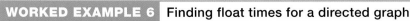

a = EST at left vertex

b = LST at right vertex

t = activity time for activity A

LST for activity $A = b - t$

float for activity $A = b - a - t$

Float time for an activity

For a non-critical activity	For critical activities
float = $LST_{right} - EST_{left}$ – activity time	float = 0

Worksheet
Critical paths and activity float times

WORKED EXAMPLE 6 | **Finding float times for a directed graph**

Determine the critical activities and float times for the non-critical activities for the project shown. Activity times shown are in days.

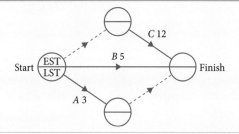

Steps	**Working**
1 First calculate ESTs and LSTs for each activity.	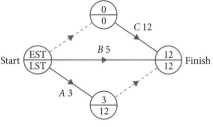
2 Identify the critical path. On the critical path EST = LST.	The critical activity is C. Dummy activities are **not** listed in the critical path.
3 Calculate float times for the non-critical activities A and B using the formula: float = $LST_{right} - EST_{left}$ – activity time	Float for activity $B = 12 - 0 - 5 = 7$ days Float for activity $A = 12 - 0 - 3 = 9$ days

Applying changes to a directed graph

a Determine the critical path and the minimum project completion time for the project shown. Activity times shown are in days.

b Extra resources are used to speed up the original project, resulting in activities *B*, *D* and *F* each being reduced by three days. Find the new critical path and the minimum project completion time.

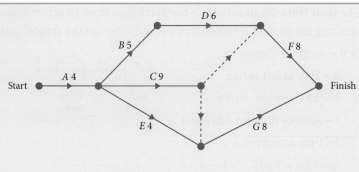

Steps	**Working**
a 1 First calculate ESTs and LSTs for each activity.	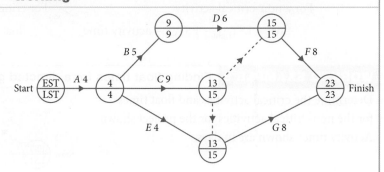
2 On the critical paths EST = LST. The minimum project completion time is EST or LST for the finish vertex.	Critical path = *A–B–D–F* Project completion time = 23 days
b 1 Reduce the times for activities *B*, *D* and *F* by three days.	Activity time for *B* = 5 − 3 = 2 days Activity time for *D* = 6 − 3 = 3 days Activity time for *F* = 8 − 3 = 5 days
2 Calculate ESTs and LSTs for each activity with the new network.	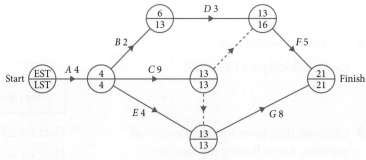 The dummy activity between the end of activity *C* and the start of activity *G* is not included in the critical path.
3 Identify the activities where EST = LST to find the critical path.	The critical path is *A–C–G* and the project completion time is 21 days.

9780170476959

© SCSA | MA2019 Q5 | Calculator-free **(11 marks)**

The network below represents a construction project. The number on each edge gives the time, in hours, to complete the activity. Each activity requires one worker.

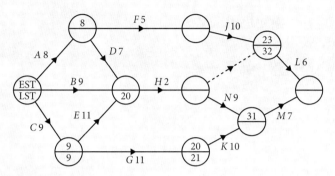

a Copy and complete the precedence table below. (2 marks)

Activity	A	B	C	D	E	F	G	H	J	K	L	M	N
Time (hours)	8	9	9	7	11	5	11	2	10	10	6	7	9
Immediate predecessor	–	–	–	A	C	A	C	B, D, E					

b Copy and complete the network showing the earliest starting time (EST) and latest starting time (LST) for each node. (Note: the first node indicates which is the EST and the LST.) (2 marks)

c Determine the critical path and the minimum completion time for the project. (2 marks)

d Calculate the float times for activities D and F. (2 marks)

e Given that the sum of all the times of the activities is 104 hours, calculate the minimum number of workers required to complete the project in the minimum completion time. (1 mark)

f What is the latest time into the project that activity F could start without affecting the minimum completion time? (1 mark)

g Explain the purpose of the dotted line on the network. (1 mark)

Video
WACE question analysis: Critical path analysis

Reading the question

- Take note of the information in the question about the scheduling problem: the numbers on the arcs representing the number of hours required for each activity.

- Highlight the type of answer required in each question part. This may require you to complete a table or calculate EST and LST values for the network.

- Even though you need to identify the number of possible paths, you are not required to list each path.

Thinking about the question

- The question requires an understanding of critical path analysis.

- You will need to know how to calculate EST and LST values for each node. You will also need to know how to find the LST and float for an activity.

- You will need to recognise a dummy activity and understand its purpose.

Worked solution (✓ = 1 mark)

a

Activity	A	B	C	D	E	F	G	H	J	K	L	M	N
Time (hours)	8	9	9	7	11	5	11	2	10	10	6	7	9
Immediate predecessor	–	–	–	A	C	A	C	B, D, E	F	G	H, J	N, K	H

correctly allocates predecessors for activity L ✓

correctly allocates all predecessors ✓

b

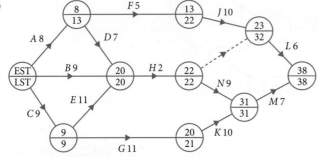

completes at least 4 nodes correctly ✓

completes all nodes correctly ✓

c At each node on the critical path, EST = LST.

Critical path is **C–E–H–N–M**. ✓ Minimum completion time is **38 hours**. ✓

d Float time for $D = 20 - 8 - 7 = $ **5** hours ✓ Float time for $F = 22 - 8 - 5 = $ **9** hours ✓

e 1 worker for activities $CEHNM = 38$ hours

1 worker for activities $AFDJL = 36$ hours

1 worker for activities $BGK = 30$ hours

The minimum number of workers required is **3**. ✓

f The latest time that activity F could start without affecting the minimum completion time is $22 - 5 = $ **17** hours. ✓

g Activity **L** depends on activities **H** and **J**. ✓

EXERCISE 9.2 Critical path analysis

ANSWERS p. 455

Recap

1 In the digraph shown, find the immediate predecessors of

a vertex G

b vertex F.

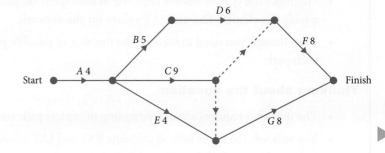

2 Copy and complete the activity table for the directed graph.

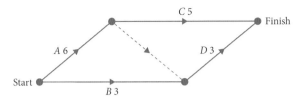

Activity	Time (hour)	Predecessor
A		
B		
C		
D		

Mastery

Use the following directed graph to answer Questions 3 and 4.

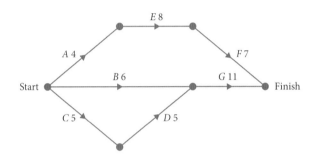

3 WORKED EXAMPLE 3 Determine the earliest starting times (ESTs) for each activity in the directed graph.

4 WORKED EXAMPLE 4 Determine the latest starting times (LSTs) for the directed graph.

5 WORKED EXAMPLE 5 Determine the earliest starting times (ESTs) and latest starting times (LSTs) for each activity in the network below and, hence, determine the critical path. Activity times shown are in hours.

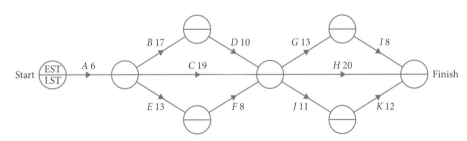

6 🔗 WORKED EXAMPLE 6 Determine the critical activities and float times for the non-critical activities for the project shown below. Activity times shown are in days.

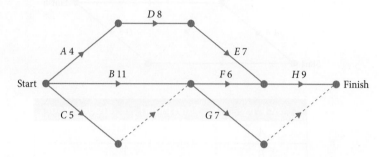

7 🔗 WORKED EXAMPLE 7

a Determine the critical path and the minimum project completion time for the project shown. Activity times shown are in days.

b Extra resources are used to speed up the original project, resulting in activities *C*, *E* and *F* each being reduced by five days. Find the new critical path and the minimum project completion time.

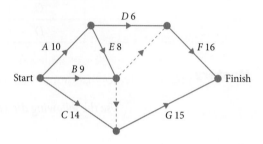

Calculator-free

8 (2 marks) The directed graph shows the sequence of activities required to complete a project. All times are in hours.

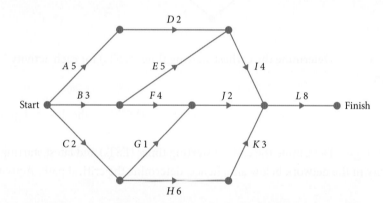

a Determine the number of activities that have exactly two immediate predecessors. (1 mark)

b There is one critical path for this project. Find the number of critical paths if the duration of activity *E* is reduced by 1 hour. (1 mark)

9 (2 marks) The directed graph below shows the sequence of activities required to complete a project. The time to complete each activity, in hours, is also shown.

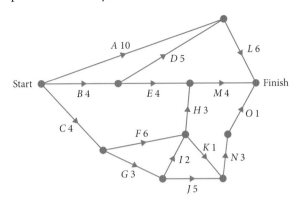

a Determine the earliest starting time for activity N. (1 mark)

b To complete the project in minimum time, some activities cannot be delayed. Find the number of activities that cannot be delayed. (1 mark)

10 © SCSA MA2016 Q2 (7 marks) A project consists of 11 activities, P to Z. The project network representing the scheduling of these activities is shown below. The times are in days.

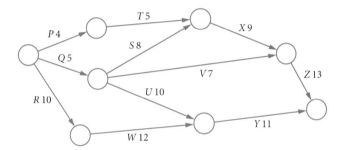

a State the critical path and the minimum completion time for this project. (2 marks)

b Determine the

   i earliest starting time for activity Y (1 mark)

   ii latest starting time for activity V (1 mark)

   iii float time for activity U. (1 mark)

c Activity W is delayed by three days. How, if at all, will this affect the critical path and minimum completion time for this project? (2 marks)

▶ **11** ©SCSA MA2018 Q6b-e (8 marks) Yana has booked three gardeners to landscape her garden. The table below shows the required activities, together with the times taken (in hours) and the immediate predecessors for each activity.

Activity	P	Q	R	S	T	U	V	W	X
Time (hours)	2.5	1.5	3	1.5	1.5	2	2.5	2	2.5
Immediate predecessors	–	–	–	P	S, Q, U	R	T, X	P	R

The network diagram, showing all tasks and durations, is shown below.

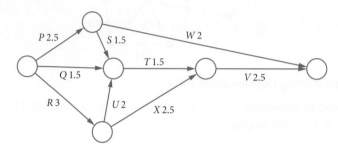

a Determine the critical path and the minimum completion time for the project. (2 marks)

b Calculate the float time for

   i activity W (1 mark)

   ii activity U. (1 mark)

c The gardeners start work at 6:30 am. They take only a half-hour break, at 12:30 pm.

   i Determine the latest starting time for activity P. (1 mark)

   ii Determine the earliest starting time for activity V. (1 mark)

d One of the gardeners becomes ill and is unable to work on Yana's landscaping job. How, if at all, will this affect the minimum completion time for this project (excluding the gardeners' break)? Explain your answer. (2 marks)

12 (3 marks) The activity network shows the sequence of activities required to complete a project. The number next to each activity in the network is the time it takes to complete that activity, in days.

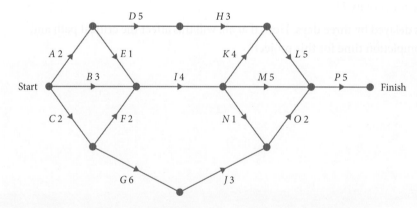

a Beginning with activity C, find the number of paths from start to finish. (1 mark)

b What is the latest starting time for activity I, in days, so that the project is completed in the shortest time possible? (2 marks)

9780170476959

13 (6 marks) A new skateboard park is to be built in Beachton. This project involves 13 activities, *A* to *M*. The directed network below shows these activities and their completion times in days.

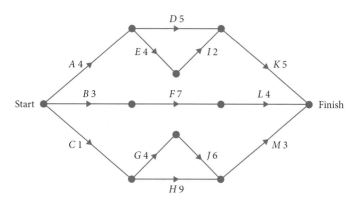

a Determine the earliest starting time for activity *M*. (1 mark)

b The minimum completion time for the skateboard park is 15 days. Determine the critical path for this project. (1 mark)

c Which activity has a float time of two days? (1 mark)

d The completion times for activities *E*, *F*, *G*, *I* and *J* can each be reduced by one day. The cost of reducing the completion time by one day for these activities is shown in the table below.

Activity	Cost ($)
E	3000
F	1000
G	5000
I	2000
J	4000

What is the minimum cost to complete the project in the shortest time possible? (1 mark)

e The initial skateboard park project in Beachton will be repeated at Campville, but with the addition of one extra activity. The new activity, *N*, will take six days to complete and has a float time of one day. Activity *N* will finish at the same time as the project.

  i Copy the network below and add activity *N*. (1 mark)

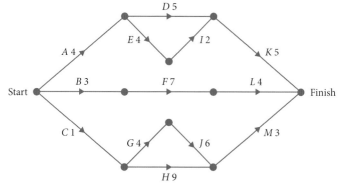

  ii What is the latest starting time for activity *N*? (1 mark)

14 (5 marks) The rides at a theme park are set up at the beginning of each holiday season.

This project involves activities *A* to *O*.

The directed network below shows these activities and their completion times in days.

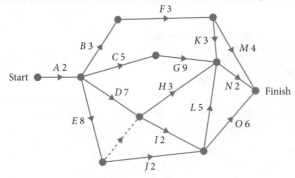

a What are the two immediate predecessors of activity *I*? (1 mark)

b The minimum completion time for the project is 19 days.

   i There are two critical paths. One of the critical paths is *A–E–J–L–N*.
      Determine the other critical path. (1 mark)

   ii Determine the float time, in days, for activity *F*. (1 mark)

c The project could finish earlier if some activities were changed. Six activities, *B, D, G, I, J* and *L*, can all be reduced by one day. The cost of this change is $1000 per activity.

   i What is the minimum number of days in which the project could now be completed? (1 mark)

   ii What is the minimum cost of completing the project in this time? (1 mark)

15 © SCSA MA2017 Q11bcd MODIFIED (8 marks) The directed graph below, consisting of 11 activities, contains information for a project in a small manufacturing company.

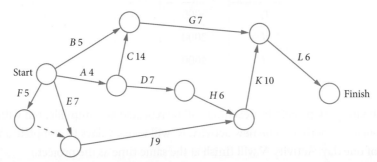

a State the critical path and the minimum completion time for this network. (2 marks)

b Determine the float time, earliest starting time, and latest starting time for activity *G*. (3 marks)

c Due to some unforeseen problems with activities *G* and *J*, **one** of these activities
will require an extra three hours to complete. Which of the activities should be chosen
for the completion time to be at a minimum? Justify your answer. (3 marks)

16 ⬛ © SCSA MA2020 Q16 (14 marks) Valeska and her sister Katrin are planning a small building project. The table below shows the required activities, together with the times taken (in days) and the immediate predecessors for each activity.

Activity	A	B	C	D	E	F	G	H
Time (days)	2	4	5	3	6	3	8	2
Immediate predecessors	–	–	A, B	B	C	C	D	D, F

a Copy and complete the project network below, showing all activities and durations. (3 marks)

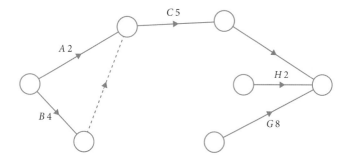

b Determine all critical activities and the minimum completion time for the project. (3 marks)

c Calculate the float times for each of the non-critical activities. (3 marks)

d If activity *H* is delayed by three days, what effect will this have on the minimum completion time and the critical activities? (2 marks)

e Extra resources become available that can be used to shorten the duration of **one** of activities *B*, *E* or *F* (on the original network) by one day. Which of these activities should be shortened and why? (3 marks)

17 (6 marks) Fencedale High School is planning to renovate its gymnasium. This project involves 12 activities, *A* to *L*. The directed network below shows these activities and their completion times, in weeks.

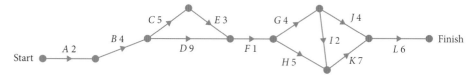

The minimum completion time for the project is 35 weeks.

a How many activities are on the critical path? (1 mark)

b Determine the latest starting time of activity *E*. (1 mark)

c Which activity has the longest float time? (1 mark)

It is possible to reduce the completion time for activities *C*, *D*, *G*, *H* and *K* by employing more workers.

d The completion time for each of these five activities can be reduced by a maximum of two weeks. What is the minimum time, in weeks, that the renovation project could take? (1 mark)

e The reduction in completion time for each of these five activities will incur an additional cost to the school. The table below shows the five activities that can have their completion times reduced and the associated weekly cost, in dollars.

Activity	Weekly cost ($)
C	3000
D	2000
G	2500
H	1000
K	4000

The completion time for each of these five activities can be reduced by a maximum of two weeks. Fencedale High School requires the overall completion time for the renovation project to be reduced by four weeks at minimum cost.

Copy and complete the table below, showing the reductions in individual activity completion times that would achieve this. (2 marks)

Activity	Reduction in completion time (0, 1 or 2 weeks)
C	
D	
G	
H	
K	

18 (5 marks) Simon is building a new holiday home for his family.

The directed network below shows the 10 activities required for this project and their completion times, in weeks.

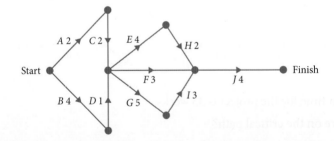

a What are the two activities that are immediate predecessors of activity *G*? (1 mark)

b For activity *D*, the earliest starting time and the latest starting time are the same. What does this tell us about activity *D*? (1 mark)

c Determine the minimum completion time, in weeks, for this project. (1 mark)

d Determine the latest starting time, in weeks, for activity *C*. (1 mark)

e Which activity could be delayed for the longest time without affecting the minimum completion time of the project? (1 mark)

9780170476959

19 (4 marks) A barn will be built on a property. This building project will involve 11 activities, *A* to *K*. The directed network shows these activities and their duration in days. The duration of activity *I* is unknown at the start of the project. Let the duration of activity *I* be *p* days.

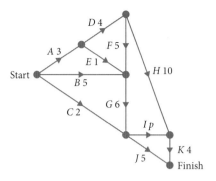

a Determine the earliest starting time, in days, for activity *I*. (1 mark)

b Determine the value of *p*, in days, that would create more than one critical path. (1 mark)

c If the value of *p* is six days, what will be the float time, in days, of activity *H*? (1 mark)

d When a second barn is built later, activity *I* will not be needed. A *dummy* activity is required, as shown on the revised directed network.

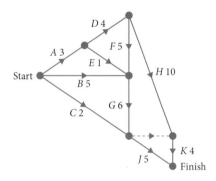

Explain what this *dummy* activity indicates on the revised directed network. (1 mark)

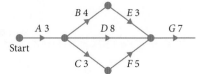

The scheduling problem

- The edges on a directed graph, or digraph, have a direction. In a weighted digraph, there is a number associated with each edge.

- An **activity table** shows the order and estimated time for each activity.

- An **immediate predecessor** is any activity that must be completed before this activity can commence.

  There are three different types of connections.

1 One preceding event
   Event B is preceded by A.

2 More than one activity with the same preceding event
   Events B and C are preceded by A.

3 An activity with more than one preceding event
   Event I is preceded by F, G and H.

- When drawing a network:

  – use a vertex to represent the start of the network

  – look for any activities that do not have any predecessors – these will be your starting activities

  – multiple predecessors to an activity will all end at the same vertex

  – an activity should not be represented by more than one arc in the network

  – two vertices can be connected by one arc only

  – a vertex indicating the completion of the project needs to be included in the network.

- A dummy activity needs to be added to a network to ensure that no two vertices are connected by multiple arcs or to maintain precedence structure.

- A dummy activity has zero time and is shown as an arc with a broken line.

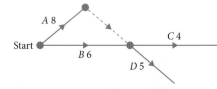

Forward scanning to determine EST

- The **earliest starting time (EST)** for an activity is the earliest time it is possible to start the activity.

- Forward scanning through a network enables us to determine the EST for every activity in the network.

  To determine the EST:

1 Draw an open circle with horizontal diameter at each vertex. At the start vertex label the top semicircle EST and the bottom latest starting time (LST).

2 Begin with the first vertex, which has an EST of 0. We will use the convention of writing the ESTs in the top semicircle of each vertex.

3 ESTs are calculated from left to right.

4 Add the activity time to the EST of the previous vertex. If more than one activity leads to the vertex, the *highest figure* obtained becomes the new EST.

5 Continue until the finish is reached.

Backwards scanning to determine LST

- Backward scanning is the process used to find the **latest starting times (LSTs)** for activities.
- The LST is the latest time an activity can start without affecting the project completion time.
- LST at the left vertex = LST at the right vertex – activity time

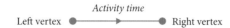

Left vertex ●———→———● Right vertex

Activity time

- Calculating LSTs at the node and for an activity:
  1 Commence LST calculations at the 'finish' vertex. At the 'finish' vertex, LST = EST.
  2 Work backwards from right to left. LSTs for the node are written in the bottom box semicircle of each vertex.
  3 To find the LST at the left vertex, work backwards, subtracting the activity time from the LST at the right vertex. Where there is more than one path to the previous vertex, take the smallest result as the LST for the node.
  4 The last LST value at the first vertex must be zero.
  5 The LST values for each activity can then be calculated using the formula:

  LST for activity = LST at right vertex – activity time

The critical path

- The **critical path** is the longest time path in the network.
- To determine the critical path:
  1 use forward scanning to determine the earliest starting times for each activity
  2 identify the path or paths that produce the final EST
  3 use backward scanning to determine the latest starting times for each activity
  4 activities for which EST = LST are on the critical path.

Activity float time

The **float time** for an activity is the maximum time an activity can be extended or postponed without affecting the project completion time.

In the diagram shown

  a = EST of activity A

  b = LST for activity A

  t = activity time for activity A

LST for activity $A = b - t$

  float for activity $A = b - a - t$

For a non-critical activity

  float = $LST_{right} - EST_{left}$ – activity time

For critical activities

  float = 0

Cumulative examination: Calculator-free

Total number of marks: 27   Reading time: 3 minutes   Working time: 27 minutes

1 (4 marks) Consider the sequence generated by the recursive rule

$$F_{n+1} = \frac{1}{2}F_n + b, F_1 = 32$$

a If $b = 4$, graph the first 5 terms of the sequence. Label the horizontal axis n and the vertical axis F_n.   (2 marks)

b Find the value of b that would produce a steady state of 12.   (2 marks)

2 (6 marks) A graph has six vertices and seven edges, as shown in the diagram.

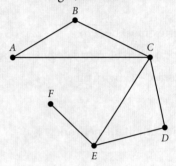

a Construct an adjacency matrix for the graph.   (2 marks)

b Explain, using your understanding of degree of vertices, why a Eulerian trail exists and, hence, state whether this trail is open or closed.   (2 marks)

c Justify whether the Eulerian trial that exists is also a Hamiltonian path.   (2 marks)

3 © SCSA   MA2020 Q5   (7 marks) A communication Wi-Fi network is to be installed to service a shopping centre connecting 12 shops located at vertices $A, B, C, D, E, F, G, H, I, J, K$ and L. The only practical connections between vertices are shown on the following network.

The number on each edge is the quoted price, in hundreds of dollars, for the direct link between the vertices.

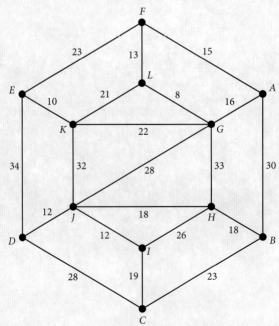

a A minimal spanning tree is to be used to determine the minimum cost of this installation.

   i Copy the network and show clearly the minimum spanning tree solution. (3 marks)

   ii Determine the minimum cost. (2 marks)

b Due to further construction at the shopping centre, edge *GJ* is now not feasible. Explain how this will change the solution for part **a**. (2 marks)

4 (6 marks) The activity network below shows the sequence of activities required to complete a project. The number next to each activity in the network is the time it takes to complete that activity, in days.

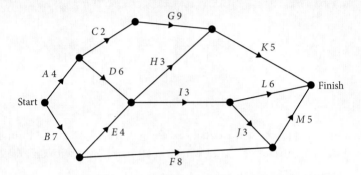

a How many activities must be completed before activity *K* can commence? (1 mark)

b What is the earliest starting time, in days, of activity *K*? (1 mark)

c What is the latest starting time of activity *L*? (1 mark)

d Find the critical path and the minimum completion time for this project, in days. (2 marks)

e What is the float time for activity *H*? (1 mark)

5 (4 marks) Roadworks planned by the local council require 13 activities to be completed. The network shows these 13 activities and their completion times, in weeks.

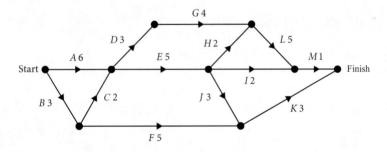

a What is the earliest start time, in weeks, of activity *K*? (1 mark)

b Find the critical paths and the minimum completion time for this project, in weeks. (2 marks)

c How many of these activities have zero float time? (1 mark)

Cumulative examination: Calculator-assumed

Total number of marks: 45   Reading time: 5 minutes   Working time: 45 minutes

1 © SCSA   MA2019 Q8def   (8 marks) Abdul has a lawnmowing business and is investigating if there is a relationship between the size of a lawn and the length of time it takes to cut the lawn. He takes a random sample of eight customers and measures the areas of their lawns and notes the times, in minutes, it takes to mow their lawns. The results are in the table below, where A is the area of the lawn in square metres and T is the time in minutes. (Note: some values are missing.)

Customer	A	B	C	D	E	F	G	H
A (m^2)		260		480	540	600	860	1180
T (min)	25	55	50	70	90	70	135	140

a The least-squares line is given by the equation of $T = 0.115A + 16.008$. Abdul charges $30 per hour. Estimate the charge for mowing a customer's lawn with an area of 500 m^2.   (2 marks)

b Explain whether the estimate determined in part **a** would be valid.   (2 marks)

c Using the least-squares line correct to three decimal places

   i calculate the residuals for Customers B and D.   (2 marks)

   ii explain the significance of the sign and the size of these residuals in reference to the least-squares line.   (2 marks)

2 (4 marks) Consider the geometric sequence 2200, 3608, 5917.12 …

a State a recursive rule for the sequence.   (2 marks)

b State the rule for the nth term   (1 mark)

c Find the 15th term to two decimal places.   (1 mark)

3 (11 marks) A travelling band, Hit the High Notes, toured Perth recently and performed four times a week at the Subiaco Arts Centre. Their manager recorded the size of audiences so that she can determine if the band should continue with the performances for a further two weeks. She recorded the data for the first three weeks, shown in the table below.

Week	Day	Time period	Size of audience	Weekly mean	Percentage of weekly mean	Seasonally adjusted size of audience
1	Thursday	1	1253		70.02	
	Friday	2	1428	A	79.80	
	Saturday	3	2320		129.65	
	Sunday	4	2157		B	
2	Thursday	5	1326		68.80	
	Friday	6	1798	1927.25	93.29	
	Saturday	7	C		132.99	
	Sunday	8	2022		104.92	
3	Thursday	9	D		75.02	
	Friday	10	1665	E	81.43	
	Saturday	11	2688		131.46	
	Sunday	12	2292		112.09	

a Determine the values of *A*, *B*, *C*, *D* and E in the table on the previous page. (5 marks)

b Determine the seasonal index for Friday, clearly showing how you derive the figure. (2 marks)

c Calculate the seasonally adjusted size of audiences for the three Fridays only and record your answers on a copy of the table on the previous page. (2 marks)

d The least-squares equation for the seasonally adjusted size of audience, *n*, and time, *t*, is

$$\hat{n} = 29.8297t + 1727$$

Predict the size of the audience for Sunday of Week 4. Comment on the validity of the prediction. (2 marks)

4 (3 marks) Emily has $8000 that she wants to invest for a certain period of time without making any withdrawals.

If she chooses to invest this money in an account earning compound interest at the rate of 7% per annum, determine

a the value of her investment after five years, if interest is paid annually (1 mark)

b the number of years required for her to triple her investment, if interest is paid quarterly. (2 marks)

5 (3 marks) The graph below shows the possible number of postal deliveries each day between the Central Mail Depot and the Zenith Post Office. The unmarked vertices represent other depots in the region. The weighting of each edge represents the maximum number of deliveries that can be made each day.

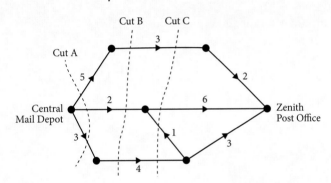

a Cut A, shown on the graph, has a capacity of 10. Two other cuts are labelled as cut B and cut C.

   i What is the capacity of cut B? (1 mark)

   ii What is the capacity of cut C? (1 mark)

b Determine the maximum number of deliveries that can be made each day from the Central Mail Depot to the Zenith Post Office. (1 mark)

6 © SCSA | MA2021 Q13 | (12 marks) A kitchen renovation project consists of a number of tasks of different durations and completed in different orders. One such renovation has this information summarised in the table below.

Task	Task description	Duration (days)	Immediate predecessor(s)
A	Prepare plans	10	None
B	Select contractor	2	None
C	Review plans	4	A, B
D	Purchase appliances	1	C
E	Remove old appliances and benches	2	C
F	Prepare electrics and plumbing	2	E
G	Construct new cupboards and benches	12	C
H	Install cupboards and benches	2	F, G
J	Install appliances	1	see part **b**
K	Tiling and splashbacks	5	J
L	Flooring	3	H
M	Complete electrical and plumbing	2	K
N	Test and handover	1	L, M

A partially completed project network is shown below for this table.

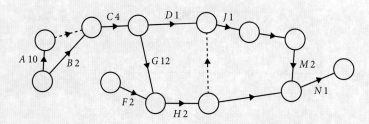

a Copy and complete the network diagram. (2 marks)

b Identify the immediate predecessor(s) of task *J*. (1 mark)

c What does the dotted line on the network indicate? (1 mark)

d Determine the critical path and the minimum completion time for the project. (2 marks)

e Determine the float time for task *D* and explain its meaning in terms of the renovation. (2 marks)

f Once task *E* (removal of old appliances and benches) begins, the kitchen cannot be used. What is the least amount of time the occupants of the house will be without a working kitchen? (1 mark)

g If task *G* was actually completed in nine days, how would this affect the critical path and minimum completion time? (2 marks)

Tasks *E* and *F* are both delayed.

h What is the maximum possible delay that does **not** affect the original minimum completion time? (1 mark)

7 (4 marks) At the Zenith Post Office all computer systems are to be upgraded. This project involves 10 activities, *A* to *J*. The directed network below shows these activities and their completion times, in hours.

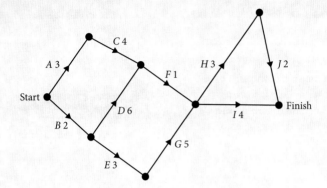

a Determine the earliest starting time, in hours, for activity *I*. (1 mark)

b The minimum completion time for the project is 15 hours. State the critical path. (1 mark)

c Two of the activities have a float time of two hours. Determine these two activities. (1 mark)

d For the next upgrade, the same project will be repeated but one extra activity will be added. This activity has a duration of one hour, an earliest starting time of five hours and a latest starting time of 12 hours. Copy and complete the following sentence by filling in the boxes.

The extra activity could be represented on the network above by an arc from the end

of activity ☐ to the start of activity ☐ . (1 mark)

Answers

CHAPTER 1

EXERCISE 1.1

1 a i *average daily screen time* and *ATAR score*

    ii The amount of screen time is likely to affect an ATAR score. However, an ATAR score is not likely to affect the amount of screen time. So, *average daily screen time* is the explanatory variable.

  b i *stress levels* (scale of 0 to 5, where 0 is none and 5 is extremely high) and *headache levels* (1 = mild, 2 = moderate, 3 = severe)

    ii The aim of the research is to see whether headache levels can be predicted from stress levels, so *stress levels* is the explanatory variable.

  c i *time taken to complete Fun Run* and *age*

    ii A person's age is likely to affect their time taken to complete a Fun Run. However, the time a person takes to complete a Fun Run will not affect their age. So, *age* is the explanatory variable.

  d i *gender* and *hand dominance*

    ii A person's gender is likely to affect their hand dominance. However, a person's hand dominance will not affect their gender. So, *gender* is the explanatory variable.

  e i *driving response times* and *levels of sleep deprivation* (1 = low, 2 = medium, 3 = high)

    ii The aim of the experiment is to see whether levels of sleep deprivation can explain driving response times, so *levels of sleep deprivation* is the explanatory variable.

  f i *English study score* and *number of television sets in the home*

    ii The number of television sets in the home is likely to affect a student's English study score. However, a student's English study score is not likely to affect the number of television sets in the home. So, *number of television sets in the home* is the explanatory variable.

2 temperature at 8 am

3 a laptop weight   **b** laptop price

EXERCISE 1.2

1 monthly electricity cost

2 E

3 a no   **b** yes   **c** yes

  d yes   **e** no

4 a The explanatory variable is *gender*. The response variable is *number of mobile phones owned*.

b

Number of mobile phones owned	Gender		Total
	Male	Female	
Only one	15	20	35
More than one	23	12	35
Total	38	32	70

5 a

Sporting club membership	Gender	
	Male	Female
Sporting club member	67%	69%
Not a sporting club member	33%	31%
Total	100%	100%

b The difference between the males and females is only 2%, suggesting that there is no association between gender and sporting club membership.

6 a

Age	Income level ($000s)		
	40–60	60–80	100+
17–26	59%	31%	10%
27–36	30%	37%	33%
37+	21%	34%	45%

b The *difference* between the ages decreases for people earning between $40 000 and $60 000 and increases for people earning more than $100 000, suggesting that there may be an association between age and income level.

7 The percentage of flats with only 1 laptop (70%) is considerably higher than the percentage of units with only 1 laptop (55%), and both percentages are considerably higher than the percentage of houses with only 1 laptop (25%). The percentages for 2, 3, 4 and more than 4 laptops for houses are all considerably greater than for both flats and units. The percentage for 2 laptops for units (35%) is considerably higher than the percentage for 2 laptops for flats (20%). So, this segmented bar chart suggests that there may be an association between type of residence and the number of laptops.

8 a

	Agree	Disagree	Undecided
Under 20	14	42	14
20–25	56	16	8
Over 25	42	5	3

b age

c i $\dfrac{16}{80} \times 100 = 20\%$

    ii

	Agree	Disagree	Undecided
Under 20	20	60	20
20–25	70	20	10
Over 25	84	10	6

d As age increases, the percentage of students who agree increases. Percentages in the Agree column are increasing with age. There are other possibilities.

9 a 85%   **b** 16%

10 a 90

b 20% of sugar traps contained more than 500 moths, whereas 10% of light traps contained more than 500 moths.

11 a

	Category A	Category B	Category C	Total
Year 7	19	11	20	50
Year 8	12	17	21	50
Year 9	13	14	23	50
Year 10	11	18	21	50
Year 11	10	15	25	50
Year 12	8	17	25	50
Total	73	92	135	300

b Year 7 = 38%, Year 12 = 16%

There is a marked drop between Year 7 and Year 12 in the percentage of students who use Category A as a mode of transport to and from school.

c **i** Generally as the students get older, the percentage using Category A as a mode of transport decreases.

ii Generally as the students get older, the percentage of students using Category C as a mode of transport increases.

iii As Category A increases, Categories B and C decrease.

12 a 1 153 000   **b** 88%

13 a The explanatory variable is the size of the house.

b 35.6%

14 a Newcastle-Sunderland and Liverpool

b

	City size	
Congestion level	Small	Large
high	4	2
medium	4	2
low	8	3
Total	16	7

c 25%

EXERCISE 1.3

1 D   **2** B

3 a *age* (years)

b *time* (min)

c 9 people

d A 50-year-old who was in the café for 20 minutes.

e 3

f They were a toddler with his/her parent/caretaker.

4 a **i** negative, linear and strong

ii The total nap time decreases as a child ages.

b **i** no association

ii There appears to be no association between the number of bad apples per tonne and the price per kilo.

c **i** negative, linear and weak

ii There is some indication that the time spent reading to a child under 8 decreases as the child gets older.

d **i** positive, linear and moderate

ii Weight tends to increase as height increases.

5 a positive association   **b** no association

c negative association   **d** positive association

e positive association   **f** negative association

6 a ClassPad

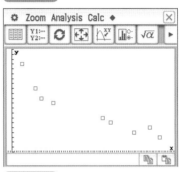

TI-Nspire

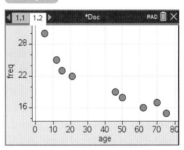

The association between age and frequency can be described as negative, linear and strong.

7 a Countries that have higher human development indices tend to have higher levels of carbon dioxide emissions.

b Male students who tend to spend more time playing sport tend to spend less time playing computer games.

8 a The association between age and the number of laps can be described as negative.

b The association between age and the number of laps can be described as linear.

c The association between age and the number of laps can be described as strong.

9 a ClassPad

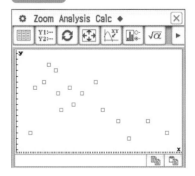

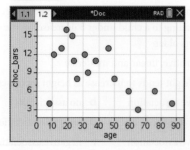

TI-Nspire

b The association between age and chocolate consumption can be described as negative, linear and moderate.

EXERCISE 1.4

1 E   **2** C

3 a The data suggests there is a weak positive linear association between *amount of exercise* and *height*.

b The data suggests there is a moderate positive linear association between *weight* and *number of hours a person spends sitting down*.

c The data suggests there is a strong negative linear association between *percentage of good peaches* and *time spent in storage*.

d The data suggests there is a moderate positive linear association between *sales* and *temperature*.

e The data suggests there is no association between *number of pets* and *temperature*.

f The data suggests there is a weak negative linear association between *number of home-cooked meals* and *income*.

4 a $r = 0.65$

b The three assumptions are that both variables are numerical, the association is linear, and there are no outliers.

c An *r* value of 0.65 indicates that there is a moderate, positive, linear association between *income* and the *number of cinema movies seen in the past year*.

5 a 0.6   **b** −0.4   **c** 0.8   **d** −0.9

e −0.5   **f** 0.2

6 a *outdoor temperature*   **b** *population*

c *household income*   **d** *age*

7 Answers will vary. Check with your teacher.

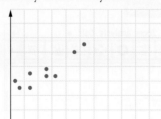

8 It's likely that there is confounding in this association. The secondary school results may have caused the university results, but we can't know because there are too many potential confounding variables that could be involved, such as innate academic ability, the capacity for hard work, economic factors or parental influence. Other answers are possible.

9 a negative

b The students in the group who spent a short amount of time playing computer games tended to be fitter.

10 a the *person's wealth* ($)   **b** lurking variable

11 a *length of stride* and *running speed*

b length of stride

c running speed

d The second researcher's study has considered other variables by controlling the factors that can be controlled, randomisation, and repeating the experiment.

e The first researcher's results may only apply specifically to Olympic-level runners who, unlike the average person, would have worked hard on stride technique over years. The results could also be influenced by other variables such as training techniques and how wind moves in large stadiums. Other answers are possible.

12 a Head lice caused people to be healthy.

b The sickness caused lice to leave.

c The people of the Hebrides had their explanatory and response variables around the wrong way in the association. They believed that the explanatory variable was the *head lice* and the response variable was the *fever*, when in fact the explanatory variable was the *fever* and the response variable was the *head lice*.

13 a 2020   **b** 0.96   **c** strong

14 a mass   **b** 0.931   **c** strong

15 a $r = 0.8054$ **b** *strong*

c linear   **d** *cost*

16 a *Tidal range* is the explanatory variable.

b
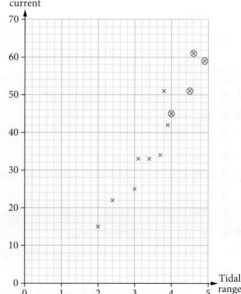

c correlation coefficient = 0.956 (three decimal places)

strong, positive, linear relationship

CUMULATIVE EXAMINATION: CALCULATOR-FREE

1 a strong and non-linear

b Correctly identifies a non-causal explanation. For example, it is a coincidence, there may be another variable/factor affecting the amount of water being used, e.g. the size of the garden.

2 a purchase price

b

Type of repair	Purchase price			
	Less than $500	From $500 to $1000	Greater than $1000	Total
Wheels and tyres	60	10	30	100
Gears and brakes	50	30	20	100
Frame and suspension	75	10	15	100

c The higher the purchase price, the lower the percentage of gears and brakes repairs that are needed.

3 a See graph (diamonds)

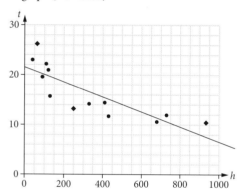

b $r = -0.8$　　**c** non-linear

d incorrect; cause not established.

CUMULATIVE EXAMINATION: CALCULATOR-ASSUMED

1 The divided bar chart does support the opinion that lunch time activity (walked, sat or stood, ran) is associated with year level. For example, the percentage who ran changed from about 78% to 40% to 10% from Years 6–8 and 8–10. (The second mark is for quoting relevant percentages.)

2 a $A = 27\,065 - (11\,985 + 9575) = 5505$

$B = 100 - (35.4 + 20.3) = 44.3\%$

b WA: $\dfrac{8026}{98\,763} \times 100 = 8.1265\%$

SA: $\dfrac{6464}{98\,763} \times 100 = 6.54\%$

WA has a higher percentage than SA.

c The number of new passenger vehicles sold is always higher than the number of sports utility vehicles sold.

d The percentages of passenger vehicles and sports utility vehicles sold in the NT is the lowest in Australia.

The percentage of other vehicles sold in the NT is the highest in the country.

3 a

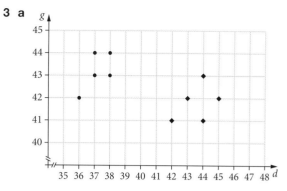

b −0.6　　**c** moderate negative linear relationship

CHAPTER 2

EXERCISE 2.1

1 a *number of applications = −9.6 × interest rate + 130*

b ClassPad

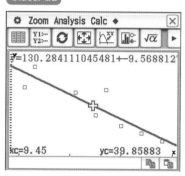

TI-Nspire

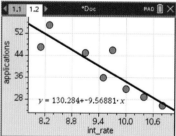

2 a

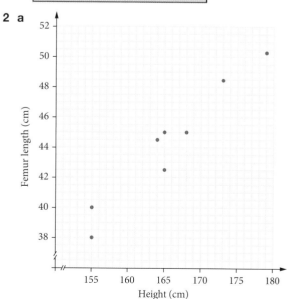

b *femur length* = 0.49 × *height* − 37.56

c

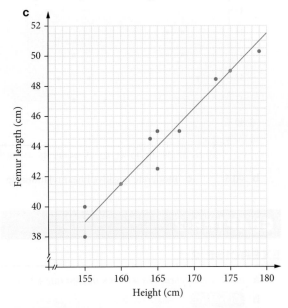

3 a The slope is −3890. The price of the car model on average decreases by $3890 for every one-year increase in age.

b The *y*-intercept is 40 000. The price of the car model was $40 000 when it was released.

4 a The slope is 0.7. This means the weight on average increases by 0.7 kg for every 1 cm increase in height.

b The *y*-intercept is −44. This means that weight is −44 kg when height is 0 cm. Negative weight is clearly impossible. This least-squares line only applies from a certain minimum height.

5 a The slope is 0.019. An increase of 1 cm in *height* is associated with an increase of 0.019 m^2 in *body surface area*.

b The *y*-intercept is −1.1. This means that body surface area is −1.1 when height is 0 cm. Negative height is clearly impossible. This least-squares line only applies from a certain minimum height.

6 a *humidity* 9 am

b *humidity 3 pm* = 0.765 × *humidity 9 am* − 1.262

c *r* = 0.871

7 a strong, positive, linear

b *male* = 0.81 × *female* + 9.69

8 a *height*

b *arm span* = 1.1 × *height* − 15.6

c *Arm span* increases by 1.1 cm for each 1-cm increase in *height*.

EXERCISE 2.2

1 A   **2** A

3 a $r^2 \approx 0.736$

b 73.6% of the variation in *shoe size* can be explained by the variation in *height*.

c 26%

d *shoe size* = 0.17 × *height* − 20

e Yes, this is an appropriate model because of the high r^2 value of 73.6%.

4 a *r* = 0.95   **b** *r* = −0.79

c *r* = 0.94   **d** *r* = −0.56

5 a While there is a moderately, strong, positive, linear association between weight and height, there is a clear outlier.

b (167, 118)

c It will increase.

6 a *average temperature* = −0.877 × *latitude* + 42.984

b The slope is −0.877. This means the average temperature decreases by 0.877 for every 1 unit increase in latitude.

7 a −0.945

b Yes, this is an appropriate model because of the high r^2 value of 89.4%.

8 a diastolic blood pressure   **b** 0.508   **c** 25.8%

9 a 81.6%   **b** 18%

10 a −0.97   **b** 93%

11 a The slope is −0.120. This means the number of errors decreases by 0.120 for every 1 minute increase in study time.

b −0.88

12 a i *weight*   **ii** −0.001

b 29%

13 a On average, *male* life expectancy increased by 0.88 years for each one-year increase in *female* life expectancy.

b 34.4 years

c 95% of the variation in *male* life expectancy can be explained by the variation in *female* life expectancy.

14 a The association between *population density* and *area* is weak, negative and linear.

b *r* = −0.38

15 a distance from the ocean

b

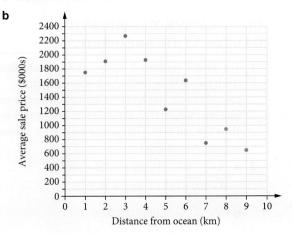

c *p* = −174.58*d* + 2331.7

d For each kilometre extra from the ocean, the selling price drops by an average of $174 580.

e i *r* = −0.84

ii strength of the linear relationship

iii strong, negative association

f $r^2 = 0.7007 \approx 70.1\%$

EXERCISE 2.3

1 strong, negative and linear

2 C

3 a 8.95 laps   **b** 5.43 laps   **c** 2.07 laps

4 a 2°C   **b** 8°C   **c** 18°C

5 a 139 cm; interpolation

   b 183 cm; extrapolation

   c The prediction for the 11-year-old involves interpolation, so it is more reliable than the prediction for a 22-year-old, which involves extrapolation.

   d 14 years old

6 a 145 grams; interpolation

   b 45 grams; extrapolation

   c 370 grams; extrapolation

   d The prediction for 25 chocolates is the most reliable because it is within the data range of 15–50; the other two predictions involve extrapolation.

   e 21 chocolates

7 a 136 cm

   b 8 years

   c **i** true   **ii** true   **iii** false

      iv true   **v** true

8 a 3 years (1 mark for calculating Australia life expectancy = 87.67… years and UK life expectancy = 84.62… years)

   b The least-squares lines of best fit equations were used to make predictions outside the available range of data.

9 a *male income*   **b** $350

   c **i** $18 250

      ii Making the required prediction involved going beyond the data used (extrapolation) to determine the line of best fit equation.

10 a **i** $1 633 380

      ii The table value is the true average value, whereas the prediction comes from the least-squares line.

   b Following this model, the selling price would eventually become less than zero.

   c Not a valid statement. Only average price increases the closer to the ocean.

EXERCISE 2.4

1 C   **2** C

3 a **i** 0%   **ii** 20%   **iii** −5%   **iv** −35%

   b **i** 0.5 metres   **ii** −0.7 metres

4 a Non-linear because it appears to be a valley shape.

   b Linear because it appears to be randomly scattered.

   c Non-linear because it appears to be a hill shape.

   d Non-linear because it does not appear randomly scattered.

5

Height (cm)	Femur length (cm)	Residual
178	50.2	−0.3
173	48.4	0.4
165	45.1	1.1
164	44.6	1.1
168	45	−0.5
165	42.6	−1.4
155	39.9	0.8
155	38	−1.1

ClassPad

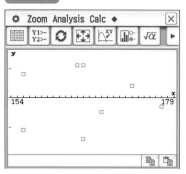

TI-Nspire

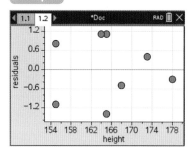

The data is probably linear because the residual values appear randomly scattered above and below the *x*-axis.

6 a The data is probably linear because the residual values appear randomly scattered above and below the *x*-axis.

   b residual = 2

   c The actual value at *x* = 11 is below the least-squares line, which means the residual must be negative. The residual plot has a positive residual value for *x* = 11.

7 a

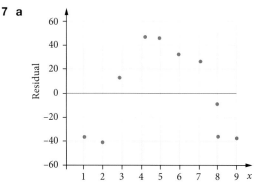

   b The residual values show a hill shape. This suggests that the association between the explanatory and response variables is *non-linear*.

ANSWERS

8 a

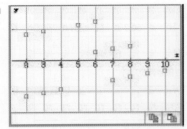

b The data is probably linear because the residual values appear randomly scattered above and below the *x*-axis.

9 a 0.8

b

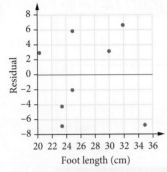

c The data is probably linear because the residual values appear randomly scattered above and below the *x*-axis.

10 a The pressure at 3 pm on average increases by 0.8894 hPa for every 1 hPa increase in the pressure at 9 am.

b 1023 hPa

c interpolation

d 3 hPa

e 93.3%

f   i The assumption is that there is a linear association between the atmospheric pressure at 3 pm and the atmospheric pressure at 9 am.

ii The residual plot has a clear pattern (a valley shape) which suggests that the association is non-linear.

11 a strong, positive, linear

b   i *apparent temperature* = 0.94 × *actual temperature* − 1.65

ii On average, when the actual temperature is 0°C, the apparent temperature is −1.65°C. (Alternative answer: When the actual temperature is 0°C, the predicted apparent temperature is −1.65°C.)

c 97% of the variation in *apparent temperature* can be explained by the variation in *actual temperature*.

d   i That the association is linear.

ii Yes, since there is no clear pattern in the residual plot.

12 a *egg density* = 18.9 × *number of male moths* − 46.8

b   i

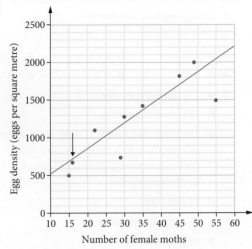

ii On average, egg density increases by 31.3 eggs/m² for each additional female moth caught.

Examples of other correct answers:

• egg density increases by 31.3 as the number of female moths increases by 1

• egg density increases by 31.3 for an increase of 1 in the number of female moths caught

• egg density increases by 31.3 for each additional female moth caught.

iii −412.5

iv 74.3%

13 a

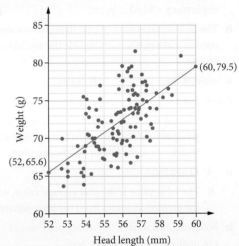

b 60.4 g

c Extrapolation, because 49.0 is outside the data range.

d 80.9 g (1 mark for correctly using the predicted value with the residual of 2.78.)

e 64.5%

f There is probably a linear association.

14 a i $y = 0.942x + 58.159$

    ii $A = 83.6$

       $B = 103.4$

    iii $0.942 \times 12 = 11.30\,\text{cm}$

    iv positive and strong

    v 94.09%

b i $C = 102.7 - 103.4 = -0.7$

    $D = 109.4 - 114.7 = -5.3$

ii

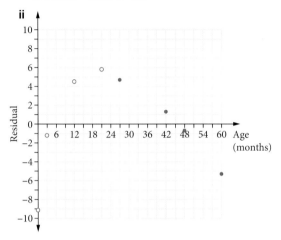

    iii A linear model is not appropriate because there is a pattern to the residuals.

CUMULATIVE EXAMINATION: CALCULATOR-FREE

1 a number of tomatoes

b $T = 224.61 \approx 224$ (225 also acceptable)

c

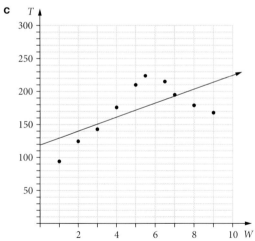

d i

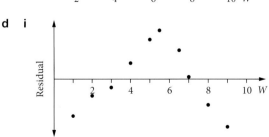

    ii A linear model is not appropriate because a pattern is evident in the residual plot.

2 a $A = 49, B = -4.0$

b

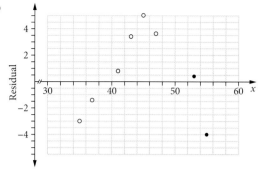

c The residual plot shows a pattern, therefore, this suggests that the least-squares line is unsuitable.

d As the correlation coefficient is low, this supports the answers to part **c**, that the least-squares line is unsuitable.

CUMULATIVE EXAMINATION: CALCULATOR-ASSUMED

1 a 19%   **b** 29 440 000

c The percentages of 15–64-year-olds in each of the three countries are similar: Australia has 67%, India has 64% and Japan has 64%.

2 a $r = 0.88$ **b** $p = 0.5w - 31.4$

c For each 1 cm increase in waistline measurement, the percentage body fat increases by 0.5.

d residual = 1.4 or 1.5

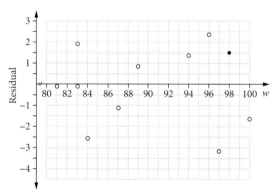

e appropriate because there is no clear pattern in the residuals

f $r^2 = d = 0.78$, hence 22% unexplained.

g i 21   **ii** Not valid, extrapolation

h The percentage body fat for player 13 is 2.6 below their predicted percentage body fat.

3 a $r^2 = 0.25$

Approximately 25% of the variation in temperature can be explained by the variation in altitude.

b $T = -0.264L + 30.94$

$r = -0.88$

c $T = -0.264 \times 22.93 + 30.94 = 24.89$

$T = -0.022 \times 9 + 24.97 = 24.77$

The prediction using latitude is more valid as the correlation coefficient is much stronger.

CHAPTER 3

EXERCISE 3.1

1 a 11th term

b nth term

c a recursive rule

d rule for the nth term

e number of terms from T_3 to T_{13}, inclusive

f sum of the first 10 terms of an arithmetic sequence

2 a i $T_{n+1} = T_n + 8$, $T_1 = 1$ **ii** 41

  b i $T_{n+1} = T_n + 6$, $T_1 = -10$ **ii** 20

  c i $T_{n+1} = T_n - 2$, $T_1 = 51$ **ii** 41

  d i $T_{n+1} = T_n - 4$, $T_1 = -7$ **ii** -27

3 a i $a = 3$, $d = 2$ **ii** $T_n = 1 + 2n$

    iii 19

  b i $a = 98$, $d = -10$ **ii** $T_n = 108 - 10n$

    iii 18

  c i $a = -9$, $d = 3$ **ii** $T_n = -12 + 3n$

    iii 15

  d i $a = 71$, $d = -7$ **ii** $T_n = 78 - 7n$

    iii 15

4 a 15th position   **b** 27th position

5 a 34th position   **b** 17th position

6 a -82, -93, -104   **b** 282, 296, 310

7 a $P_{n+1} = P_n + 6$, $P_1 = 10$

  b $P_n = 4 + 6n$

  c 22nd term

8 a $D_n = 2 + 3n$ **b** 68 disposals

  c 153 disposals   **d** 803 disposals

9 a $T_n = 87 - 3n$ **b** 69%

  c after the 11th weekly assessment

10 a

  b i $T_n = -3n + 11$ **ii** -502

11 a fourth year: 22 cm   **b** $T_n = 30 - 2n$

    fifth year: 20 cm

  c 15th year   **d** 270 cm

12 a $T_n = 90 - 6n$ **b** 48 L

  c 504 L   **d** 630 L

EXERCISE 3.2

1 a $T_{n+1} = T_n - 6$, $T_1 = 33$

  b $T_n = 39 - 6n$

  c 13th position

2 405

3 a i $a = 200$, $r = 1.1$ **ii** $T_{n+1} = 1.1T_n$, $T_1 = 200$

    iii 10% increase

  b i $a = 30$, $r = 0.7$ **ii** $T_{n+1} = 0.7T_n$, $T_1 = 30$

    iii 30% decrease

  c i $a = 3$, $r = 2.5$ **ii** $T_{n+1} = 2.5T_n$, $T_1 = 3$

    iii 150% increase

4 a 1231.5, 2155.1, 3771.5   **b** 42.3, 31.7, 23.8

5 a i 2, 6, 18   **ii** $a = 2$, $r = 3$

    iii $T_n = 2(3)^{n-1}$ **iv** 1458

  b i 1200, 600, 300   **ii** $a = 1200$, $r = \dfrac{1}{2}$

    iii $T_n = 1200\left(\dfrac{1}{2}\right)^{n-1}$ **iv** 18.75

  c i 30, 36, 43.2   **ii** $a = 30$, $r = 1.2$

    iii $T_n = 30(1.2)^{n-1}$ **iv** 89.58

6 a $T_n = 2(4)^{n-1}$ **b** 8th position   **c** 8192

7 a 45   **b** 24   **c** $\dfrac{27}{2}$ **d** $\dfrac{32}{9}$

8 a $C_n = 20(1.071)^{n-1}$

  b

  c Geometric growth; the growth is increasing and it is not a straight line.

9 a 1.27   **b** $R_n = 253(1.27)^{n-1}$

  c $658 million   **d** 2031

10 a $Q_{n+1} = 0.95Q_n$, $Q_0 = 400$ **b** $Q_n = 380(0.95)^{n-1}$

  c $Q_n = 400(0.95)^n$ **d** 176 quokkas

11 a $A_{n+1} = 0.75A_n$, $A_1 = 320$

  b $A_n = 320\left(\dfrac{3}{4}\right)^{n-1}$ **c** $\dfrac{405}{4}$

12 a $T_{n+1} = \dfrac{2}{3}T_n$, $T_0 = 54$

  b $T_n = 36\left(\dfrac{2}{3}\right)^{n-1}$ or $T_n = 54\left(\dfrac{2}{3}\right)^n$

c $T_5 = 36\left(\dfrac{2}{3}\right)^4 = 36 \times \dfrac{16}{81} = \dfrac{64}{9}$ or

$T_5 = 54\left(\dfrac{2}{3}\right)^5 = 54 \times \dfrac{32}{243} = \dfrac{64}{9}$

13 a $r = \dfrac{30\,256}{22\,579} \approx 1.34$

b $T_n = 15 \times 1.34^{n-1}$ or $T_n = 11.19 \times 1.34^n$

c 2026

d $5\,372\,640

14 a i 215.38 units per 100 L

ii Thursday 17 December

b i $C_{n+1} = 1.03C_n, C_1 = 200$

ii 268.78 units per 100 L

EXERCISE 3.3

1 C

2 geometric sequence; 133.49

3 a 1, 8, 29, 92　　　**b** 26, 6, −4, −9

c 20, 40, 70, 115　　**d** 70, 70, 70, 70

4 a $k = 45$　　**b** $k = -28$　　**c** $k = -1000$

5 a i

M_n axis graph with points:
(10, 22.7)
(20, 29.5)
(30, 31.4)
(40, 31.8)
(50, 32.0)

ii levels out to a steady state

b i

M_n axis graph with points:
(10, 57.1)
(20, 67.8)
(30, 85.3)
(40, 113.8)
(50, 160.1)

ii increasing

6 a 32　　　**b** 35　　　**c** 15

7 a 5　　　**b** 11

8 a 18.92

b i 11.1　　**ii** 27.75　　**iii** −2.96

9 a $L_{n+1} = 0.96L_n + 30, L_1 = 980$

b 750 birds　　**c** 44 birds　　**d** 750

e −60; removing 60 birds at the end of each year

10 a $T_{n+1} = 0.5T_n + 20, T_1 = 32$

b approaching a steady state of 40 turtles

c 40

11 a i Only 92.5% of the crocodiles remain in the river each year.

ii 20

b

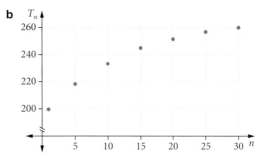

c The population increases, then levels out towards a steady state.

d The population of crocodiles will settle to approximately 267.

CUMULATIVE EXAMINATION: CALCULATOR-FREE

1 a i $T_{n+1} = \dfrac{1}{2}T_n, T_1 = 256$

ii $T_n = 256\left(\dfrac{1}{2}\right)^{n-1}, \dfrac{1}{64}$

b i 11, 18, 25　　　**ii** 5, 7.5, 11.25

c the 37th term

2 a $T_{n+1} = T_n + 4000, T_1 = 14\,000$

b $T_n = 32\,000\left(\dfrac{3}{4}\right)^{n-1}$

c Program A because viewer numbers are increasing.

d 42 000 viewers

CUMULATIVE EXAMINATION: CALCULATOR-ASSUMED

1 a 24

b

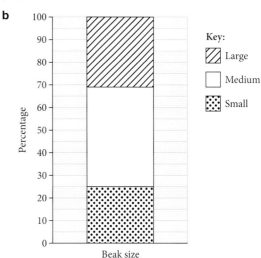

The three sections can be in any order.

c Yes, the information does provide support for the contention that *beak size* is associated with *sex*. The second mark requires a statement similar to one of the following:

- 50% of males had large beaks, which was higher than females, with 7%.
- 48% of males had medium beaks, which was higher than females, with 38%.
- 2% of males had small beaks, which was lower than females, with 55%.

2 a The response variable is *evening congestion level*.

b *Evening congestion level* = 0.922 × 60 + 8.48 = 63.8%

c From the scatterplot, when the *morning congestion level* is 47%, the actual *evening congestion level* is 50%.

The residual value to one decimal place is −1.8.

d 85% of the variation in the evening congestion level can be explained by the variation in the morning congestion level.

3 a $T_{n+1} = 1.09T_n$, $T_0 = 660$　**b** $T_n = 660 \times 1.09^n$

c 21 months　　　　**d** 362

4 a 10%　　　　　　**b** 47

c Deborah will always have 42 mealworms in the long run.

d i 45　　　　**ii** 27

e 14

CHAPTER 4

EXERCISE 4.1

1 a i 5 vertices (*A, B, C, D, E*), 8 edges (*AB, AE, BC, BD, BE, CD, DE, DE*)

ii

Vertex	A	B	C	D	E	Sum
Degree	2	4	2	4	4	16

degree sum = 2 + 4 + 2 + 4 + 4
　　　　　　= 16
　　　　　　= 2 × 8
　　　　　　= 2 × number of edges

b i 5 vertices (*A, B, C, D, E*), 5 edges (*AB, AE, BC, BD, BE*)

ii

Vertex	A	B	C	D	E	Sum
Degree	2	4	1	1	2	10

degree sum = 2 + 4 + 1 + 1 + 2
　　　　　　= 10
　　　　　　= 2 × 5
　　　　　　= 2 × number of edges

2 a No, as edge *CF* does not exist in graph *H*.

b

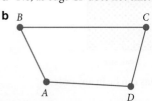

Other answers are possible.

3 a i *L* has a loop, as there is a road from Lucton to Lucton, without visiting other towns.

ii There are multiple edges *MN, LM, LN, KN* and *LK*, as there are two different routes for each of the mentioned journeys.

b

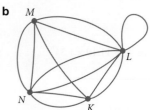

c

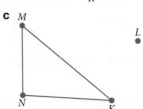

4

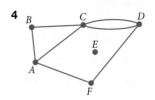

5 a graph *B*　　　　**b** graphs *A, B, C, D, E*

6 a degree sum = 2 + 3 + 2 + 3 + 2 = 2 × 6 = 12

b i *D, E*　　　　**ii** *A, C, E*

iii degree sum = 2 × 8 = 16

7 a Many possible answers, with *A* as the central vertex.

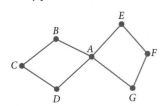

b not possible

Given that vertices *B, D, E, G* (or equivalent) are required to be connected to *A* to have a degree of 2, then *A* will always have a degree greater than 2.

8 a graph *H*　　　　**b** graph *L*

9 a i Alex　　　　**ii** Cameron, Dale

b isolated vertex

10 a Appletown, East Swan

b i

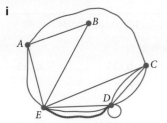

ii A driver leaving Dalton is able to return to Dalton without visiting any other suburbs.

11 a *G*. It is an isolated vertex as there are no borders shared with other states.

b *F*. Western Australia borders two other states/territories so must have degree 2. C and F have degree 2. As *C* is connected to *B*, which represents NSW as it is the only state with degree 4, and WA is not connected to NSW, Western Australia is represented by vertex *F*.

c *D*. deg(D) = 1

d degree sum = 20. The degree sum of 20 is twice the number of shared borders (10) between any two states and/or territories in Australia.

EXERCISE 4.2

1 E **2** C

3 a i Simple, as no loops or multiple edges.

ii Not complete, as every vertex is not directly connected to every other vertex.

b i Not simple, as there are multiple edges *FH*.

ii Not complete, as it does not meet the conditions of simple.

c i Simple, as no loops or multiple edges.

ii Complete, as every vertex is directly connected to every other vertex. It is K_4.

d i Not simple, as there is a loop at *A*.

ii Not complete, as it does not meet the conditions of simple.

4 a two bridges

b one bridge

c no bridges

d six bridges

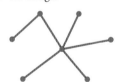

5 a i Other answers are possible.

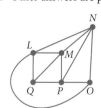

ii $v = 6, f = 7, e = 11; 6 + 7 - 11 = 2$ so Euler's formula holds true.

b i Other answers are possible.

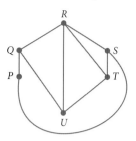

ii $v = 6, f = 5, e = 9; 6 + 5 - 9 = 2$ so Euler's formula holds true.

c i Other answers are possible.

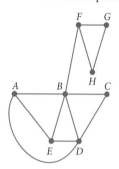

ii $v = 8, f = 6, e = 11; 8 + 6 - 11 = 3 \neq 2$ so Euler's formula does not hold true.

d i Other answers are possible.

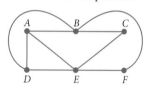

ii $v = 6, f = 5, e = 9; 6 + 5 - 9 = 2$ so Euler's formula holds true.

6 a 5 **b** 14 **c** 4

7 Not bipartite, as the 'triangular' subgraph containing *A*, *B* and *E* is not bipartite.

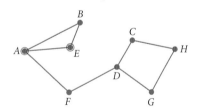

8 a i 6 **ii** 10

iii

n	2	3	4	5	6
Edges	1	3	6	10	15

$$\frac{n(n-1)}{2}$$

b 3

9 a $f = 4$

b $v = 5, f = 3, e = 6; 5 + 3 - 6 = 2$, so Euler's formula holds true.

10 a graphs *A*, *B*, *D* **b** graphs *B*, *D*

11 a no two edges cross

b $v = 3, e = 4, f = 3, 3 + 3 - 4 = 2$, so the graph satisfies Euler's formula.

12 a $f = 3$, $v = 7$, $e = 8$; $7 + 3 - 8 = 2$, so the graph is planar.

b i He cannot remove $Tap - B$, $B - C$ or $C - D$.

  ii a bridge

13 a i $5 + 5 - e = 2$, $e = 8$

  ii Other versions of this graph are possible.

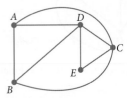

b i minimum = 4, maximum = 10

  ii minimum = $n - 1$

  iii complete

14 a i 2 edges

  ii 8 edges

  iii If connected, $e = 7$, so 5 edges are required. If not connected, Euler's formula does not hold and so, 3 edges are required.

b i Plan Q

EXERCISE 4.3

1 graph A

2 true

3 a D

b B and F; both vertices have an out-degree of 0.

4 R–M: 27 minutes

  R–N–M: 28 minutes

  R–N–P–M: 25 minutes

  R–Q–P–N–M: 39 minutes

  R–Q–P–M: 26 minutes

  ∴ shortest time is 25 minutes along R–N–P–M.

5 10 km; Home–A–D–School

6 10 km $(4 + 1 + 3 + 2)$

7 a 3

b 1000 m; Entrance–$P1$–$P3$

c Yes, the path from Entrance → $P1$ → $P3$ still exists as a possible path in that direction.

8 $CDEAB$; 23 minutes

9 $AQSPSRTUB$; time = 360 minutes

10 a 1.4 km   **b** $ETUTSV$; total = 2 km

11 $BDEFHJ \Rightarrow 22$ ∴ 2200 metres

12 a Shortest path is $ACFH$; distance = 44 km

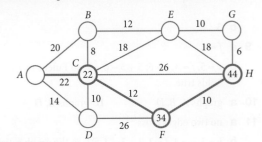

b $22 + (26 - x) < 44 \Rightarrow x > 4$, also $x < 26$, so $4 < x < 26$. The direct road between C and H can be reduced by between 4 and 26 kilometres

EXERCISE 4.4

1 B (10 km = 4 + 1 + 3 + 2)

2 B (23 minutes = 7 + 2 + 2 + 2 + 3 + 7)

3 a

	A	B	C	D	E
A	0	1	0	1	0
B	1	0	0	1	0
C	0	0	0	1	1
D	1	1	1	0	1
E	0	0	1	1	0

b

	A	B	C	D
A	1	1	0	0
B	1	0	3	0
C	0	3	0	0
D	0	0	0	0

c

	A	B	C
A	0	2	3
B	2	0	2
C	3	2	0

4 a i Not a simple graph as there are multiple edges BC (two) and a loop at D.

  ii Other versions are possible.

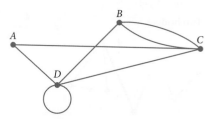

b i Not a simple graph as there are multiple edges AE (three) and a loop at B.

  ii Other versions are possible.

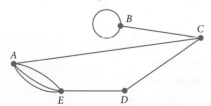

5 a

	A	B	C	D	E
A	0	1	0	1	0
B	0	0	1	0	0
C	0	0	0	0	1
D	0	1	0	0	0
E	0	1	0	0	0

b The entry in the cell EE would change from 0 to 1.

6 $\begin{bmatrix} 1 & 0 & 3 \\ 0 & 0 & 2 \\ 1 & 1 & 0 \end{bmatrix}^3 = \begin{bmatrix} 7 & 3 & 18 \\ 2 & 0 & 10 \\ 6 & 5 & 3 \end{bmatrix}$

$\begin{bmatrix} 1 & 0 & 3 \\ 0 & 0 & 2 \\ 1 & 1 & 0 \end{bmatrix}^4 = \begin{bmatrix} 25 & 18 & 27 \\ 12 & 10 & 6 \\ 9 & 3 & 28 \end{bmatrix}$

9780170476959

7

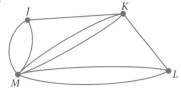

$$
A = \begin{array}{c} \\ J \\ K \\ L \\ M \end{array}
\begin{array}{cccc} J & K & L & M \\ \end{array}
\left[\begin{array}{cccc}
0 & 1 & 0 & 2 \\
1 & 0 & 1 & 2 \\
0 & 1 & 0 & 2 \\
2 & 2 & 2 & 0
\end{array} \right]
$$

$$
A^2 = \begin{array}{c} \\ J \\ K \\ L \\ M \end{array}
\begin{array}{cccc} J & K & L & M \\ \end{array}
\left[\begin{array}{cccc}
5 & 4 & 5 & 2 \\
4 & 6 & 4 & 4 \\
5 & 4 & 5 & 2 \\
2 & 4 & 2 & 12
\end{array} \right]
$$

4 routes:

$M–J–K$

$M–J–K$

$M–L–K$

$M–L–K$

8

$$
\begin{array}{c} \\ A \\ B \\ C \end{array}
\begin{array}{ccc} A & B & C \\ \end{array}
\left[\begin{array}{ccc}
1 & 1 & 1 \\
1 & 0 & 1 \\
1 & 1 & 0
\end{array} \right]
$$

9

$$
\begin{array}{c} \\ A \\ B \\ C \\ D \\ E \\ F \end{array}
\begin{array}{cccccc} A & B & C & D & E & F \\ \end{array}
\left[\begin{array}{cccccc}
0 & 1 & 0 & 0 & 0 & 0 \\
0 & 0 & 1 & 0 & 0 & 1 \\
0 & 0 & 0 & 1 & 1 & 0 \\
0 & 0 & 0 & 0 & 1 & 0 \\
1 & 1 & 0 & 0 & 0 & 0 \\
0 & 0 & 1 & 0 & 0 & 0
\end{array} \right]
$$

10 a There are no one-way road connections, as the adjacency matrix is symmetrical about the leading diagonal, meaning it is an undirected graph.

b

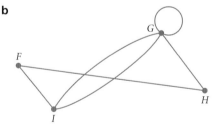

Other versions are possible.

11 a

$$
\begin{array}{c} \\ A \\ B \\ C \\ D \\ E \end{array}
\begin{array}{ccccc} A & B & C & D & E \\ \end{array}
\left[\begin{array}{ccccc}
0 & 1 & 0 & 1 & 0 \\
1 & 0 & 1 & 1 & 1 \\
0 & 1 & 0 & 1 & 0 \\
1 & 1 & 1 & 0 & 1 \\
0 & 1 & 0 & 1 & 0
\end{array} \right]
$$

b A child cannot be friends with themselves.

12 a A '0' means that two individuals cannot communicate directly with each other.

b $f = 1, g = 0$

13 Other versions of the graph possible.

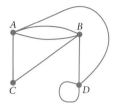

14 a There is no doorway between the area(s).

b $X = 1, Y = 2$

c By calculating the sum of the row or column corresponding to the specific area.

d

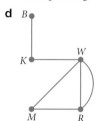

e two (two via R)

15 a

$$
A = \begin{array}{c} \\ W \\ X \\ Y \\ Z \end{array}
\begin{array}{cccc} W & X & Y & Z \\ \end{array}
\left[\begin{array}{cccc}
1 & 2 & 2 & 1 \\
2 & 0 & 1 & 1 \\
2 & 1 & 0 & 1 \\
1 & 1 & 1 & 1
\end{array} \right]
$$

b

$$
A^3 = \begin{array}{c} \\ W \\ X \\ Y \\ Z \end{array}
\begin{array}{cccc} W & X & Y & Z \\ \end{array}
\left[\begin{array}{cccc}
36 & 31 & 31 & 26 \\
31 & 19 & 20 & 20 \\
31 & 20 & 19 & 20 \\
26 & 20 & 20 & 18
\end{array} \right]
$$

$a_{33}^3 = 19$; there are 19 different routes from Y to Y using three edges.

16 a

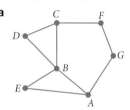

b i The matrix is symmetrical about the leading diagonal.

ii 4

iii $BCF, BCD, BDC, BEA, BAE, BAG$

CUMULATIVE EXAMINATION: CALCULATOR-FREE

1 a i $A_1 = 4, A_2 = 19, A_3 = 34, A_4 = 49, A_5 = 64$

ii $A_n = 4 + 15(n - 1)$ or $A_n = 15n - 11$; $A_{50} = 739$

b i long-term decreasing (to steady state of 0)

ii $B_{n+1} = 0.75B_n, B_1 = 100$

2 a i

ii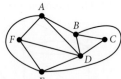

iii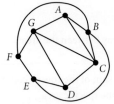

b $v = 6, f = 6, e = 10; 6 + 6 - 10 = 2$, so the graph verifies Euler's formula.

3 a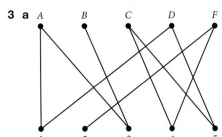

b i yes **ii** yes

c A simple graph has no loops or multiple edges.

CUMULATIVE EXAMINATION: CALCULATOR-ASSUMED

1 a i $P_5 = 132$ **ii** $P_{14} = 61$

b The long-term population of birds on the island reaches a steady state of 50.

2 a rainfall $= -2.68(35) + 131 = 37.2\,\text{mm}$

b i 80.81% of the variation in the rainfall data can be explained by the variation in the percentage of clear days data

ii $r = \pm\sqrt{0.8081} = -0.899$
∴ a strong negative linear correlation

3 a Entrance–P–S; 45 metres

b Entrance–L–T–M; $35 + 35 + 40 = 110$ metres

c The path from G–R–Café will be 55 metres in length, which is equal to the path G–T. Hence, it will not reduce the shortest path in part **b**.

4 a i The adjacency matrix is not symmetrical about the leading diagonal.

or

There are two ways from B to C but only one way from C to B (or two ways from D to C but only one way from C to D).

ii A simple graph contains no loops or multiple edges. There is a loop from C to C. There are two edges AB and DC.

b i Here are three 2-stage routes from C to D.

ii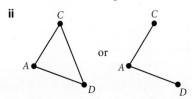

CHAPTER 5

EXERCISE 5.1

1 a Closed trail: no repeated edges, a repeated vertex, starts and finishes at the same vertex

b Cycle: no repeated edges, no repeated vertices (except the first and last), starts and finishes at the same vertex

c Path: no repeated edges, no repeated vertices, starts and ends at different vertices

d Walk only: a repeated edge

e Open trail: no repeated edges, a repeated vertex, starts and ends at different vertices

2 a Path: no repeated edges, no repeated vertices, starts and ends at different vertices

b Cycle: no repeated edges, no repeated vertices (except the first and last), starts and finishes at the same vertex

c Closed trail: no repeated edges, a repeated vertex (E), starts and finishes at the same vertex

d Walk only: a repeated edge (DE is the same as ED)

e Walk only: a repeated edge (FE)

f Open trail: no repeated edges, a repeated vertex (E), starts and ends at different vertices

3 a A–B–A, A–C–A, A–A–A

b Not a closed trail nor a cycle, as it contains a repeated edge

4 a It is a walk with no repeated edges. It starts and finishes at different vertices.

b 6

c 44 minutes

d A–B–C–E–A; 31 minutes.

5 a i 8 **ii** open trail

b i 7 **ii** (open) path

c The teenagers cannot leave through H because of the direction of the arc HI and HG.

d I–G–E–D–I; length 4

e I–G–E–D–A–B–C–D–I; length 8

EXERCISE 5.2

1 D

2 D (closed trail: a repeated vertex, no repeated edges, starts and finishes at the same vertex)

3 Neither, as all vertices have an odd degree.

4 a $\deg(A) = 3$, $\deg(B) = 2$, $\deg(C) = 3$, $\deg(D) = 2$. Two odd vertices, so semi-Eulerian.
A–B–C–A–D–C, Other answers are possible.

b $\deg(A) = 4$, $\deg(B) = 2$, $\deg(C) = 4$, $\deg(D) = 2$, $\deg(E) = 4$, $\deg(F) = 2$. No odd vertices, so Eulerian.
A–B–C–F–E–D–A–E–C–A. Other answers are possible.

5 There are two vertices of odd degree (V and W), and so an open Eulerian trail exists, but not a closed Eulerian trail. So, the skate is not possible.

6 5 ways; remove edge *BF*, *DF*, *EF*, *DE* or *DE*.

7 Graph *H*; it has two odd vertices (*D* and *F*).

8 a The graph has more than two odd vertices.

b

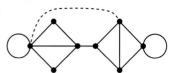

9 a

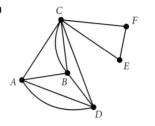

b The graph now has no odd vertices and so a closed Eulerian trail is possible.

10 a Any of the following possibilities:

Tap–*B*–*C*–*D*–*G*–*F*–*D*–*E*–*F*

Tap–*B*–*C*–*D*–*G*–*F*–*E*–*D*–*F*

Tap–*B*–*C*–*D*–*F*–*E*–*D*–*G*–*F*

Tap–*B*–*C*–*D*–*E*–*F*–*D*–*G*–*F*

Tap–*B*–*C*–*D*–*F*–*G*–*D*–*E*–*F*

Tap–*B*–*C*–*D*–*E*–*F*–*G*–*D*–*F*

b open Eulerian trail (semi-Eulerian trail)

11 a The graph contains 4 vertices of odd degree, meaning a closed Eulerian trail does not exist. The inspector will need to drive along the same road twice.

b 2 roads

c 108 km

EXERCISE 5.3

1 D

2 C

3 a **i** *K*–*L*–*M*–*N*   **ii** *K*–*L*–*M*–*N*–*K*

b **i** *A*–*B*–*C*–*D*   **ii** *A*–*B*–*C*–*D*–*A*

c **i** *B*–*A*–*D*–*E*–*F*–*C*   **ii** *B*–*A*–*D*–*E*–*F*–*C*–*B*

4 a

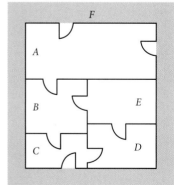

b Hamiltonian cycle

5 a *G*–*A*–*B*–*C*–*D*–*F*–*E*–*G*. Other answers are possible.

b *D*–*L*–*E*–*F*–*G*–*H*–*I*–*J*–*K*–*D*. Other answers are possible.

6 *D*–*G*–*F*–*C*–*E*–*A*–*B*

7 *E*–*P*–*Q*–*R*–*S*–*V*–*U*–*T*–*E*

8 a Hamiltonian cycle

b *B*–*C*–*G*–*J*–*H*–*F*–*E*–*D*–*B*; 4800 m

9 *A*–*D*–*C*–*B*–*E*–*G*–*H*–*F*; 70 km

10 a factory–*T*–*S*–*Q*–*R*–*S*–*U*–factory

b Town *S* is visited twice.

CUMULATIVE EXAMINATION: CALCULATOR-FREE

1 a There is the presence of an outlier that weakens the relationship between the data sets.

b $-0.05(20) + 57.6 = 56.6\%$. This prediction is not reliable because of the weakness of the regression model.

c **i** The correlation coefficient would become more negative, i.e. closer to −1.

   ii The gradient of the least-squares line would become more negative.

   iii The *y*-intercept would become more positive, i.e. closer to 100%.

2 a graph *B*; all vertices have a degree of zero.

b Yes, a Hamiltonian path exists given that all vertices can be visited once with no repeated vertices, no repeated edges and a different first and last vertex.

3 a *D*–*E*–*C*–*A*–*B*–*D*. Other answers are possible.

b The graph is semi-Eulerian, as it contains exactly two odd vertices.

4 a None

b Plan *P* and Plan *R*

5 a

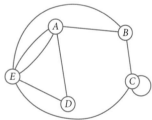

b **i** Neither, as there are three vertices of odd degree (*B*, *C*, *E*)

   ii Semi-Hamiltonian, as a path can be found that does not repeat vertices and edges, visits all vertices but does not start and finish at the same vertex, e.g. *A*–*D*–*E*–*B*–*C*.

CUMULATIVE EXAMINATION: CALCULATOR-ASSUMED

1 a $T_5 = 39.54°C$

b $T_9 = 30.53$, $T_{10} = 29.84$. After 10 minutes.

c The surface temperature of the egg tends towards a long-term steady state of 28.57°C.
$k = 0.65k + 10 \Rightarrow 0.35k = 10 \Rightarrow k = 28.57$

2 a $120

b Queens Garden and Riverville

c $v = 6$, $e = 11$, $f = 7$; $6 + 7 − 11 = 2$ and so Euler's formula holds true.

3 a i Lizards eat insects.

    ii A bird cannot be eaten by any of the animals in the food web.

b
$$N = \begin{array}{c} \\ \\ \\ \\ \end{array} \begin{bmatrix} 0 & 1 & 1 & 1 \\ 0 & 0 & 0 & 0 \\ 0 & 1 & 0 & 0 \\ 0 & 1 & 1 & 0 \end{bmatrix} \begin{array}{c} I \\ B \\ L \\ F \end{array}$$

with column headings I B L F

c Any given animal species in the food web does not eat another member of its own species.

4 a 86 km; G–O–N–M **b** town K

CHAPTER 6

EXERCISE 6.1

1 ClassPad

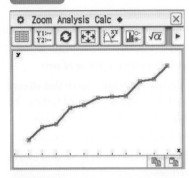

TI-Nspire

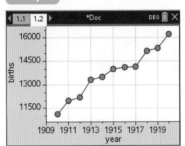

2 a Likely to show seasonality because more would be bought in winter when there is more rain.

b Likely to show seasonality because more would be bought in summer.

c Likely to show seasonality because more people would have the time to shop for them on weekends than weekdays.

d Not likely to show seasonality because milk is consumed regularly every day.

e Likely to show seasonality because the occupancy rates would be higher during the weeks of the holiday periods.

f Not likely to show seasonality because salaries are paid regularly throughout the year.

g Likely to show seasonality because fruit is harvested in certain months.

h Likely to show seasonality because AFL matches are only played during certain months of the year.

3 a no trend   **b** decreasing   **c** increasing

4 a i ClassPad

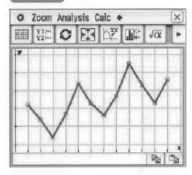

TI-Nspire

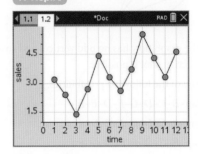

    ii increasing trend

    iii no possible outlier or irregular trend

    iv 4 seasons in a period

b i ClassPad

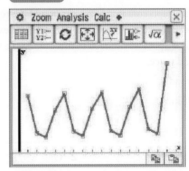

TI-Nspire

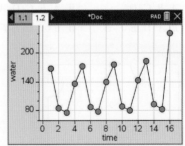

    ii no trend

    iii a possible outlier – Dec 2021

    iv 4 seasons in a period

c **i**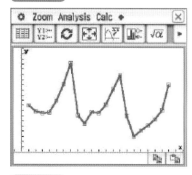

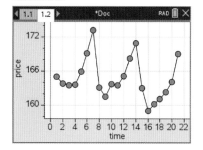

ii decreasing trend

iii no possible outlier or irregular trend

iv 7 seasons in a period

EXERCISE 6.2

1 **a** no trend

   b no trend with a possible outlier

   c decreasing trend

   d increasing trend

2 **a** $3057   **b** $2824

3 **a**

Year	Number of road deaths	Three-point moving means
1990	225	
1991	240	$\dfrac{225 + 240 + 201}{3} = 222$
1992	201	$\dfrac{240 + 201 + 192}{3} = 211$
1993	192	$\dfrac{201 + 192 + 185}{3} = 192.67$
1994	185	$\dfrac{192 + 185 + 160}{3} = 179$
1995	160	$\dfrac{185 + 160 + 172}{3} = 172.33$
1996	172	$\dfrac{160 + 172 + 127}{3} = 153$
1997	127	$\dfrac{172 + 127 + 132}{3} = 143.67$
1998	132	$\dfrac{127 + 132 + 101}{3} = 120$
1999	101	$\dfrac{132 + 101 + 100}{3} = 111$
2000	100	

b The smoothed number of road deaths in 1994 is 179. There was not enough data to calculate the smoothed number of road deaths in 2000.

c

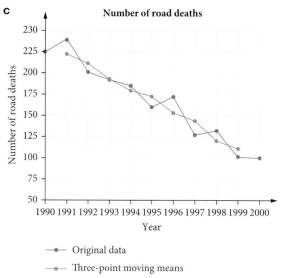

d The graph of the smoothed data indicates a decreasing trend.

4 a

Year	Number of textbook sales	Four-point moving means	Four-point moving means with centring
2010	2250		
2011	2230		
		$\dfrac{2250 + 2230 + 2000 + 2010}{4} = 2122.5$	
2012	2000		2090
		$\dfrac{2230 + 2000 + 2010 + 1990}{4} = 2057.5$	
2013	2010		2153.75
		$\dfrac{2000 + 2010 + 1990 + 3000}{4} = 2250$	
2014	1990		2255.625
		$\dfrac{2010 + 1990 + 3000 + 2045}{4} = 2261.25$	
2015	3000		2383.625
		$\dfrac{1990 + 3000 + 2045 + 2989}{4} = 2506$	
2016	2045		2632.25
		$\dfrac{3000 + 2045 + 2989 + 3000}{4} = 2758.5$	
2017	2989		2627.25
		$\dfrac{2045 + 2989 + 3000 + 1950}{4} = 2496$	
2018	3000		2505.375
		$\dfrac{2989 + 3000 + 1950 + 2120}{4} = 2514.75$	
2019	1950		2423
		$\dfrac{3000 + 1950 + 2120 + 2255}{4} = 2331.25$	
2020	2120		2243.375
		$\dfrac{1950 + 2120 + 2255 + 2297}{4} = 2155.5$	
2021	2255		
2022	2297		

b There is not enough data to calculate the smoothed number of sales for 2011. The smoothed number of sales for 2019 is 2423.

d The graph of the smoothed data indicates an increasing trend until 2016 and a decreasing trend from 2017 to 2022.

c

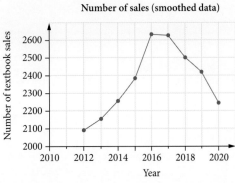

Number of sales (smoothed data)

5

	A	B	C
1	Time period	Gym users	3-point moving averages
2	1	1788	
3	2	1420	1436.00
4	3	1100	1403.33
5	4	1690	1396.00
6	5	1398	1366.00
7	6	1010	1342.33
8	7	1619	1311.33
9	8	1305	1307.33
10	9	998	1296.33
11	10	1586	1268.00
12	11	1220	1225.00
13	12	869	

6 a

Time period	4-monthly sales ($'000)	3-point moving average
1	20	
2	27	24
3	25	24.7
4	22	25.7
5	30	25.7
6	25	26.3
7	24	27.3
8	33	29
9	30	29.3
10	25	30.3
11	36	30.3
12	30	31.3
13	28	32
14	38	33
15	33	34.3
16	32	35.3
17	41	36
18	35	

b

Time period	No. of patrons each quarter	4-point centred moving average
1	94	
2	120	
3	103	103
4	97	101.5
5	90	99.875
6	112	98.375
7	98	97
8	90	95.5
9	86	93.625
10	104	91.5
11	91	88.75
12	80	85.875
13	74	82.875
14	93	80
15	78	77.625
16	70	74.75
17	65	
18	79	

7 a $A = 63$, $B = 36.5$, $C = 30.2$

b The 4-point moving averages are more suitable because the:

- data is collected quarterly and the peaks always occur in December quarter; or
- 4-point centred smoothed data has a decreasing trend while the 5-point smoothed data has fluctuations.

8 a 21   **b** 19

9 a To smooth out the data or to identify the trend.

b $A = 816$, $B = 880.8$, $C = 888.5$

c The most appropriate is the 5-point moving average. The data has a 5-point cycle.

or

The values in the 5-point moving average column are continually decreasing.

10 a & b i

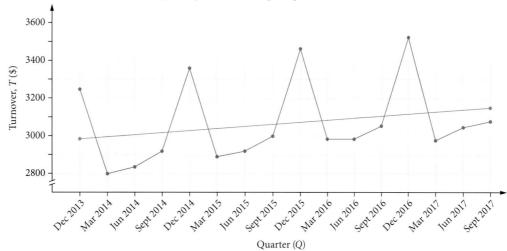

Quarterly retail turnover per capita, Australia, 2013–2017

b ii There is an increasing/upwards trend. The high points are in the December quarter and the low points are in the March quarter.

c $3053.90

1 12

2 increasing trend

3 12-point centred moving average. It has 12 seasons (months) in one period (year).

4 a 7 months   **b** October   **c** January

   d 12   **e** 1

   f

Jan	250%
Feb	210%
Mar	160%
Apr	70%
May	20%
Jun	20%
Jul	10%
Aug	0%
Sep	60%
Oct	90%
Nov	120%
Dec	190%

   g 100%   **h** 150%

5 a

Q1	Q2	Q3	Q4
1.91	1.39	0.17	0.52

   b

Q1	Q2	Q3	Q4
191%	139%	17%	52%

   c Q2

6 a

Year	Q1	Q2	Q3	Q4
Seasonal index	1.437	1.581	0.303	0.679

   b

Year	Q1	Q2	Q3	Q4
2023	7.655	5.063	3.300	4.418
2024	6.263	5.696	9.900	1.473
2025	2.784	5.696	3.300	10.309

   c

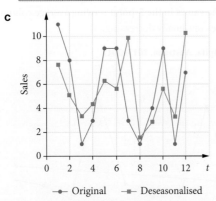

     —•— Original   —■— Deseasonalised

7 a 0.7 is the missing seasonal index.

   b 43%   **c** $70 000

8 a 96.04%   **b** $3179.31   **c** $2985.30

9 a $A = 90, B = 158, C = 63.4\%$

   b i 0.625 (or 0.590)   **ii** 140

10 a $A = 884, B = 1153.5, C = 87.66\%$

   b

Quarter	March	June	September	December
Seasonal index	0.9846	0.8774	0.9826	1.1554

   c $E = 1062$

1 1.29   **2** $240 400   **3** $197 800

4 a *deseasonalised number of sales*
    = −3.69 × quarter number + 39.11

   b

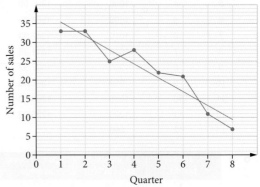

Deseasonalised number of sales 2022–2023

During 2022–2023, the sales of the costume decreased on average by 3.69 per quarter.

   c 6 costumes

   d 5 costumes

5 a *sales = −0.127 × month + 211.123*

   b $207

   c $178

6 a $A = 10, B = 128.6, C = 14.25$

   b Winter 74.6; Spring 115.3

   c $\dfrac{13}{0.958}$

   d Winter, as the seasonal index (74.6) is the lowest

   e $y = -0.24n + 14.88$

   f 10 sprinkler systems

   g The number of systems installed is declining since the gradient of the least-squares line is negative.

7 a 5-point moving average

   b $A = 68.18, B = 20.8, C = 15$

   c SI = 1.5262 (152.62%)

   d

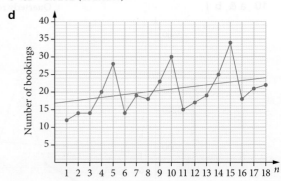

   e i 41.12 (approximately 41)

     ii Not valid, since it is extrapolation.

CUMULATIVE EXAMINATION: CALCULATOR-FREE

1 a 36% of the variation in their birth weight can be explained by the variation in the gestation period.

  b 42.2 grams

  c It may not be valid (even though it is interpolation) as the correlation is moderate.

  d There is not enough evidence to say there is a causal relationship between higher birth weight and a shorter gestation period as there may be other factors involved.

2 a 7, 3, –1

  b $T_n = 11 - 4n$

  c the 23rd term

3 a Other answers are possible.

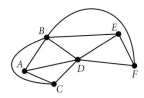

  b 7 faces; $v - e + f = 2 \Rightarrow 6 - 11 + 7 = 2$ and so Euler's formula holds true.

  c 4 edges: AE, AF, CE, CF

  d 3 edges. All vertices have odd degree and so there are three pairs of odd degree vertices. One edge connecting each pair needs to be removed so that all vertices have even degree, e.g. BC, AD and EF.

CUMULATIVE EXAMINATION: CALCULATOR-ASSUMED

1 a

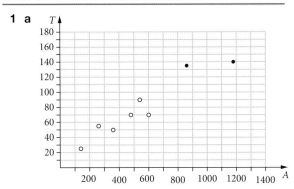

  b Equation of least-squares line is $T = 0.115A + 16.008$. The coefficient of determination is 0.904.

  c The time taken to mow the lawn increases by 0.115 minutes per square metre.

2 a 5.5%   **b** 2873 people

3 a $S–V–Y–W$; 9 minutes

  b $S–T–U–X–Z–W–V–Y$; 21 minutes

c

$$A = \begin{array}{c} \\ S \\ T \\ U \\ V \\ W \\ X \\ Y \\ Z \end{array} \begin{array}{c} \begin{array}{ccccccccc} S & T & U & V & W & X & Y & Z \end{array} \\ \left[\begin{array}{ccccccccc} 0 & 1 & 0 & 1 & 0 & 0 & 0 & 0 \\ 0 & 0 & 1 & 0 & 0 & 0 & 0 & 0 \\ 0 & 0 & 0 & 0 & 0 & 1 & 0 & 0 \\ 0 & 1 & 0 & 0 & 0 & 0 & 1 & 0 \\ 0 & 1 & 0 & 1 & 0 & 0 & 0 & 0 \\ 0 & 0 & 0 & 0 & 1 & 0 & 0 & 1 \\ 0 & 0 & 0 & 0 & 1 & 0 & 0 & 0 \\ 0 & 0 & 0 & 0 & 1 & 0 & 0 & 0 \end{array} \right] \end{array}$$

  d S, T, U, V, Y

  e $a_{62}^3 = 2$. There are two ways to travel from store X and finish at store T, by visiting two other stores. $X–Z–W–T$ and $X–W–V–T$.

4 a $A = 178.65$, $B = 183$

  b There are 4 seasons in a year; therefore, a 4-point centred moving average is used.

  c $\hat{M} = -3.65T + 165$

  d The gradient of the line is negative, indicating that the trend of the rainfall pattern is decreasing.

  e $C = 192.80$, $D = 52.4$, $E = 23.8$, $F = 158.65$, $G = 173.21$

  f $\dfrac{252.28 + 300.61 + 173.21}{3} = 242.03$

  g $H = 113.54$

  h $t = 17$ for Summer 2021/22.

  $R = -5.6754(17) + 202.00 = 105.5182$

  The predicted rainfall for Summer 2021/22 will be 106 mm. The prediction is not reliable as it is more than one period out of the recorded data.

CHAPTER 7

EXERCISE 7.1

1 a Bank 1 = 7.68%, Bank 2 = 7.78%, Bank 3 = 7.74%, Bank 4 = 7.7%

  b Georgio should choose Bank 2 because it pays the higher effective rate of interest and will therefore pay more interest.

  c Bank 1 with 7.68% interest would earn Georgio the least interest.

  d The nominal and effective interest rates for Bank 4 are the same because the rate compounds annually.

2 a 9.38%   **b** 11.61%   **c** 12.36%   **d** 6.18%

3 a Quarterly, as the effective rate of interest is the lowest, he will pay less interest.

  b $4.06   **c** $3153.81

4 a A: 3.77%, B: 3.76%, C: 3.75%

  b When nominal interest rates are equal, the compounding period with most periods in one year will always be the best investment (have the largest effective interest rate).

  The best loan rate is C and the best investment rate is A.

5 a 6.98%   **b** $83.76

6 a The lowest effective rate of interest is 7.72%.

b 7.58% p.a. compounding six-monthly

7 difference = effective interest rate – nominal interest rate
$$= 5.5\% - 5.4\% = 0.1\%$$

8 Option B: effective interest rate
$$= \left(1 + \frac{0.054}{365}\right)^{365} - 1 = 0.055\,48 = 5.55\%$$

Therefore, Option A is better since 5.64% is higher.

9 a 4.39%

b If the interest is compounded annually.

c Account C, as it has the highest effective interest rate.

10 a 7.8% p.a.

b Aussie Bank has the higher effective interest rate because it has the same nominal interest rate as Power Bank but a higher number of compounding periods.

c It is 8%. The nominal and effective interest rates for Power Bank are the same because the rate compounds annually.

d Bank of Western Australia: 8.03%, Aussie Bank: 8.16%. Jillian should choose Aussie Bank since it has a higher effective interest rate.

EXERCISE 7.2

1 E   **2** C

3 a Answers can vary slightly depending on when values are rounded.

i $5018   **ii** $1138   **iii** the third year

n	Depreciation	Value after n years ($)
0		12 000
1	$\frac{16}{100} \times 12\,000 = 1920$	$12\,000 - 1920 = 10\,080$
2	$\frac{16}{100} \times 10\,080 = 1613$	$10\,080 - 1613 = 8467$
3	$\frac{16}{100} \times 8467 = 1355$	$8467 - 1355 = 7112$
4	$\frac{16}{100} \times 7112 = 1138$	$7112 - 1138 = 5974$
5	$\frac{16}{100} \times 5974 = 956$	$5974 - 956 = 5018$

b $V_0 = 12\,000$, $V_{n+1} = 0.84V_n$

c 84%

d geometric decay

e about $1000

4 a $V_0 = 25\,000$

$V_1 = 0.6V_0 = 0.6 \times 25\,000 = 15\,000$

$V_2 = 0.6V_1 = 0.6 \times 15\,000 = 9000$

b 40%   **c** 5 years

5 a $V_n = 0.75^n \times 200\,000$ **b** $20\,023

c 6 years   **d** $8899

6 a $V_0 = 64\,000$, $n = 2$, $V_2 = 40\,960$, $r = ?$

Solve $40\,960 = \left(1 - \frac{r}{100}\right)^2 \times 64\,000$

b $7000

7

n	Depreciation	Value after n years ($)
0		4000
1	600	3400
2	510	2890

8 a $V_0 = 32\,000$, $V_{n+1} = 0.89V_n$

b $V_0 = 134\,000$, $V_{n+1} = 0.915V_n$

9 a

n	0	1	2	3
Value of car after n years ($)	13 000	11 050	9393	7984

b 15% per year

c $T_n = 13\,000 \times 0.85^n$

d $3542

e $n = 4.265$. Julie will sell her car at the end of the fifth year.

10 a $V_1 = 0.9 \times 60\,000 = 54\,000$

$V_2 = 0.9 \times 54\,000 = 48\,600$

b 10%   **c** 11 years

11 a 15%   **b** $5484.23

12 a $3000 - 2760 = $240

$\frac{240}{3000} = 0.08 = 8\%$

b $S_0 = 3000$, $S_{n+1} = 0.92S_n$

EXERCISE 7.3

1

n	Depreciation	Value at the start of year n ($)
0		100 000
1	$\frac{15}{100} \times 100\,000 = 15\,000$	$100\,000 - 15\,000 = 85\,000$
2	$\frac{15}{100} \times 85\,000 = 12\,750$	$85\,000 - 12\,750 = 72\,250$
3	10 837.50	61 412.50
4	9211.88	52 200.62

  a $52 200.62

  b $15 000

  c $47 799.38

2 a $62 098.75

  b 43 months

  c **i** 15 quarters

    ii 2 months

3 a 7.4%

  b $135 000

4 a $N = 156$, C/Y or CpY $= 52$

  b $N = 10$, P/Y or PpY $= 2$

  c $N = 48$, PV $= -15\,000$

  d I% $= 3.6$, PV $= -4000$, FV $= 5000$

  e $N = 6$, PV $= -70\,000$, C/Y or CpY $= 1$

  f $N = 4$, I% $= -12.5$, FV $= 18\,600$

5 $5995

6 a **i** $6039.75

    ii Time required to double investment is 10.7 years or 10 years 9 months.

  b Option B: effective interest rate = 5.55%. Therefore, Option A is better since 5.64% is higher. See worked solutions for ClassPad use of the effective interest function.

7 a $21 315   **b** $16 782.37

  c 10.14%

8 a $184 388.30   **b** $15 360

9 a 3.60%   **b** 0.06%

10 a 24.02%   **b** $1154.70

11 a $6089.34   **b** 14.7%

EXERCISE 7.4

1 B

2 C

3 a $V_0 = 30\,000$, $V_{n+1} = 1.0385 V_n - 2300$

  b The recurrence relation models a combination of linear decay and geometric growth.

  c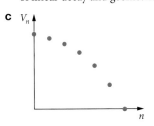

4 a $15 000

  b $465

  c $V_0 = 15\,000$

$V_1 = 1.006 V_0 - 465$
$\quad = 1.006 \times 15\,000 - 465$
$\quad = 14\,625.00$

$V_2 = 1.006 V_1 - 465$
$\quad = 1.006 \times 14\,625.00 - 465$
$\quad = 14\,247.75$

$V_3 = 1.006 V_2 - 465$
$\quad = 1.006 \times 14\,247.75 - 465$
$\quad = 13\,868$

  d 7.2%   **e** 7 months

5 a $r = \dfrac{2850.00}{380\,000.00} \times 100 = 0.0075 \times 100 = 0.75\%$

  b 9% per annum compounding monthly

  c

Month	Opening balance	Interest	Repayment	Closing balance
1	380 000.00	2850.00	3854.00	378 996.00
2	378 996.00	2842.47	3854.00	377 984.47
3	377 984.47	2834.88	3854.00	376 965.35

6 a $A = \$2664.24$, $B = \$2664.24$, $C = \$106.57$, $D = \$20.81$

  b $9979.19

7 a $V_1 = 1.0036 \times 80\,000 - 288 = 80\,000$

$V_2 = 1.0036 \times 80\,000 - 288 = 80\,000$

The value of the loan stays at the principal value $80 000 for all compounding periods.

  b $200 000   **c** $6825

8 a $225 000

  b $V_0 = 225\,000$, $V_{n+1} = 1.003\,V_n - 675$

9 a $\dfrac{98.67}{16\,000} \times 100 = 0.616\,6875\%$

$0.616\,6875 \times 12 = 7.400\,25\% \approx 7.4\%$

  b

Month	Opening balance	Interest	Repayment	Closing balance
1	16 000.00	98.67	600.00	15 498.67
2	15 498.67	95.58	600.00	14 994.24
3	14 994.24	92.46	600.00	142 486.71
4	14 486.71	89.33	600.00	13 976.04

  c $T_{n+1} = \left(1 + \dfrac{0.074}{12}\right) T_n - 600, \quad T_0 = 16\,000$

  d 30 months

  e $1540.99

  f Wendy could borrow $21 984.27.

10 a 1.5% per month

  b $A = \$219.91$, $B = \$14\,380.32$

  c 44 months

  d $21 961.26

  e He would never pay off the loan. Using this suggestion the repayment would equal the interest per month, so he only ever pays interest and the balance never decreases.

  f The payment required per month is $786.98.

11 a

Number of years (n)	0	1	2	3
Amount owing ($)	15 000	15 600	16 224	16 872.96

   ii $A_{n+1} = 1.04A_n$, $A_0 = 15\,000$

  b $T_{n+1} = 1.04T_n - 2400$, $T_0 = 15\,000$

  c **i** $B_{n+1} = 1.0033B_n - 200$, $B_0 = 15\,000$

    ii 87 months   **iii** $17 262.09

12 a **i** $643.85   **ii** $317 428.45

  b $S_0 = 320\,000$, $S_{n+1} = 1.003S_n - 1600$

EXERCISE 7.5

1 D

2 a 2% per month   **b** $2346.98

3 a 5.2%   **b** 3.8%   **c** 6.6%   **d** 20.9%

  e The payments are made to the bank, so the money is moving away from the person.

4 a **i** $154.66   **ii** $9279.60   **iii** $1279.60

  b **i** $2097.64   **ii** $755 150.40   **iii** $455 150.40

  c **i** $1514.18   **ii** $60 567.20   **iii** $25 567.20

  d **i** $442.63   **ii** $115 083.80   **iii** $40 083.80

5 a **i** $36 157.44   **ii** 38%   **iii** 102

    iv $195.05   **v** $1030.05

  b **i** $32 815.84   **ii** 18%   **iii** 272

    iv $221.69   **v** $471.69

6 a 4.5%   **b** $3517.50   **c** 3.2%

  d $9250.00

7 a $413.00   **b** $7132.42   **c** $127 207.81

8 a $1001.71   **b** $3568.12

9 a $807.23   **b** 47 months

  c **i** $43 234   **ii** $281.02

10 150

EXERCISE 7.6

1 B   **2** C   **3** 14 years

4 a $817.14   **b** $52 269.08

5 a $161.50   **b** $371.48   **c** $716.93

6 4.1%   **7** 21 years

8 a Loan amount reduces at a quicker rate or less interest is paid

  b **i** $316 386.79   **ii** $301 279.69

  c Option 1 best by 301 422.81 − 301 279.69 = $143.12

9 a $225   **b** $16 801   **c** $622.75

10 a $107.50   **b** $250

  c $420.40. Using a finance solver:

    $N = 12$, $I\% = 12.9$, $PV = 3776.15$, $PMT = -330$

    $FV = -90.400\,655\,97$, $P/Y = C/Y = 12$

    So, the last payment of $330 must be increased by $90.40 to repay the loan.

11 a **i** 300 months   **ii** $3694.25

  b $2265.04

12 a **i** $65 076.22   **ii** $4676.22

  b $28 204

EXERCISE 7.7

1 D   **2** B

3 a $V_0 = 45\,000$, $V_{n+1} = 1.005V_n - 250$

  b 10.4%   **c** $65.39

4 a $380 000.00   **b** $3854.00

  c 0.75% per month   **d** 9% p.a.

  e $2834.88   **f** $1004.00

  g $376 965.35

5 17 years   **6** $3355.08

7 a $V_0 = 175\,000$, $V_{n+1} = 1.012\,V_n - 3500$

  b $98 467

8 a $365 428.00   **b** $39 214.20   **c** $4271.49

9 a $a = 200\,000$, $R = 1.0075$, $d = -2000$

  b 186

  c **i** 93 months   **ii** $2282.19

  d **i** 166 months   **ii** $678.05

10 a $3000   **b** 18 months

EXERCISE 7.8

1 B   **2** C

3 a $V_1 = 1.006 \times 31\,000 - 186 = 31\,000$

    $V_2 = 1.006 \times 31\,000 - 186 = 31\,000$

  The value of investment stays at the principal value $31 000 for all compounding periods.

  b $53 571.43   **c** $13 650

4 a 6.3%   **b** $8102   **c** 8.7%

  d $45 292

5 a $150 000   **b** 3.75%   **c** $440 678

6 a 3.75%   **b** $20 000

7 a ClassPad

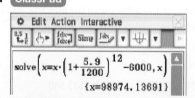

  TI-Nspire

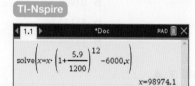

  b Quarterly payment required = $4283.77

8 a 6.3%   **b** $80 000   **c** $35 208

9 a **i** Annuity   **ii** $700 420.20   **iii** $681 970.53

  b $10 278.03

10 a

P	Q	X	Y
$15 000	$648 000	$645 312	$12 260.93

  b 7.6%   **c** 16.5 years   **d** $5690.52

  e Effective interest rate is 7.82%; therefore, his belief is not true.

EXERCISE 7.9

1 C **2** D

3 a $V_0 = 27\,000$, $V_{n+1} = 1.006V_n + 310$

  b 7.2% **c** $839.76

4 a $r = \dfrac{40.00}{500.00} \times 100 = 0.08 \times 100 = 8\%$

  or $r = \dfrac{83.20}{1040.00} \times 100 = 0.08 \times 100 = 8\%$

  b 8% p.a.

  c

Year	Opening balance	Interest	Payment	Closing balance
1	500.00	40.00	500.00	1040.00
2	1040.00	83.20	500.00	1623.20
3	1623.20	129.86	500.00	2253.06

5 a 8 months **b** $534.81

6 $1 014 606.61

7 a $500.00 **b** $660.87

8 a six months **b** $7059

9 a $307 794.50 **b** $18 496.02 **c** $1854.06

10 a 4.39%

  b if the interest is compounded annually

  c Account C, as it has the highest effective interest rate.

  d $r = \dfrac{125}{25\,000} \times 12 \times 100 = 6\%$

  e $c = 1.005$, $d = 250$

  f **i** After two years, Elena has $34 536.98, so she will need an extra $463.02.

    ii extra $18.21/month

11 a 6.54%

  b $11 276.52 (1 mark for balance after 3 years $9964.628…)

12 a $1560 **b** $805.65

13 a $45.33

  b $V_0 = 2500$, and $V_{n+1} = 1.0034V_n + 150$

  c 5.87%

14 a $986.25 **b** $655 539.45

  c $5082.39 **d** $4097.12

CUMULATIVE EXAMINATION: CALCULATOR-FREE

1 a The graph is semi-Eulerian, as it has exactly two vertices of odd degree.

  b Because all edges of the graph are bridges, meaning they would have to be crossed twice to start and finish at the same vertex. This means that some vertices will have to be visited more than once.

2 $V_0 = 20\,000$, $V_{n+1} = 1.0002V_n - 1223.13$

3 $V_1 = 1.01 \times 2000 - 20$

  $V_1 = 2020 - 20 = 2000$. As $V_1 = V_0$, this is a perpetuity.

CUMULATIVE EXAMINATION: CALCULATOR-ASSUMED

1 a 0.53

  b **i** 56.9%

    ii 56.9% of the variation in *height* is explained by the variation in *age*.

2 a 13 **b** 15

3 a 160 m

  b There are exactly two vertices of odd degree, meaning that an open Eulerian trail exists but not a closed one.

  c 1250 m. The open Eulerian trail would start and finish at the two vertices of odd degree, giving the distance of 1180 m. The shortest distance from these two vertices is 70 m, so 1180 + 70 = 1250.

  d 9 edges

4 a Brianna's method of fitting a line of best fit through the deseasonalised data is more appropriate in modelling the sales data, as it accounts for the trend in the data and removes the effect of seasonality.

  or

  The five-point moving average method used by Allan is not suitable as quarterly sales data implies four seasons in a cycle, therefore a five-point moving average method will not model the data well.

  b $x = 105$

  c Quarters 1 and 3. The seasonal indices for both quarters are above 100%.

  d **i** $t = 52$ for 4th quarter 2022

    $s = -1.9585(52) + 105 = 3.158$

    Deseasonalised sales for 4th quarter 2022

    $= \dfrac{3.158}{0.78} \times 1000 = \$4048.72$

    ii Although the sales data is extrapolated, it is considered as reliable because the prediction is within one cycle beyond the data collected.

  e Based on Brianna's analysis, I would not recommend that Zen Supermarket acquire Chaos Grocers. The regression line fitted to the deseasonalised data suggests a downward trend in sales and if the trend continues, the quarterly sales of Chaos Grocers will go below $0, making it not feasible.

5 a **i** $T_{n+1} = \left(1 + \dfrac{1.8}{1200}\right)T_n - 420$, $T_0 = 14\,999$

    ii Andrew still owes $10 189.43

  b Andrew owes $5408.99 ($5409) after two years.

  c 38 months **d** $52.63 **e** $16 792.63

6 a **i**

Number of quarters money is invested	1	2	3	…	16
Value of investment ($)	4536.45	4573.20	4610.24	…	5120.00

    ii $T_0 = 4500$, $T_{n+1} = T_n\left(1 + \dfrac{0.0324}{4}\right)$

b Solve $5120 = 4500\left(\dfrac{1+x}{365}\right)^{1460}$.

$x \approx 3.23\%$ p.a.

Therefore, increasing the compounding period to daily reduces the required interest rate.

7 a **i** $33 164.78

  ii It will take an extra 53 months to pay off the loan.

b 17.3%

c He should have chosen the reduced interest rate as he would have paid less for the car.

or

He should have chosen to increase his repayments because he would have paid off the loan sooner and it would have cost only an extra $907.

CHAPTER 8

EXERCISE 8.1

1 a This graph is not a spanning tree. It has an edge (*CI*) that isn't in the original graph.

b This graph is not a spanning tree. It has a cycle.

c This graph is not a spanning tree. It is not connected.

d This graph is a spanning tree. It has 9 vertices and 8 edges.

e This graph is a spanning tree. It has 9 vertices and 8 edges.

f This graph is not a spanning tree. It has a loop.

2

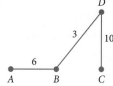 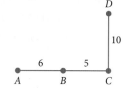

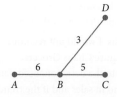

total weight of minimum spanning tree = 14

3

total weight = 20

4 64 m

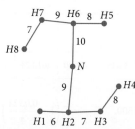

5 a

	A	B	C	D
A		④	6	
B	4		5	③
C	6	5		2
D		3	②	

minimum spanning tree *AB, BD, CD*

weight = 9

b

	A	B	C	D	E	F
A		6	1	3		5
B	6				④	7
C	①			4		
D	③		4		6	
E		4		6		⑤
F	⑤	7			5	

minimum spanning tree *AC, AD, AF, BE, EF*.

weight = 18

c

	A	B	C	D	E	F
A		10		④	5	
B	10		⑦		9	9
C		7		⑦	8	8
D	4		7		2	
E	5	9	8	②		
F		9	⑧			

minimum spanning tree *AD, BC, CD, CF, ED*.

weight = 28

d

	A	B	C	D	E	F	G
A		9				⑤	3
B	9		7				④
C		⑦		10			
D			10		9		⑥
E				9		8	⑦
F	5				8		
G	③	4		6	7		

minimum spanning tree *AF, AG, BC, BG, DG, EG*

weight = 32

6 a 4 edges   **b** 5

7 Option A is the only tree. Each of the other options contains a cycle.

8 a

	A	B	C	D	E	F	G	H
A	–	8					6	3
B	8	–	4			3	4	
C		4	–	6		3		
D			6	–	5	4		
E				5	–	3		7
F		3	3	4	3	–		4
G	6	4				4	–	8
H	3				7		8	–

b

	A	B	C	D	E	F	G	H
A	=	8					6	3
B	8	=	4			3	(4)	
C		4	=	6		(3)		
D			6	=	5	(4)		
E				5	=	(3)		7
F		(3)	3	4	3	=		4
G	(6)	4					=	8
H	(3)				7		8	=

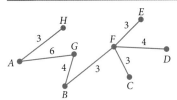

9 a

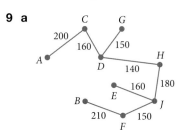

b Min length
= 200 + 160 + 150 + 140 + 180 + 160 + 150 + 210
= 1350 metres

10 a *EP*(0.4), *PQ*(0.3), *QR*(0.2), *RS*(0.4), *VS*(0.5), *ST*(0.2), *UT*(0.3). The minimum length is 2.3 km.

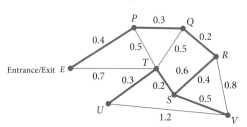

b The minimum length will decrease by 0.1 km (as *SU* would be used instead of *TU*).

11 $x = 9$

12 $k = 5$

13 a $300   **b** $920   **c** *N* and *P*

d   i

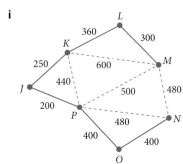

    ii $120

14 a minimum spanning tree

b

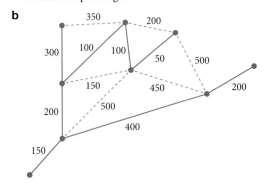

EXERCISE 8.2

1 a 3 edges   **b** 15

2 70

3 a incoming flow = 900 vehicles per hour

   b outgoing flow = 600 vehicles per hour

   c maximum flow = 600 vehicles per hour

4 44

5 a 65

   b

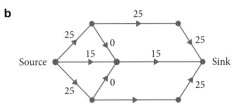

6 a 14   **b** 13

7 $a + b + c + e$

8 a 22   **b** 18

9 a Cut *A* = 25 + *x*, cut *B* = 23 + *x*, cut *C* = 24 + *x*, cut *D* = 32, cut *E* = 27

   b 25 litres per minute

10 a 21   **b** cut *D*

11 a 23   **b** 10

12 350

13 a *FJH* – 13   *FJH* – 13

     FCABEH – 8   *FGH* – 10

     FGH – 10   *FGDEH* – 6

     FGDEH – 6   *FCDEH* – 5

     FCDBEH – 10   *FCABEH* – 8

     FCDEH – 2   *FCDBEH* – 7

     Total is 49   Total is 49

     Maximum flow is 4900 cars per hour.

   b

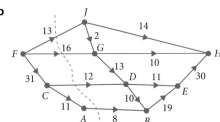

c **i** One of *FG*, *CD*, *FJ* or *AB*

    ii By increasing any one of *FG*, *CDJ* or *AB* by 3 (300 cars per hour), the new maximum flow would be 5200 cars per hour.

14 **a** $x + 19 = 57$, so $x = 38$.

  b Listing 1. *SBCT*-19, *SAT*-22, *SACT*-12

               gives a total of 53 L per minute

       Listing 2. *SACT*-12, *SAT*-22, *SABCT*-4, *SBCT*-15

               gives a total of 53 L per minute. Alternatively, a minimum cut can be made through *AT*, *AC* and *BC*, which gives $22 + 12 + 19 = 53$.

  c Using Listing 1:

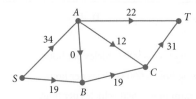

      Using Listing 2:

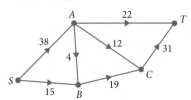

  d Using Listing 1:

      SBDT = 6

      SABDCT = 3

      Therefore, maximum flow = $53 + 9 = 62$

      New maximum flow = 62 litres per minute

      This is an increase of 9 litres per minute.

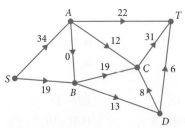

      Using Listing 2:

      SBDT = 6

      SBDCT = 3

      Therefore, max flow = $53 + 9 = 62$

      New maximum flow = 62 litres per minute

      This is an increase of 9 litres per minute.

      or

      Minimum cut through AT, CT and DT gives 62 as before. This is an increase of 9 litres per minute.

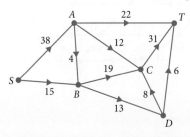

15 **a** *ABEF* – 100

     ADF – 100

     ACF – 350

     ACEF – 200 gives a total of 750, i.e. 750 litres per minute. Other paths are possible.

  b

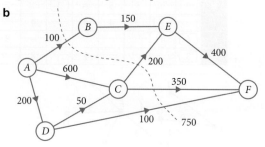

  c

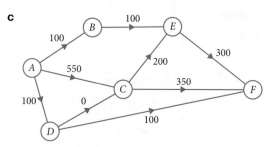

  d 550 litres per minute

  e $550 - 480 = 70$

     Therefore, increase arc *DF* from 100 to 170.

16 **a** Cut 1: $12 + 17 + 26 = 55$ L/min

     Cut 2: $12 + 24 + 26 = 62$ L/min

     Cut 3: $11 + 24 + 26 = 61$ L/min

  b **i** The flow into a node must equal the flow out of that node.

      ii $y = 8$, $z = 22$

  c *DT*, *ET*, *BE* and *DC*

  d 48 L/min

  e **i** Increase along *SBCT* by 2 to get 50 L/min

      ii Minimum cut through *DT*, *CT* and *ET* = 50 on Diagram 1

         or

         DT, *CT* and *ET* are now at full capacity after the increase in flow from part **e i**.

17 **a** 37

  b **i** *A–B–E–C–D*

      ii

Group	Maximum group size	Path taken from *A* to *D*
1	17	answered in part **b i**
2	11	*A–F–E–D*
3	7	*A–G–F–B–C–D*
4	2	*A–B–E–D*

18 **a** 43   **b** 22   **c** 7

19 **a** 26 trucks   **b** 23 trucks   **c** 8 trucks

EXERCISE 8.3

1 B **2** D

3 task 1 – Christen, task 2 – Cathy, task 3 – Celeste, task 4 - Cheryl

4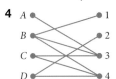

Allocation $A3, B1, C4, D2$

5

	1	2	3	4
A	18	7	0	8
B	0	4	2	3
C	8	0	4	0
D	3	0	5	6

Allocation $A3, B1, C4, D2$

Time = 7 + 6 + 5 + 5 = 23 hours

6 $A1, B3, C2$ for a minimum distance of 42 km

7 $A2, B3, C1, D4$; total maximum = 144

8 $A1, B4, C2 = 57$ or $A3, B1, C2 = 57$

9 a 8

b i

		Delivery driver			
		John	Kerry	Liam	Max
Deliveries	1	4	0	9	5
	2	0	1	7	11
	3	8	4	0	0
	4	0	3	3	1

ii The number of lines (3) is not the same as the number of drivers and tasks (4).

c

Delivery driver	John	Kerry	Liam	Max
Delivery	2	1	3	4

minimum total delivery time = 109 minutes

10 a

$$
\begin{array}{c}
 & J1 & J2 & J3 & J4 \\
A & 4 & \boxed{7} & 7 & \boxed{8} \\
B & \boxed{2} & 6 & 3 & \boxed{4} \\
C & 3 & \boxed{4} & \boxed{5} & 2 \\
D & \boxed{9} & 7 & \boxed{4} & 6
\end{array}
$$

b Job 1 goes to Ben, Job 2 goes to Adam, Job 3 goes to Demi, Job 4 goes to Cate.

11 a

$$
\begin{array}{c}
 & A & B & C & D \\
S1 & 4 & 5 & 2 & 12 \\
S2 & 10 & 4 & 8 & 2 \\
S3 & 3 & 4 & 6 & 14 \\
S4 & 9 & 3 & 5 & 3
\end{array}
$$

b

Salesperson	1	2	3	4
Machine	B	C	D	A

c $S1 – A, S2 – B, S3 – C$ and $S4 – D$ gives a total of 17 sales.

Solution to part **b** gives a total of 36 sales. Greater by 19.

12 a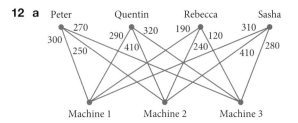

b The number of columns does not equal the number of rows, i.e. not a square.

c

Worker	Peter	Quentin	Rebecca	Sasha
Machine	1	3	4	2

Maximum total length of shade cloth
= 300 + 320 + 410 = 1030 metres

13 14 minutes

14 a George

b

Person	Position
Harriet	drums
Ian	saxophone
Keith	keyboard

15 a 17

b Four lines are required before allocating four tasks to four people and there are only three.

c

	Worker			
Task	Julia	Ken	Lana	Max
W	0	0	4	0
X	2	2	0	10
Y	1	3	0	0
Z	7	0	3	5

d W – Julia, X – Lana, Y – Max, Z – Ken

16 a

Student	Sport
Blake	tennis
Charli	football
Huan	basketball
Marco	athletics

b Anita – 400, Imani – 200, Jordan – 100, Lola – 300

17 a 2, 0, 7, 5

b She is the only child with a zero in the column for Concert 2.

c 3, 4, 2, 1 or 3, 1, 2, 4

d 56

18 a Colin must plan Tour 2.

b 43 minutes

CUMULATIVE EXAMINATION: CALCULATOR-FREE

1 a $T_{n+1} = T_n - 3, T_1 = 18$

b year 7

c 103 cm

2 a path **b** cycle **c** open trail

3 a Laura should sign up for the compounding period with the lowest effective annual rate of interest. This is because the lower the effective annual rate, the less interest she will need to pay at the end of the year.

Look at the table for the 5% and pick the lowest. Therefore, Laura should sign up for the loan with interest compounded quarterly (5.095%).

b To determine the interest Laura would earn by investing, multiply 200 by the effective interest rate from the table for 3.5% and quarterly.

$$\text{Interest} = 200 \times \frac{3.553}{100}$$
$$= 3.553 \times 2$$
$$= 7.106$$

(note the 100 and divide by 100 cancel)

≈ $7.11 rounded to the nearest cent.

c To determine the value of the investment, we need to determine the interest and add this to the original amount, using the effective interest rate from the table for 5.5% compounded monthly.

$$10\,000 \times \frac{5.641}{100} = 100 \times 5.641 = 564.1$$

(dividing by 100 cancels two zeroes from 10 000)

564.1 + 10 000 = $10 564.10

4 a

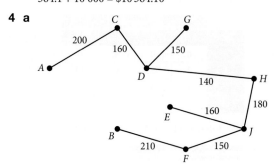

b 1350 metres

5 a

Type	Type 1	Type 2	Type 3	Type 4
Workshop	C	D	A	B

Total minimum annual cost is $149 000

b

$$\begin{bmatrix} 12 & 18 & 10 & 14 \\ 12 & 21 & 17 & 11 \\ 10 & 18 & \boxed{16} & \boxed{17} \\ \boxed{8} & \boxed{22} & \boxed{20} & \boxed{4} \end{bmatrix}$$

c

Type	Type 1	Type 2	Type 3	Type 4
Workshop	A	B	D	C

CUMULATIVE EXAMINATION: CALCULATOR-ASSUMED

1 a strong negative

b The residual plot indicates that the association between the *mean height* and the *mean age* of women is non-linear.

2 a

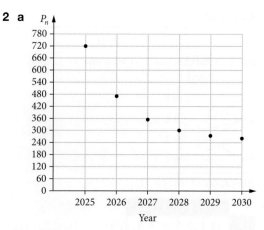

b 250 fish

3 a There exist vertices of odd degree, i.e. not all vertices are of even degree.

b Any one of the following:
- start at *C* and finish at *D*
- start at *D* and finish at *C*
- start at *D* and finish at *H*
- start at *H* and finish at *D*
- start at *G* and finish at *H*
- start at *H* and finish at *G*.

c 335 m

4 a $\text{monthly interest rate} = \dfrac{\text{annual interest rate}}{12}$

$= \dfrac{15\%}{12} = 1.25\%$

b $C = 11\,496.88 \times 1.25\% = 143.71$

$D = 11\,496.88 + 143.71 - 400 = 11\,240.59$

5 a Cut 1 = 58, Cut 2 = 62

b 25 + 29 = 54, which is the flow out of the source

c $x = 1$, $y = 5$, $z = 20$

d maximum flow = 56

e *ET* will increase the flow by 2 (*PBCET*)

DG will increase the flow by 2 (*PBDGT*)

or cut 1 = minimum cut with edge *ET* or *DG*, i.e. an increase of 2

Therefore, either proposal will increase the flow by 2.

6 a i 45

ii The maximum flow through the network is ≤45.

b $SABT = 10$ $SCDT = 14$

c One possibility:

Using answers from part **b**:

$SABT = 10$, $SCDT = 14$, $SACEDT = 3$, $SAET = 4$, $SET = 9$

maximum flow = 40 kilolitres per hour

or

Another possibility:

$SABT = 10$, $SACDBT = 5$, $SACDT = 5$, $SCDT = 4$, $SCEDT = 7$, $SEDT = 1$, $SET = 8$

maximum flow = 40 kilolitres per hour

d

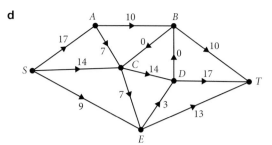

e

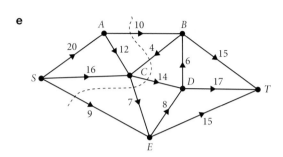

7 a

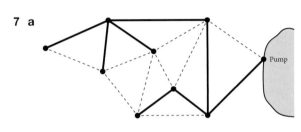

b minimum spanning tree

CHAPTER 9

EXERCISE 9.1

1

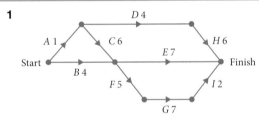

2

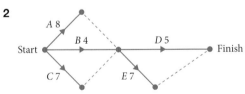

3

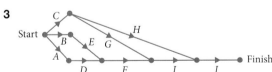

4 a I, J and K **b** 2

5 a B, E and D **b** 7

6

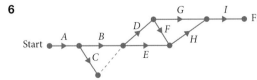

7

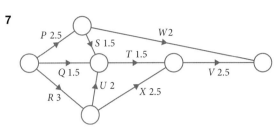

8

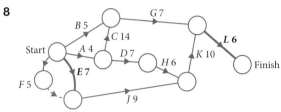

9

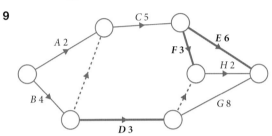

10 a

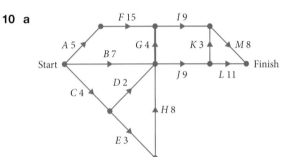

b B, C, D, E and H

EXERCISE 9.2

1 a C and E **b** C and D

2

Activity	Time (hours)	Predecessor
A	6	–
B	3	–
C	5	A
D	3	A, B

3

Activity	EST
A	0
B	0
C	0
D	5
E	4
F	12
G	10

4

Activity	EST	LST
A	0	2
B	0	4
C	0	0
D	5	5
E	4	6
F	12	14
G	10	10

5

Activity	EST	LST
A	0	0
B	6	6
C	6	14
D	23	23
E	6	12
F	19	25
G	33	35
H	33	36
I	46	48
J	33	33
K	44	44

Critical path is A–B–D–J–K.

6 critical activities A, D, E and H

float for non-critical activities

float for activity C = 8 − 0 = 8 days

float for activity B = 2 − 0 = 2 days

float for activity F = 13 − 11 = 2 days

float for activity G = 21 − 11 = 10 days

7 a critical path A–E–F, project time 34 days

b critical path AEG, project time 28 days

8 a 2   **b** 3

9 a 12   **b** 4

10 a critical path QSXZ, minimum completion time 35 days

b **i** EST for activity Y is day 22.

ii LST for activity V is day 5.

iii float time for activity U is 9 days.

c critical path changes to RWY

Minimum completion time is now 36 days (an extra day).

11 a The critical path is RUTV. The minimum completion time is 9 hours.

b **i** 4.5 hours   **ii** 0 hours

c **i** 7:30 am

ii Adding a break of half an hour means the project takes 9.5 hours (finishing at 4 pm).

Therefore, V must start no later than 1:30 pm.

d Minimum completion time is now 10 hours (a delay of an hour).

One gardener can do all tasks on the path (R, U and V).

A second gardener will need to do P, S and Q. As these take a total of 5.5 hours, T will have to be delayed by half an hour. The second gardener can then do T and W, in that order, which would also add an extra half an hour.

12 a 4   **b** 4

13 a 11 days   **b** A–E–I–K

c activity H   **d** $2000

e **i**

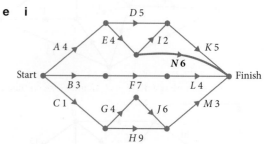

ii 9 days

14 a D and E

b **i** A–E–I–L–N   **ii** 6 days

c **i** 17 days   **ii** $4000

15 a ADHKL. Minimum completion time is 33 hours

b float time = 2 hours, earliest starting time = 18 hours, latest starting time = 20 hours

c If G is increased by 3 hours, new completion time is 34 hours (ACGL).

If J is increased by 3 hours, new completion time is 35 hours (EJKL).

Therefore, choose G to ensure minimum completion time is the smallest.

16 a

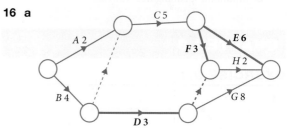

b There are two critical paths: BCE and BDG, so the critical activities are B, C, E, D and G. Minimum completion time = 15 days

c A has a float time of 2 days, F has a float time of 1 day, H has a float time of 1 day

d H has a float time of 1 day so a delay of 3 days would increase the minimum completion time to 17 days. The critical activities would now be B, C, F and H (i.e. the new critical path is BCFH).

e Shorten B, which will reduce the minimum completion time by 1 day, as it is on both critical paths.

Shortening F will have no effect since it is not a critical activity.

Shortening E would only be effective if D or G were reduced as well, as E is only on one critical path.

17 a 8   **b** 12   **c** *J*   **d** 29

    e $C - 0, D - 1, G - 2, H - 1, K - 1$

18 a Activities *C* and *D*

    b Activity *D* must be on the critical path.

    c 17 weeks   **d** 3 weeks   **e** Activity *F*

19 a 18   **b** 1   **c** 7

    d The *dummy* activity indicates that activity *K* cannot proceed unless activities *H*, *G* and *C* are completed. Activities *C* and *G* are immediate predecessors of *K*.

CUMULATIVE EXAMINATION: CALCULATOR-FREE

1 a

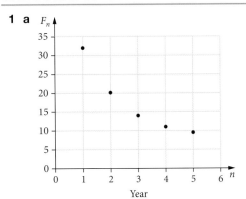

    b 6

2 a

$$M = \begin{array}{c} \\ A \\ B \\ C \\ D \\ E \\ F \end{array} \begin{array}{cccccc} A & B & C & D & E & F \\ \left[\begin{array}{cccccc} 0 & 1 & 1 & 0 & 0 & 0 \\ 1 & 0 & 1 & 0 & 0 & 0 \\ 1 & 1 & 0 & 1 & 1 & 0 \\ 0 & 0 & 1 & 0 & 1 & 0 \\ 0 & 0 & 1 & 1 & 0 & 1 \\ 0 & 0 & 0 & 0 & 1 & 0 \end{array} \right] \end{array}$$

    b There are exactly two odd vertices and so an open Eulerian trail exists.

    c An open Eulerian trail is *F–E–D–C–B–A–C–E*. Other answers are possible. This is not a Hamiltonian path as vertex *C* is repeated.

3 a i

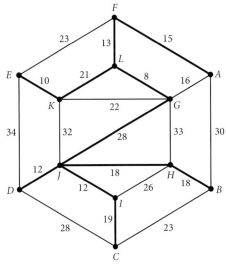

    ii Cost = $17 400

    b Now includes edge *AB*. Total cost = $17 600.

4 a 7   **b** 15 days

    c 16 days   **d** *BEIJM*, 22 days

    e 3 days

5 a 14 weeks

    b *ADGLM* and *AEHLM*, 19 weeks

    c 7

CUMULATIVE EXAMINATION: CALCULATOR-ASSUMED

1 a $T = 73.508$

    estimated charge: $36.75

    b Estimate would be valued since it is interpolation.

    c i Residual for customer B is 9.092.

       Residual for customer D is −1.208.

    ii The change in sign indicates the residuals are above and below the least-squares line. The size indicates that the residual for D is closer to the line than the residual for customer B.

2 a $T_{n+1} = 1.64 T_n$

    for $n = 1, 2, 3 \ldots$

    b $T_n = 2200(1.64)^{n-1}$

    for $n = 1, 2, 3 \ldots$

    c 2 239 940.87

3 a $A = 1789.50, B = 120.53, C = 2563, D = 1534,$ $E = 2044.75$

    b $\dfrac{79.8 + 93.29 + 81.43}{3} = 84.84$

c

Week	Day	Time period	No. of audience	Weekly mean	Percentage of weekly mean	Seasonally adjusted No. of audience
1	Thursday	1	1253	A	70.02	
	Friday	2	1428		79.80	1683.17
	Saturday	3	2320		129.65	
	Sunday	4	2157		B	
2	Thursday	5	1326	1927.25	68.80	
	Friday	6	1798		93.29	2119.28
	Saturday	7	C		132.99	
	Sunday	8	2022		104.92	
3	Thursday	9	D	E	75.02	
	Friday	10	1665		81.43	1962.51
	Saturday	11	2688		131.46	
	Sunday	12	2292		112.09	

d Sunday Week 4 is when $t = 16$

$29.8297 (16) + 1727 = 2204.28$

The predicted number of audiences for Sunday Week 4 is 2204. This prediction is reliable because it is within 1 period (week) of the data set.

4 a value of investment after 5 years = $P(1 + r)^n$
(or financial function in CAS)

$= 8000 \times (1 + 0.07)^5 = \$11\,220.41$

b Number of years required to triple investment:

Using the quarterly interest rate (1.75% or 0.0175)

$8000 \times (1 + 0.0175)^n = 3 \times 8000$ (use SOLVER in CAS)

$\Rightarrow n \approx 63.326$ quarters (not years)

$\dfrac{63.326}{4} = 15.83$ but must round to 16 years (this

is a type of question where must always round up anyway otherwise the investment will not have tripled yet)

5 a   i 9   **ii** 13

b 7

6 a

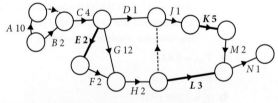

b Immediate predecessors of task *J* are *D* and *H*.

c *H* is the predecessor for *J*

d *ACGHJKMN*. Minimum completion time is 37 days.

e Float time = 13 days. This means task *D* can be delayed by up to 13 days.

f 15 days

g Critical path stays the same. Minimum completion time is now 34 days.

h 8 days

7 a 10

b *B–E–G–H–J*

c *A* and *C*

d end of activity *E* to the start of activity *J*

Glossary and index

3-point moving average smoothing A smoothing technique that involves finding the average of consecutive sets of three data points. (p. 225)

4-point centred moving average smoothing A smoothing technique that involves finding the average of consecutive sets of four data points involving the process of centring the averages. (p. 226)

5-point moving average smoothing A smoothing technique that involves finding the average of consecutive sets of five data points. (p. 225)

6-point centred moving average smoothing A smoothing technique that involves finding the average of consecutive sets of six data points involving the process of centring the averages. (p. 228)

activity (directed graphs) Interconnected steps represented by an arc in a directed graph and shown by a line with an arrow. (p. 391)

activity table A table showing the order and estimated time for a series of activities. (p. 391)

adjacency matrix An $n \times n$ matrix in which the entry in the ith row and jth column is the number of edges joining the vertices i and j, and n is the number of vertices in the graph. In an adjacency matrix, a loop is counted as 1 edge. For a directed graph, the entry in the ith row and jth column is the number of directed edges (arcs) joining the vertices i and j in the direction i to j. (pp. 171, 364)

adjacent vertices Any two vertices connected by an edge. (p. 144)

allocation problem *See* **assignment problem**.

amortisation table A table that shows the step-by-step calculations of how a loan is reduced. (p. 285)

annuity A type of investment where a sum is invested, interest is compounded at a fixed rate and withdrawals are made at regular intervals, usually until the value of the investment is $0. (p. 305)

annuity investment An investment that involves making an initial deposit followed by additional regular payments into an account earning a fixed rate of compound interest. (p. 318)

annuity table A table that shows step by step calculations for each time period of an annuity. (p. 307)

appreciation The increase in value of items over time. (p. 267)

arc The directed edge of a digraph. (p. 165)

arithmetic decay (linear decay) Occurs when the terms of a sequence decrease by a constant amount. This is represented by a negative d value. (p. 89)

arithmetic growth (linear growth) Occurs when the terms of a sequence increase by a constant amount. This is represented by a positive d value. (p. 89)

arithmetic sequence A sequence where there is a common difference between consecutive terms. (p. 89)

asset An item purchased by businesses to help them function. (p. 267)

assignment problem The process of finding the best way to match the elements in two groups, such as a group of workers to a set of tasks, to optimise a stated objective such as minimising cost, distance or time. (p. 363)

average percentage method A method for calculating a seasonal index where the data for each season is expressed as percentages of the average. (p. 237)

balance The value of an investment or loan at any time. (p. 279)

bipartite graph A graph with vertices that can be separated into two distinct sets, e.g. set X and set Y, so that each edge of the graph only connects a vertex from set X to set Y. (pp. 159, 363)

bivariate data Data associated with two related variables. (p. 3)

bridge An edge of a graph that keeps it connected. (p. 155)

capacity of a cut The total of all the flow capacities passing across a cut in the direction from source to sink. (p. 354)

causation A relationship between an explanatory and a response variable where the change in the explanatory variable actually causes a change in the response variable. Knowing that two variables are associated, no matter how strongly, is not sufficient evidence by itself to conclude that the two variables are causally related. (p. 26)

centring The extra step of taking a 2-point moving average of the smoothed values when smoothing with an even number of points. (p. 226)

circuit *See* **closed trail**.

closed Eulerian trail A trail that includes every edge in a graph, but only once and starts and finishes at the same vertex. An Eulerian trail may include repeated vertices. (p. 195)

closed path A path that starts and finishes at the same vertex. (p. 191)

closed trail A trail that starts and finishes at the same vertex. (p. 190)

closed walk A walk that starts and finishes at the same vertex. (p. 190)

coefficient of determination (r^2) In a linear model between two variables, proportion of the total variation that can be explained by the linear relationship existing between the two variables, usually expressed as a percentage. (p. 49)

coefficients The values a and b in the least-squares line equation are called the 'coefficients' of the equation. This is a general term for these values in an equation. (p. 43)

coincidence A high level of correlation between two variables occurring by chance without any underlying cause. (p. 26)

common ratio The constant number used in geometric sequences to generate new terms. It is denoted by r. Each new term is found by multiplying the previous term by r. (p. 100)

common response A high level of correlation between two variables where a third factor is causing the correlation. (p. 26)

complete bipartite graph A bipartite graph in which every vertex in the first distinct set of vertices is connected to every vertex in the second distinct set, and vice versa. (p. 159)

complete graph A simple graph in which every vertex is connected to every other vertex by an edge. (p. 155)

compound interest Interest that is added to the principal, where the interest for the next time period is calculated using this new balance. (p. 262)

compounding period The length of the time period that elapses before interest compounds. (p. 262)

confounding A high level of correlation between two variables where it's unclear how the variables are related and there may be an unknown factor involved. (p. 26)

confounding variable An unknown factor that may be causing a high level of correlation between two variables. (p. 26)

connected graph A graph in which any vertex can be reached using a sequence of edges, starting from any other vertex. If removed, it leaves the graph disconnected. (p. 155)

correlation coefficient (r) A measure of the strength of the linear relationship between a pair of variables $-1 \leq r \leq 1$. Calculating r may be performed by using appropriate technology. (p. 23)

critical path The longest time path between the start and finish. (p. 402)

critical path analysis A step-by-step project management technique that is used to examine every activity in a project and how each activity affects the project completion time. (p. 399)

cut A line in a network that stops all flow from source to sink. (p. 353)

cycle A closed walk in which all the edges and vertices are different, except for the first and last vertex. (p. 191) *See also* **closed path.**

degree (of a vertex) The number of edges that enter or exit from the vertex; thus, loops are counted twice. (p. 144)

degree sum (of a graph) The sum of the degrees of all the vertices in a graph. (p. 144)

depreciation The decrease in value of assets bought by a business over time. (p. 267)

deseasonalisation The process of using seasonal indices to remove the seasonal component of time series data, using the formula

deseasonalised value = $\dfrac{\text{actual value}}{\text{seasonal index}}$ (p. 237)

digraph *See* **directed graph.**

directed graph (digraph) A graph containing vertices and edges, where each edge has an indicated direction between the two vertices it connects. (p. 165)

divided column graph A graph where two or more percentage segmented bar charts are shown on the same axes. (p. 9)

dummy activity An activity of zero time, shown as a directed edge with a broken line, added to a network to ensure that no two vertices are connected by multiple edges or to maintain precedence structure. (p. 393)

earliest starting time (EST) The earliest time it is possible to start an activity. (p. 399)

edge A line connecting two vertices in a graph. (p. 144)

effective interest rate The interest after compounding has been taken into account, which allows us to compare different rates. (p. 262)

Eulerian graph A connected graph that contains a closed Eulerian trail. (p. 195)

Eulerian trail A trail that includes every edge in a graph, but only once. (p. 195)

Euler's formula For connected planar graphs, $v + f - e = 2$, where v = the number of vertices, f = the number of faces, and e = the number of edges. (p. 157)

explanatory variable A variable that we expect to predict or explain another variable. (p. 3)

extrapolation Occurs in the context of fitting a linear relationship between two variables, when the fitted model is used to make predictions using values of the explanatory variable that are outside the range of the original data. (p. 60)

face The region of a planar graph that is enclosed or bounded by edges, including the outer infinitely large region. (p. 157)

finance solver An application that solves finance problems. (p. 275)

first-order linear recurrence relation A recurrence relation where the only term used to recursively generate T_{n+1} is its immediately prior term T_n and where the relationship between T_{n+1} and T_n is linear. The form is $T_{n+1} = bT_n + c$, $T_1 = a$. (p. 116)

float time The maximum time an activity can be extended or postponed without affecting the project completion time. Activities on the critical path all have float times of zero. (p. 403)

future value The new reduced value of an asset being depreciated at any point in time or the balance of loans and investments at any point in time. (p.267)

geometric decay Occurs when the terms of a sequence decrease by a constant ratio. This is represented by an r value between 0 and 1. (p. 100)

geometric growth Occurs when the terms of a sequence increase by a constant ratio. This is represented by an r value greater than 1. (p. 100)

geometric sequence A sequence where there is a common ratio, r, between consecutive terms. Each new term is equal to its preceding term multiplied by this constant ratio. (p. 100)

gradient *See* **slope.**

graph A diagram consisting of a set of points, called vertices, that are joined by a set of lines called edges. (p. 144)

Hamiltonian cycle A Hamiltonian path that starts and finishes at the same vertex. (p. 202)

reducing balance depreciation Depreciation where the future value of an asset is reduced every year by a fixed percentage of its value in the preceding year. (p. 267)

reducing balance loan A loan where interest is calculated on the amount still owing after each repayment is made. (p. 282)

reducible interest rates *See* **nominal interest rates**.

re-seasonalisation The process of using seasonal indices to find the original non-seasonalised value using the formula
actual value = deseasonalised value × seasonal index (p. 240)

residual The difference between the observed value and the value predicted by a statistical model (e.g. by a least-squares line). (p. 65)

residual plot A scatterplot with the residual values shown on the vertical axis and the explanatory variable shown on the horizontal axis. Residual plots are useful in assessing the fit of the statistical model (e.g. by a least-squares line). (p. 69)

response variable A variable whose changes we expect to be predicted or explained by another variable. (p. 3)

scatterplot A graph used to compare two numerical variables where the explanatory variable is plotted on the x-axis and the response variable is plotted on the y-axis. (p. 16)

seasonal adjustment A term used to describe a time series from which periodic variations due to seasonal effects have been removed. (p. 236)

seasonal indices Values used to make seasonal adjustments seasonal series data. (p. 235)

seasonal variation A regular rise and fall in the time series each year. (p. 219)

semi-Eulerian graph A connected graph that contains an open Eulerian trail (p. 195)

semi-Eulerian trail *See* **open Eulerian trail**.

semi-Hamiltonian graph connected graph contains an open Hamiltonian path. (p.

sequence An ordered list of numbers, separated by commas. (p. 89)

shortest path problem A problem involving finding the shortest distance, shortest time or least cost a starting vertex to an end vertex by following a sequence edges and vertices. (p. 165)

simple graph A graph without any loops or multiple edges between adjacent vertices. (p. 154)

sink The final vertex of a flow network. (p. 353)

slope The measure of the steepness of a line. The coefficient in the equation for the least-squares line $y = a + bx$. (p. 42)

smoothing A technique for levelling out fluctuations in time series data to produce a smoother graph, which allows us to see trends more clearly. (p. 223)

source The first vertex of a flow network. (p. 353)

spanning tree A tree subgraph that includes all the vertices of the original graph. (p. 341)

steady state The eventual value of a sequence whose terms converge to the same number. In a first-order recurrence relation, this occurs when $T_{n+1} = T_n$. Not all sequences reach a steady state. (p. 120)

subgraph A graph formed using a subset of the vertices and edges in a larger graph. (p. 146)

term A value in a sequence, often represented by T_n, where n is the position of the term. (p. 89)

time series Bivariate data where the explanatory variable is time measured at equally spaced intervals. (p. 217)

time series plot The graph of a time series with time plotted on the horizontal axis. (p. 217)

trail A walk that does not contain repeated edges. In a trail, vertices can be repeated. (p. 190)

traversability problem A problem involving the tracing of a graph to find Eulerian trails. (p. 196)

traversable A graph for which there exists either an open or a closed Eulerian trail, i.e. a semi-Eulerian or an Eulerian graph. (p. 196)

tree A connected graph with no loops, multiple edges or cycles. (p. 341)

trend The term used to describe the general direction of a time series (increasing/decreasing) over a long period of time. (p. 219)

trend line A line of best fit for time series. (p. 245)

trend line forecasting A line fitted to a time series to make predictions about the future. (p. 245)

two-way frequency table A frequency table used to explore the association between two categorical variables, each with at least two categories. The row and column totals represent the total number of observations in each row and column. (p. 5)

undirected graph A graph containing edges that do not indicate a direction between two adjacent vertices. (p. 154)

vertex (plural **vertices**) A point in a graph or network diagram. (p. 144)

walk A sequence of connected vertices such that from each vertex there is an edge to the next vertex in the sequence. Walks can include repeated edges and vertices. (p. 190)

weight (of a graph) A quantity given to an edge in a weighted graph that represents information, such as distance, time and cost, between adjacent vertices. (p. 165)

weighted graph A graph in which each edge is labelled with a quantity, called the weight, to represent extra information between adjacent vertices, such as distances, time or costs. (p. 165)

y-intercept The y value of a point where the line crosses the y-axis. (p. 42)

9780170476959